CRIME AND PUNISHMENT
IN
EARLY MARYLAND

CRIME AND PUNISHMENT
IN
EARLY MARYLAND

BY

RAPHAEL SEMMES, LL. B., PH. D.

Author of
Captains and Mariners of Early Maryland

BALTIMORE
THE JOHNS HOPKINS PRESS
1938

PRINTED IN THE UNITED STATES OF AMERICA
BY J. H. FURST COMPANY, BALTIMORE, MARYLAND

Man's inhumanity to man
Makes countless thousands mourn.

ROBERT BURNS

CONTENTS

CHAPTER   PAGE

   I. The Majesty of the Law 1

  II. Trial, Punishment and Imprisonment . . . 21

 III. Houses, Clothing and Theft 41

 IV. Livestock and Hog Stealing 58

  V. Servant Discipline and Punishment . . . 80

 VI. Homicide, Assault and Suicide 119

VII. Drunkenness, Profanity and Witchcraft . . 145

VIII. Adultery, Fornication and Bastardy . . . 174

 IX. Defamation 207

  X. Sickness, Chirurgery and Burials 232

BIBLIOGRAPHICAL CHAPTER NOTES 259

INDEXES 319

CHAPTER I

THE MAJESTY OF THE LAW

It was the policy of the proprietary officials to uphold the dignity of their official position at all times. This was true whether one held the rank of governor, councillor, burgess or judge. Captain Samuel Tilghman, commander of the good ship Golden Fortune, was arrested " for sundry contemptuous and scandalous words uttered by him against his Lordship's rule and government." When Colonel Nathaniel Utie was thought to have uttered words similar in nature to Captain Tilghman's, he, too, was arrested. Both men were compelled to appear before the provincial court to explain their remarks.[1]

On another occasion, George Butler, clerk of the provincial court, was charged with having abused Charles Calvert, uttering " most scandalous and opprobrious speeches " which reflected not only upon his Lordship, but also upon his father " of noble memory." In a humble petition to Charles Calvert, Butler said that as he was aware of his error and foolish remarks, he begged the proprietor's pardon. He further asked his Lordship to consider his forlorn and miserable condition and that of his wife, then " big with child." The sheriff was ordered to bring Butler before the members of the council. After he had again apologized for his utterances, an order was given for Butler's release from prison.[2]

Councillors, too, resented any remarks which reflected on their character or official position. It was for this reason that Captain John Price, one of his Lordship's councillors, objected to the slanderous remarks which Luke Gardiner had made about him. According to Price, Gardiner had spread abroad a report that the captain kept a dog in order to kill his neighbour's livestock so that he would have " the whole range himself." The judges of the provincial court considered this statement as " a disrespectful expression to one of his Lordship's council." When Gardiner admitted that he should not have

1

made such remarks, the members of the court, " in hopes of his more respective carriage hereafter," pardoned Gardiner.[3]

John Saxon, doorkeeper of the assembly, became involved in difficulty when he " in most scurrilous language and behaviour " mocked and abused Colonel William Stevens, another member of the council. After Saxon had humbly begged the pardon of Stevens, the proprietary officials decided to overlook his remarks. Giles Brent, also a councillor and at one time acting governor of Maryland, complained against Edward Lennin for having " defamed him publicly " in the house of Edward Commins and elsewhere. Brent asked that Lennin should be brought " to such punishment as the fault shall be found to deserve." Just what were the slanderous remarks made by Lennin or what punishment, if any, he received the records do not disclose.[4]

Members of the lower house of the assembly, burgesses or delegates, as they were called, also maintained the dignity of their official position. Accordingly, when James Lewis was accused of having " abused " John Vanhack, one of the members of the lower house, Lewis was compelled to go before the delegates and upon his knees ask the forgiveness of the house and of Vanhack in particular. William Bretton, clerk of the assembly, on another occasion, reviled a member of that body calling him " a factious fellow." After Bretton made a humble submission to the house and apologized to the delegate whom he had offended, adding that he did not make the remark out of any " abusive intent," but because of " some distemper " at the time, the assembly decided to pardon Bretton.[5]

In two other cases, those who had insulted the members of the assembly, did not escape so easily. When, in one instance, Edward Husbands, who was a chirurgeon by profession, threatened, menaced and cursed the delegates, he was ordered whipped on his bare back with twenty lashes " by the hand of the common hangman." Similar punishment was meted out to Edward Erbery. He spoke of the members of the lower house as " pitiful rogues and puppies " and of " the turdy, shitten assembly." Erbery also referred to Charles Calvert as " a rogue." As the delegates decided that such remarks were " a scandal " to the governor, to them, and to the whole province, they ordered that Erbery should be brought before them. When taxed with having made these slanderous statements, Erbery said that he was drunk at the time he made them. This did not seem a sufficient excuse to the members of the lower house. Accordingly, they referred the case to the upper house to take what action they thought necessary. The councillors at once ordered that Erbery in the presence of the members of the assembly should be tied to an apple tree and publicly whipped upon his bare back with thirty-nine lashes. After receiving this punishment, Erbery was again brought before the members of the assembly to ask their forgiveness.[6]

From the instances given, it is evident that governor, councillor and burgess alike did everything they could to maintain the dignity and prestige of their office. But what of the judges of the provincial court and of the justices of the peace in the county courts. What did they do to uphold the majesty of the law?

Those who today attend a session of an English court cannot help but be impressed with dignity surrounding all the judicial proceedings. Some of this is due to the gowns and wigs worn by the judges which seem to enhance the dignity of the court. It is interesting to note that the proprietary courts followed this English custom. About thirty years after the founding of the colony, Cecil Calvert, the Maryland proprietor, sent the following instructions to his son, Charles Calvert, then acting as governor of the province:

That you seriously take into your consideration to find and speedily to propose unto us some convenient way of and for the making of some visible distinction and distinctions between you, our Lieutenant General, our Chancellor, Principal Secretary, General Officers, Councellors, Judges and Justices, and the rest of the people of our said Province, either by the wearing of habits, medals, or otherwise.

It was, no doubt, because of these instructions that Charles Calvert and his councillors soon afterwards ordered that every judge of the provincial court must appear in court at the days appointed " for their setting " with his ribbon and medal under penalty of a fine for every time they failed so to appear.[7]

The wearing of a distinctive dress was not the only regulation adopted by the early Maryland courts to lend prestige to their official position. During one session of the assembly, one of the questions discussed was the need of having a law to punish " contempts in court or out of court by not standing bare-headed before courts and magistrates." The commissioners, or justices of the peace, of Kent county adopted a regulation of this kind. As they reasoned, it was " the custom of England, grounded upon the word of God that due respect be given to magistrates." The county justices therefore ordered that " no man presume, except a member of the court, to stand with his hat on his head in the presence of the court, whilst the court is sitting, or use any incivil language, upon pain of such fine, or other punishment, as the court shall think fit." It was not long after this rule was put into practice that Henry Carline appeared at a meeting of the Kent county court with his hat on his head " contrary to civil respect to magistrates." When Carline refused to remove his hat at the request of the justices, the latter fined him three hundred pounds of tobacco for his " contempt." [8]

The insolent behaviour of the Quakers was especially resented as whenever they appeared in court they would " presumptuously stand covered." Not only would members of this sect do this, but they would also refuse to testify under oath, act as jurors, or hold any office in the colony, nor would they comply with military regulations. For these reasons the governor and council instructed the justices of the peace in every county that whenever any Quakers came into their county, they should be seized and then whipped " from constable to constable until they be sent out of the province." [9]

One ran the risk of punishment if he had had too much to drink when he appeared in court. At a session of the provincial court Thomas Bushnell, " for his misbehaviour in court being much in drink," was fined five hundred pounds of tobacco and cask. Nor would their honours tolerate any profanity in the court room. John Cornelius was fined on two separate occasions for swearing in the presence of the court and William Empson was made to pay a fine for the same offense.[10]

A man by the name of Henry Higges acted disrespectfully towards the justices of the Cecil county court. It was said that Higges coming into the court room with his hat on presumed to sit on the same bench with the justices. Later, he called for drinks, swore " God damn them " that he had come to drink a bowl of punch with them. Higges had to beg the commissioners' pardon and promise " amendment." Swithen Wells, clerk of the Cecil county court, was dismissed from office because of the insolent language he used towards the presiding justice of that court. It seems that Wells instead of reading from a law book, as he had been requested to do, threw the book at the justice telling him to read it himself. But this was not the only difficulty in which Wells was involved. Later, when intoxicated, he appeared before the same justices, called them fools and " God damned them all." For these " grand affronts and indignities " to the commissioners, Wells was ordered arrested, but he escaped punishment by apologizing to the justices and pleading his drunkenness as an excuse.[11]

Ralph Rawlings insulted the commissioners of the Talbot county court by saying that if he bribed them with a bottle of drams he could have " his business done as he pleased." Outraged by this " scandalous reproach," the Talbot justices ordered Rawlings taken outside the court house, stripped to his waist, and given thirty lashes with a whip " well laid on." John Davis also became involved in difficulty with the Talbot commissioners, when, on one occasion, " with malicious and wicked intent " he scandalized the proprietary officials by saying that the Talbot county court had not done him justice in a case " betwixt him and Jonathan Sibery." At his trial Davis acknowledged " his fault and humbly upon his knees put himself upon the favour of the court." Upon this humble submission, the court decided to pardon Davis " his default." [12]

An amusing case of contempt of court occurred in Kent county. Mary Baxter has left us an account of the incident. It appears that Thomas Hynson upon entering the room where the court held its sessions remarked: " Now we are in his Majesty's dog-house." Mrs. Baxter reproved Hynson, telling him that his Majesty " did not use to keep dogs in any court house where justice was administered." When John Winchester made a remark similar in nature to Hynson's, Mrs. Baxter also reprimanded him, telling Winchester that it was a house for men and not dogs. Neither Hynson or Winchester appear to have been brought to account for their disrespectful language. Perhaps the reason for this was that in Hynson's case he himself had at one time been a commissioner of the Kent county court.[13]

Because of his temper Captain Robert Vaughan, of Kent county, was always getting into trouble. On one occasion, Vaughan insulted the justices of the Kent court by bending his fist over their heads and by using " opprobrious " epithets. The captain also swore at the clerk of the court. For his contemptuous attitude Vaughan was fined nearly a thousand pounds of tobacco. This large sum seems to have had a chastening effect on the captain, as he now sent the following lengthy petition to the commissioners of the Kent county court:

Gentlemen: My dear respects to you being remembered, I would entreat you to take into your serious consideration the griefs and sorrows, which I have sustained, and do sustain, through my great oversight, caused by my infirmity, I committed in court, in using very unfitting language, which I can confess, I am very sorry for. But I hope, the measure of grief is no further good, than it makes way for joy; for a bad action salved up with a free forgiveness is as not done; and as a bone once broke is stronger after well setting, so is love after reconcilement. I know how easy it is to detest some faults in others, which we flatter in ourselves; for, in the best men, nature is partial in itself.

Therefore, it is good to sentence others' frailities with the remembrances of our own. But, God willing, it shall be my duty hereafter to keep myself from all violent passion, which causeth discontent, knowing that all things happen from Him whom Himself came.

I pray look upon the crosses and wants that God hath laid upon me at present, withal, having lost almost all my hogs, and the greater part of my cattle, being wanting and dead. Yet, methinks, I see some men ready to add to my afflictions, rather than to yield me any comfort in my sorrow.

The end, for which I write these lines unto you, is that you would be pleased to remit the fines that were laid upon me for my offences committed; and it shall be understood, a thankfulness from him that is,

Your poor friend to love you,

ROBERT VAUGHAN.

Kent, the 1st of
April, 1653.

After considering this petition, the Kent commissioners decided to remit the fine which they had imposed on Captain Vaughan for the contempt which he had shown towards them as justices. Vaughan was warned that he must be careful to live up to the promise he had made in his petition about being more respectful in his attitude towards the commissioners in the future.[14]

It was rather hard, however, for Captain Vaughan to live up to any promise about keeping his temper, or, as he said in his petition, to keep himself "from all violent passion." Apparently it was impossible for him to control his almost ungovernable temper. This was not the first time that Vaughan had become involved in difficulty because of his temper. Several years before this he had insulted the governor of Maryland and had accused him of " partiality " in the administration of justice. When called to account for these remarks, the captain admitted that his statements were false, " rashly done," and that he was sorry he had made them. Vaughan promised " amendment " and begged the governor to withdraw his action against him. The governor finally agreed to pardon Vaughan.[15]

There are two other incidents which show Captain Vaughan's irascible nature. These occurred when he was acting as commander of Kent Island. In one case he complained to the commissioners of the island against a man by the name of William Laut. It was Vaughan's contention that—

William Laut hath uttered divers reviling and upraiding speeches against his person and his authority, full of insolency, arrogancy and pride, inciting and animating thereby those people committed to his charge to sedition and rebellion, and the lessening of his power and authority, for which acts of his, Captain Robert Vaughan requires the said Laut may be censured and punished, as the foulness thereof shall by this court be found to deserve.

The Kent justices, after listening to this imposing indictment, informed Laut that if he would say that he was sorry for what he had said they would take no action against him. The Kent commissioners realized, no doubt, that Vaughan's temper might make him exaggerate the remarks attributed to Laut.[16]

The other instance showing Vaughan's testy nature was when he became angry with one Walter Smith because of some testimony which the latter had given in a case involving Joan Hood. Vaughan had Smith seized and laid him " neck and heels." While thus bound the captain told Smith that he had better " clear the oath that he took concerning Joan Hood." Smith replied that he had already cleared it as well as he could. Not satisfied with this answer, Vaughan gave Smith a kick and called for more weight to put on his neck.[17]

Other illustrations of the zealous way in which the county commissioners upheld the dignity of their position can be given. Joseph Wickes, one of the justices of the Kent county court, complained against John Bowles " for several divers scandalous words in a very unseemly manner the said Bowles hath spoken against him." The fellow commissioners of Captain Wickes at once ordered the sheriff of the county to take Bowles into his custody and keep him " secure," until he found sufficient security for his good behaviour during the next twelve months. Bowles was compelled to appear at every court session during that time.[18]

Disborough Bennett, another of the justices of the Kent county court, appeared before his fellow commissioners to complain against William Dowland " for that Dowland came to his house and abused him in a very unseemly manner." The Kent county court took the same action as they had in the Bowles case, except that Dowland was put on his good behaviour " during the court's pleasure." [19]

William Smith made some slanderous remarks about James Ringgold, one of the commissioners of the Talbot county court. When, later, on his bended knees in court Smith asked the justice's pardon, Ringgold said that he would " pardon him the twenty stripes" which the county commissioners had ordered him to receive.[20]

The cases which have so far been considered were those in which persons by word, or action, acted disrespectfully towards the justices or judges of a court. But what if some one attempted by force to hinder the execution of justice? By the terms of an early law, any person striking any officer, juror, or witness, in the presence of the court, or striking any one with a drawn weapon in the presence of the court, or striking any magistrate while in the performance of his duties, might be punished by losing a hand, or being burned in the hand, or could be put " to any other corporal shame, or correction, not extending to life," or be fined, as the court should deem fit. Furthermore, any person who struck a sheriff of a county, or any other officer, in the execution of his office, could be fined, or " censured," as the members of a court should think fit. Colonists were required to aid and assist a sheriff in the performance of his duties and " to acknowledge, obey, and respect him in the quality of sheriff, as they will answer the contrary at their peril." [21]

There is no record of a case in early Maryland where a person was accused of striking some one in the presence of the court, or striking a magistrate or justice of the peace. George Mee, however, became involved in difficulty when he struck Nicholas Keeting near the door of the room in which the provincial court was sitting. This was considered " an abuse and affront " to the court for which Mee was fined five hundred pounds of tobacco. Upon Mee's " humble submission " to the governor, this fine was remitted. In Kent county, Mrs. Thomas Bradnox acted very disrespectfully in the presence of Captain Joseph Wickes, one of the Kent county commissioners. It appears that on one occasion, Sarah Taylor, a servant, came to see Wickes because of the cruel treatment she had received at the hands of Captain and Mrs. Bradnox, her master and mistress. Mrs. Bradnox accompanied Sarah to hear what her servant had to say to the magistrate. When Sarah aired her complaints, Mrs. Bradnox, notwithstanding the presence of Justice Wickes, struck her servant with " a rope's end." Captain Wickes claimed by doing this in his presence Mrs. Bradnox " broke the peace." [22]

2

There are several instances in the early records of a person attempting to interfere with a sheriff, or constable, in the performance of their duties. Henry Morgan, sheriff of Kent county, said that when he was sent to the house of Thomas Munday to carry out an order of a commissioner of the county, Munday " did with violence present a gun against him . . . bidding him go from his house," and that afterwards Munday struck him. When the sheriff's charge against Munday was shown to be true, the provincial court ordered that Munday should be tied to a post, and with " a good pokicory wand, to have forty good stripes about the shoulders." This beating was to be inflicted by Sheriff Morgan, whom Munday had offended.[23]

John Holliday and Edward Coppage (or Coppedge) were two other men who refused to cooperate with a sheriff in the performance of his duties. Coppage failed to comply with the request of the sheriff of Kent county when he not only refused to assist him " in a boat upon the country's service," but also insulted the sheriff by using " some contumelius words or speeches." The provincial court lost no time in taking action. The county sheriff was ordered to strip Coppage to the waist and give him twenty lashes on his bare back " for his contempt in disobeying lawful authority and power and for his upraiding the sheriff in words." Holliday, it seems, when called upon by the sheriff of Anne Arundel county to aid and assist him " in what then actually concerned his office," refused to help him. For " his obstinacy and contempt " in disobeying the sheriff's request, Holliday was given the same punishment that Coppage had received. Holliday had been asked to aid and assist the sheriff in apprehending Thomas Thurston, a Quaker. When, still later, the sheriff of Anne Arundel made the same request of Holliday, again the latter refused to cooperate. Probably Holliday was himself a Quaker. Whatever the reason for his actions, the provincial court, instead of ordering another whipping for Holliday, imposed a fine of five hundred pounds of tobacco upon him.[24]

Mistress Margaret Brent, acting as his Lordship's attorney, brought an action against Edward Commins for having in contempt of his Lordship's authority " wittingly and knowingly taken certain persons under execution " out of Sheriff George Manners' custody. At his trial Commins said that he did not consider the persons in question under execution " because they had their liberties in going to and again up and down the county." Sheriff Manners testified, however, that Commins had declared that he would take the persons out of his custody " for there was no law in the province." Commins was " censured " and fined two thousand pounds of tobacco for interfering with the sheriff in the performance of his duties.[25]

Richard Willan, sheriff of St. Mary's county, had a long list of grievances to present against Thomas Innis. In " a humble petition " to the judges of the provincial court, Willan recounted how he had faithfully performed the duties of sheriff, enduring and undergoing many dangers and hazards. As a result his health had been impaired, his vigor and strength were gone, and he was both sick and lame. In this weak physical condition Sheriff Willan went on board a New England vessel then anchored near St. Mary's. It appears that Thomas Innis happened to be on board of this boat. Innis, for some undisclosed reason, bore a grudge against Willan. Accordingly, as soon as he espied the sheriff he set upon him " with scoffs, contempts, hateful and provoking complaints . . . blows, boxes and buffets." The result of this assault was that the sheriff's nose was punched in and his teeth knocked out. Innis was about to throw Willan overboard had not the sheriff called for help and " required the peace of the Lord Proprietor to be kept." Afterwards Innis boasted how he had attacked the sheriff and defied his authority.

All this was more than Willan could endure. He asked that Innis should be punished for the lack of respect he had shown, for " the greatness of his place, as being sheriff . . . and the lowness of his condition, as being weak, and not able to defend himself." Only by punishing Innis, said Sheriff Willan, would his honour as sheriff and that of the Lord Proprietor, whom he represented, be vindicated. Unfortunately a torn and illegible record prevents us from knowing what was the final determination of this case.[26]

In addition to the above cases, all of which involved some

interference with a sheriff in the performance of his duties, there are two instances where a colonist refused to cooperate with a constable. Henry Gott, a constable of Kent county, presented Thomas Dickes " for contempt against the government." It was Gott's contention that Dickes questioned his authority by refusing to assist him in the execution of his office and also that Dickes would not let him inspect his supply of corn. Dickes' defence was that he had refused to aid the constable because of " some provoking words " which this official had used to him. This plea was held insufficient and Dickes had to pay a fine of one hundred pounds of tobacco for his contempt of authority.[27]

In the other case, Richard Bennett was presented " for contempt of government " by Thomas Bassett, constable of New Town hundred. Bassett claimed that Bennett refused to assist him in arresting Robert Ford, a man thought guilty of hog killing. William Young was one of the witnesses called upon to testify. Young said that when Bassett had ordered Bennett in the name of the Lord Proprietor to help him arrest Ford for killing a hog, Bennett had said: " Thomas, why are you so hasty? Where is your warrant? " To this the constable had replied: " My oath is my warrant." Young added that he could not swear that Bennett had refused to go with the constable to apprehend Ford. In this case Bennett was finally pardoned by the governor. This was due to the fact that this was Bennett's first offence of this kind, and that he had promised " amendment for the future." [28]

So far all of the cases which have been considered are where some colonist interfered or refused to cooperate with a sheriff, or constable, in the performance of their duties. There are also several instances of where the sheriff, himself, or the constable did not properly perform the duties required of them. Sheriffs were appointed in each county by the governor. The commissioners of each county submitted a list containing three names. The governor then chose one of these men to serve as sheriff for the period of one year. After his appointment the sheriff was required to take the following oath:

You shall swear that well and truly you will serve the Lord Proprietor of this province in the office of sheriff, and do the Lord's profit in all things that belongeth to you to do by way of your office as far forth as you may; you shall truly and rightfully treat the people of your sherifwick, and do right as well to poor as to rich, in all that belongeth to your office; ye shall truly return and truly serve all writs and warrants lawfully directed to you, to your cunning; and in all things ye shall well and faithfully discharge the office of sheriff committed unto you, to your skill and power. So help you God.

Before entering upon his duties, the sheriff was required to put up a bond as a guarantee that he would " perform and execute the office of sheriff truly, faithfully, and diligently, both to the Lord Proprietor and to all the people, according to his oath . . ." [29]

There are several cases of where an attempt was made to hold a sheriff responsible for his actions. The inhabitants of Cecil county complained to the proprietary officials against Edward Inglish, sheriff of that county, because of the arrogant way in which he had acted in the performance of his duties. No action, however, appears to have been taken against Inglish. The people of the same county, on another occasion, sought to hold William Peirce, a sheriff, responsible for the actions of his deputy sheriff. It appears that the deputy sheriff refused to arrest some disturbers of the peace when ordered to do so by the county commissioners. Peirce, in his own defence, said that he was sick at the time and knew nothing of the failure of his deputy sheriff to properly perform his duties. When Sheriff Peirce added that he would appoint a new deputy, the governor and council dismissed the charge against him.[30]

Thomas Sprigge, sheriff of Calvert county, permitted Richard Newell, a man who had been convicted of killing some of Cuthbert Fenwick's hogs, to escape. For " his connivance and willful permittance " of the escape of the prisoner, Sprigge was fined two thousand pounds of tobacco. The sheriff was also told that he must produce Newell at the next meeting of the county court so that he could have his back lashed for having killed some hogs. If he failed to do so, the sheriff would have to pay an additional fine of two thousand pounds of tobacco. In another case the judgment of the court was much less severe on the sheriff. This was when Hannah Lee claimed that while German Gillett was " under execution " for a tobacco debt, and in the custody of the Sheriff William Evans, of St. Mary's county, the latter had allowed Gillett to go free before he had paid the sum due her. Hannah asked that the sheriff should " make satisfaction to the debts value." Evans pleaded that James Jolly, an innkeeper, had taken Gillett away, " out of his custody, and out of this county." Instead of holding Sheriff Evans responsible for Gillett's debt to Mrs. Lee, the provincial court simply ordered Jolly to bring Gillett back again into the county and deliver him again into the sheriff's custody.[31]

Constables in each hundred of a county were appointed by the county commissioners. These men, besides executing all warrants received by them, must take an oath that they would arrest all those who committed a breach of his Lordship's peace, raise " hue and cry " and pursue murderers, thieves, fugitive servants, and also do what they could to suppress drunkenness. Justices of the peace, sheriffs, constables or coroners, who refused to take their oaths of office after their appointment might be fined or imprisoned.[32]

There are two instances of where constables refused to take the required oath of office. For his " contempt " in refusing to serve as constable after he had been appointed to this office, John Rigby was imprisoned. In another similar case, when William Elliott was tendered the oath as constable of one of the hundreds in Kent county, he refused to take it " for conscience's sake." As Elliott does not appear to have ever been summoned before the Kent county commissioners to explain his refusal to take the oath of office, he may never have been punished or required to serve as constable. He was, no doubt, a Quaker.[33]

Several constables were charged with having failed to perform the duties required of them. It was said that John Hall, a constable, had without authority released Thomas Thurston, a Quaker, from prison. Hall was fined one thousand pounds of tobacco for allowing Thurston to go free. When Edward Lloyd, a member of the governor's council, intervening in Hall's behalf, said that he was sure that the constable had not

acted " out of any malice," the governor decided to remit half
of the fine. John Bagby, another constable, was accused of
failure to perform his duty in that when he was commanded to
search for and arrest Thurston, he had refused to do so. Bagby's
defence was that as another man had been appointed to succeed
him as constable he did not think that he still held the office of
constable at the time he received the order to arrest Thurston.
In view of this plea, the court decided to dismiss the charge
against Bagby.[34]

The question arose whether Joseph Spernon, a constable, was
guilty of a failure to properly perform his duties. It appears
that Spernon, after he had captured two runaway servants, instead of taking them to the nearest justice of the peace for com-
mittment, had taken them to his own house from which they
had later made their escape. After all the evidence was heard
the constable was finally exonerated.[35]

Contempt for judicial authority could be shown by a colonist
in other ways besides acting disrespectfully in the presence of
the court, or refusing to assist or aid a sheriff, or constable, in
the performance of their duties. For example, Thomas Gerrard
was warned by the members of the provincial court that he
must appear in court to answer a suit instituted by Captain
Thomas Cornwallis, or his refusal to do so would be considered
" a contempt of the government." Again, when Walter Pake
(or Peake) refused to appear in court to defend a suit of
defamation which Paul Simpson had entered against him, the
judges of the same court declared Pake's action " a contempt
to this court." The sheriff was ordered to arrest Pake and keep
him in safe custody until he gave security for his appearance at
the next meeting of the court.[36]

Mistress Margaret Brent, herself, was reprimanded by the
provincial court because in a suit involving a house and one
hundred acres of land, she had left the colony and " wilfully
refused to appear." As the judges considered this " a slighting
and contempt of this court and government," they decided to
hear the case notwithstanding Margaret's absence. In another
lawsuit John Robinson, carpenter and barber-chirurgeon, was
sued by Margaret Brent. When the sheriff said that he had

warned Robinson to answer the suit " upon peril of judgment "
and Robinson had refused to come to court, the judges declared
Robinson " in poenam contumacie," and allowed Mistress
Brent's claim.[37]

In other words, when any one was sued in the provincial
or county courts, he must either appear in person, or by attorney, to defend the case. Failure to do so would probably
result in a judgment being entered against him. Moreover, if
any one instituted a suit against another, he must either appear
in person, or by attorney, to prosecute the case, otherwise he
might be nonsuited. In one instance, where a man was arrested
" in an action of debt," committed to prison, and then the
plaintiff failed to prosecute the suit, the Talbot county court
commissioners not only released the defendant from jail, but
also compelled the plaintiff to pay the cost of his imprisonment.[38]

When speaking of attorneys in early Maryland we must not
think of them in the modern sense of the word. During the
proprietary period, few of the men who acted as attorneys had
studied law at the Inns of Court in London before coming
to Maryland. Matthew Ward, of Cecil county, was admitted
to Gray's Inn and Thomas Bland had studied at the Inner
Temple for six years, but most of the men who represented
litigants had not had any prior legal training in England before
coming to the colony. The same was true of most of the
judges. For this reason it is quite remarkable that the provincial
courts followed so closely the procedure and practice of the
English courts during the same period. In seventeenth-century
Maryland, appointing an attorney meant designating some one
who acted as an agent, or attorney in fact, rather than as an
attorney at law for a client. And yet—even these attorneys in
fact sometimes displayed a surprising knowledge of law. Indeed, the average layman of those early days knew more about
the law than one of today.[39]

Generally speaking, if an attorney could show " his deputation under handwriting," he was nearly always permitted to
represent his client. In the Charles county court, however, we
find the parties to the litigation asking permission of the county
justices to choose attorneys to plead their cases. One man said

that "through simplicity" he was unable to argue his own
case and therefore needed an attorney. Another man said that
as he was an Irishman who found it very difficult "to deliver
himself in our English tongue," he, too, wanted an attorney to
represent him in court. The justices granted both requests.
But this, as has been stated, was not the usual practice. Frequently the power of attorney was entered in the court records.
The following is a typical form:

Know all men by these presents that I, . . ., of the Province of
Maryland, have constituted and ordained Mr. . . . to be my lawful
attorney:

To answer to all such suits now depending in law, in the
county of St. Mary's: and all suits now depending, or shall be
be called upon in the provincial court, in the province of Maryland. In
all manner of actions of what nature soever as well to answer, as to
require judgment against all persons that shall prove to stand indebted
to me; or any other way summoned at the courts aforesaid, at my suits,
promising to allow and ratify all what shall be done, by my said attorney. Allowing this, my letter of attorney, to be in the forementioned
premises, of so much force and virtue, as any can be made, although
any clause, or clauses, should be omitted herein required, as witness
my hand the . . . day of . . . 16—

Signed: [40]

In one instance a man, who was appointed to act as an attorney of another, was given the power "in his stead to appoint
any that he shall think convenient," and in other cases, to appoint one or more attorneys to act with him during the trial of
the case. If a man undertook to act as an attorney of a non-
resident, and the suit was lost, then the attorney was liable
to pay on behalf of him whom he had represented all the costs,
debts, damages, etc., to the party recovering against his client.
There were a few persons who could not act as attorneys. A
wife could not act in this capacity for her husband. Women
could, however, represent other persons in court and could
themselves institute suits and defend actions brought against
them. In one case the plaintiff objected to the defendant's attorney on the ground that he was a servant. The plaintiff maintained that the servant must show a certificate that he had been

freed, as a servant could not act, he said, as an attorney except
for his master, mistress, or overseer, and not even in that case
" unless by the admittance of the adverse party." The court
upheld the plaintiff's contention.[41]

What did an attorney receive for his services? Anthony
Rawlins gave Walter Gess (or Guest) an axe for his trouble in
representing him as his attorney. As Gess had once been a
servant and was so ignorant that he could not write his own
name, the payment of an axe for his services was not as ridiculous as it sounds. In another case Richard Fouke sued John
Meekes for his services as an attorney. It appears that Humphrey Haggate had previously represented Meekes as his attorney. Accordingly, Fouke asked satisfaction not only for his
predecessor's trouble, but also " for himself and his horse " in
attending three sessions of the court as the attorney of Meekes.
For all the trouble he had taken, Fouke said that he had asked
only two thousand pounds of tobacco which Meekes had refused to pay. Fouke added that he left Meekes " and his conscience with his ungrateful actions to the silent determination
of the court." Unfortunately, there is no record of the court's
deciding this matter either silently or in words.[42]

As many attorneys were in the habit of demanding excessive
fees of their clients, the proprietary officials felt that something
should be done. Also there were too many persons who were
acting as attorneys. To break up these practices, a law was
passed which provided that thereafter only those could act as
attorneys in the chancery and provincial courts who had been
" admitted, nominated and sworn " by the governor, and that
no one could practice in the county courts who had not been
appointed by the county commissioners. The fees which attorneys could ask of their clients were also regulated. For his
services in representing a client in the chancery court, an attorney could ask eight hundred pounds of tobacco, for prosecuting or defending a case in the provincial court four hundred
pounds of tobacco could be demanded, while for the same
services in a county court two hundred pounds of tobacco could
be asked which was later reduced to sixty pounds. Any attorney
asking more than these fixed amounts was not only liable

to pay a fine of two thousand pounds of tobacco, but he might also be thereafter disbarred from acting as an attorney. As tobacco was worth during the proprietary period about two pence per pound, this meant that for representing a client in the chancery court the attorney received about six pounds sterling, while for the same services in a county court he received about ten shillings. When, as sometimes was the case, tobacco was worth only a penny a pound the attorney's compensation in terms of English money was only half as much. In fixing the fees which an attorney might charge, no mention was made about the time it might take to reach a decision or the amount involved in the suit.[43]

There is only one instance in the early records of where a man was disbarred from practicing as an attorney. This was when Benjamin Randall, of Kent county, petitioned the governor and council that:

Whereas your petitioner hath been permitted to practice as an attorney in the county court of Kent county in which employment he hath well and honestly behaved and demeaned himself, notwithstanding which the commissioners of the said county have denied your petitioner the liberty of pleading without any just cause shown for the same to the ruin of your petitioner, being never culpable or convicted of any crime which might debar him of the liberty of a subject.

Your petitioner doth humbly beg your Lordship's order and license to plead in the said county court until he shall be legally convicted of any crime that may disable him from such practice and your petitioner heard to make his own defence. And he shall pray, etc.

As a result of this petition, the proprietary officials informed the Kent county commissioners that they must allow Randall to practice before their court unless they could show grounds for his disbarment.[44]

Before closing this chapter it is interesting to note that while, as we have seen, justices of the county courts were quick to resent any disrespectful language used in their presence, they, too, were subject to regulations which they must observe. Certain days were designated on which the commissioners of each county must meet to give their attention to matters relating to orphans. Failure of any justice to appear on the days appointed

would make him liable to pay a fine of from one hundred to two
hundred pounds of tobacco. The proceeds from these fines were
used for building stocks, whipping posts and pillories. For
" the well governing " of the county courts, magistrates would
pass rules and regulations, which they and the litigants appearing before them, must observe. These were posted at the court
house door. The county justices could impose a fine not exceed-
ing one hundred pounds of tobacco on any one who failed to
observe these court regulations.[45]

The county commissioners appear to have lived up to the
regulations which they adopted. In Talbot county the fellow
justices of Thomas South fined him one hundred pounds of
tobacco for not coming to court, while two other men, also
members of this court, were each fined one hundred pounds of
tobacco for leaving the court before a session had ended.
Anthony Mayle, another Talbot commissioner, was fined five
hundred pounds of tobacco for using abusive words in the
presence of the court.[46]

During one session of the Charles county court, when there
were not enough commissioners present to warrant the holding
of a court, three absent justices were fined " for their default
in not appearing upon the court days." In Somerset county, the
commissioners declared that after their court was in session no
member of the court could depart, " without leave of the court,
upon penalty of ten pounds of tobacco for every hour being
absent." [47]

CHAPTER II

___

TRIAL, PUNISHMENT AND IMPRISONMENT

In order to obtain some idea of the trial of a man accused of crime during seventeenth-century Maryland, let us attend a meeting of the provincial court held at St. Mary's on the third day of January, 1666. Members of the court present, included Charles Calvert, the Governor, Philip Calvert, the Chancellor, and Thomas Truman, Baker Brooke, and William Evans. It was "nine of the clock" in the morning when these men assembled. The court was called to order by the sheriff's deputy, or crier, making the following proclamation:

All Justices of the Peace, Coroners, Stewards of Leets and Liberties, and other officers that have taken any inquisitions, indictments, or recognizances, whereby you have let any man to bail, put in your records thereof forthwith that we may proceed.

Whereupon William Calvert, his Lordship's Attorney General, presented the following indictment:

Let it be inquired for the Right Honourable, the Lord Proprietor, whether Pope Alvey, of Newtown hundred in the county of St. Mary's, cooper, the nineteenth day of December in the year one thousand, six hundred and sixty five, at Breton's Bay in Newtown hundred in the county aforesaid, a certain cow of black colour to the value of two pounds, ten shillings sterling, of the goods and chattels of Colonel William Evans then and there being found, feloniously took, stole, killed, and carried away against the peace of the said Lord Proprietor, his rule and dignity.

Thereupon the crier made proclamation again, saying—

You good men that be impanelled to inquire for the Right Honourable, the Lord Proprietor, for the body of this province, answer to your names every man at his first call upon pain and peril that shall fail therein.

Then the members of the grand jury were summoned and they replied as their names were called. Sixteen men answered the roll call, including Abraham Rowse, who acted as the foreman of the grand jury. Rowse, after placing his hand on a bible, took the following oath:

You, as foreman of this inquest for the body of this province, shall diligently inquire and a true presentment make of all such matters and things as shall be given you in charge. His Lordship's counsel, your fellows and your own you shall keep secret. You shall present no man for envy, hatred or malice, neither shall you leave any man unpresented for love, fear, favour, or affections, or hope of reward, but you shall present things truly as they come to your knowledge, according to the best of your understanding—So help you God.

The other members of the grand jury were then sworn as follows:

The same oath which your foreman hath taken on his part, you and every of you on your behalfs shall well and truly observe and keep—So help you God.

The crier now warned all persons to keep silent " whilst the charge is in giving to the grand jury upon pain of imprisonment." Thereupon the indictment which William Calvert, the Attorney General, had presented was shown to the members of the grand jury. Immediately afterwards evidence was produced which tended to support the indictment. This was in the form of depositions, or testimony, which had been sworn to before John Jarboe, a justice of the peace. Apparently, when Alvey had been apprehended by the constable for stealing catttle, he was brought before Jarboe to whom he had admitted, under oath, that he had killed an animal which he thought belonged to Colonel William Evans because of the markings on the ears.

The deposition of James Pattison, the constable who had arrested Alvey, was then introduced in evidence. According to the constable, when he went to see Alvey, the latter had denied that there was any meat in his house. Upon investigation, however, Pattison said that he had found some fresh beef " newly killed and the hide of a beast hidden under corn husks in the hen house without ears . . . and young meat salted and drying upon tobacco stalks in the loft." When asked to explain how he had acquired this meat, Alvey told Pattison that he had bought part of it and some he had killed at the request of the owner. Walter Pake and Gregory Rowse, who had assisted Pattison in searching Alvey's house, " made oath verbatim with the constable's declaration."

The grand jury next considered the sworn testimony of Henry Aspinall, a son-in-law of Walter Pake. Aspinall testified that he went to see Alvey because he thought he had killed one of his own cows. When questioned about the cows, Alvey told Aspinall that if he had killed some of his cattle he would pay him for them. Finally, the deposition of Daniel Hammond was introduced for consideration by the grand jury. Although a youth of only seventeen years, Daniel was married to one of Alvey's daughters. This lad told how after Alvey had killed a cow he had asked him to help dress it. After they had taken the meat into the house, Daniel said that his father-in-law had hidden the hide under some husks in a hen house. The boy further testified about the markings of the ears of the slaughtered animal, stating that one ear had a hole in it and that the other was cropped and slit. This testimony was important as this was the way in which Colonel Evans' cattle were marked.

After deliberating for some time, the grand jury informed the members of the provincial court that they had reached a decision and that while they were willing that the judges should change the form of the " billa vera," or true bill, which they had found, they would not permit any change " in the substance." In other words, the members of the grand jury had come to the conclusion that the facts warranted Pope Alvey being presented, or indicted, on the charge of stealing and killing a cow which belonged to Colonel William Evans.

The grand jury was then dismissed, and the crier now commanded—

All persons to keep silent for now they will proceed to the pleas of the Crown, to the arraignment of the prisoner upon life and death, and that all persons that have any evidence to give against the prisoner at the bar, draw near and give your attendance upon penalty of forfeiting your recognizances.

The clerk of the court then told the sheriff to bring Alvey " to the bar." While the cooper was facing the members of the court, the clerk continued:

Thou are here indicted by the name of Pope Alvey for that thou, on the 19th day of December, etc., etc. [The indictment was read in full].

After he had finished reading the indictment, the clerk added:

What sayest thou, art thou guilty of the felony whereof thou stands indicted, or not guilty?
    Alvey: Not guilty.
    Clerk: How wilt thou be tried?
    Alvey: By God and the Country.
    Clerk: God send thee a good deliverance.

When Alvey asked to be tried "by God and the Country" he was using an expression which meant that he wished to be tried before a jury of twelve men, chosen from the country, or neighbourhood, in which he lived. These men would decide if upon the facts offered in evidence Alvey was guilty, or not guilty, of the offence with which he was charged. Such a jury was known as a petit, or petty jury, to distinguish it from a grand jury whose task, as we have seen, was to determine whether the facts warranted finding a bill of indictment, or true bill, against the prisoner. In accordance with the request which Alvey had made for a trial by jury, which he could make as a matter of right, the clerk of the provincial court asked the sheriff for a panel, or list, of jurors who should hear the case against Alvey. The sheriff at once made his return of twelve men impanelled to serve upon the petty jury. Captain Thomas Manning acted as the foreman of this jury. Addressing these men, the clerk said:

You good men that be impanelled to inquire between the Right Honourable, the Lord Proprietor, and the prisoner at the bar answer to your names every man at the first call upon pain and peril that shall fail therein.

After the jurymen had answered the roll call, the clerk told Alvey to hold up his hand at the bar. After the prisoner had done this the clerk continued:

These good men that were last called and have appeared are those that shall pass between the Lord Proprietor and you upon your life and your death. Therefore if you will challenge them, or any of them, you may challenge them as they come to the Book to be sworn before they be sworn and you shall be heard.

When Alvey did not challenge any of the men chosen to act as members of the jury, the court crier then made this announcement:

If any man can inform his Lordship's council, the attorney general, or this inquest to be taken between the Lord Proprietor and the prisoner at the bar of any treason, murder, felony, or other misdemeanor committed or done by the prisoner at the bar, let them come forth and they shall be heard: the prisoner stands at the bar upon his deliverance.

None came forward to give any information of this character against the prisoner. The twelve jurors were then sworn, each man " severally " by this following oath:

You shall well and truly try and true deliverance make between his Lordship, the Lord Proprietor, and the prisoner at the bar, whom you shall have in charge, according to your evidence—so help you God.

After the clerk had " called them over," and the sheriff had counted them to make sure there were twelve men present, the clerk, addressing the jurymen, asked:

Are you all sworn or not?
Jurors: Yes.

After the prisoner again had been told to hold up his hand at the bar, the clerk, still directing his remarks to the members of the jury said:

Look upon the prisoner you that be sworn and hearken to his cause. You shall understand that he stands indicted by the name of Pope Alvey. [The indictment was then read.] Upon this indictment he hath been arraigned. Upon his arraignment he hath pleaded not guilty, and for his trial hath put himself upon God and the Country, which Country are you, so that your charge is to inquire whether he be guilty of this felony whereof he stands indicted, or not guilty.

If you find him guilty, you shall inquire what lands, tenements,

3

goods and chattels, he had at the time of the felony committed, or at any time sithence.

If you find him not guilty, then you shall inquire if he did fly for it (tried to escape), or not. . . . You shall inquire what goods and chattels he had at the time he did fly for it, or at any time sithence. If you find him not guilty, nor that he did fly for it, say so and no more, and hear your evidence.

The same witnesses were now called whose depositions had already been considered by the grand jury. These included, as we have seen, Pattison, the constable, Aspinall and others. Apparently, all these men reaffirmed what they had previously stated in their depositions.

After hearing all the evidence the jury withdrew. The sheriff was instructed to provide for a room for them " without having meat or drink " until they had reached a verdict. When the jurors had finally agreed, they returned to the court room, where each one answered as his name was called. When the clerk asked them if they had reached a decision, the jurymen replied in the affirmative and added that the foreman would speak for them. After prisoner Alvey had been told to hold up his hand for the third time, the clerk of the court, addressing the jurors, said:

Look upon the prisoner, you that be sworn. What say you? Is he guilty of the felony whereof he stands indicted, or not guilty?

The foreman of the jury now came forward with their verdict written on the back side of the indictment. It read:

Guilty, and the cow worth eleven pence and no more.

It will be recalled that Alvey was indicted for stealing and killing a cow belonging to Colonel William Evans worth two pounds, ten shillings. The jury, however, now declared the animal was worth only eleven pence, a very low estimate of the cow's value. In rendering this verdict, the jurors were taking into consideration the distinction between grand and petit, or petty, larceny. At common law, at this time, larceny was divided into these two classes according as the value of the property stolen was over or under twelve pence.

This attempt of the jurymen to protect Alvey by returning a verdict making him guilty of petit and not grand larceny did not succeed. They were ordered by the provincial court judges to retire in order to reconsider their verdict and " to have a special care in what they did." Not long afterwards the jurors returned to the court room and again they were asked if they had agreed upon their verdict to which they replied in the affirmative. This time, however, the word, " Guilty," alone appeared written on the back of the indictment. Nothing was stated as to the value of the cow.

The members of the jury were now asked if the prisoner Alvey owned any lands, goods, or chattels at the time he committed the felony or at any time since.

> Foreman: None to our knowledge.
> Clerk: Did he (Alvey) fly (i. e. try to escape)?
> Foreman: No.
> Clerk: Hearken to your verdict as the court recordeth it—
> You say Pope Alvey is guilty of the felony whereof he stands indicted?
> Foreman: Yes.
> Clerk: And so you say all?
> Foreman: Yes.

The petty jury was now dismissed and the clerk instructed the sheriff " to set the prisoner to the bar." The presiding judge of the court now addressed Alvey:

> You do remember that you have been indicted for felony by you done and committed. Upon your indictment you have been arraigned and have pleaded, ' Not Guilty,' and for your trial you have put yourself upon God and the Country, which Country hath found you guilty. What can you now say for yourself, why, according to the law, you should not have judgment to suffer death.

Alvey at once replied that he claimed " the benefit of clergy." In making this request, the prisoner sought to take advantage of a privilege, originally claimed only by a member of the medieval church when accused of crime before a temporal court. It meant that the ecclesiastic claimed exemption from trial by a temporal court and asked to be tried by an ecclestiastical court, which could not inflict the death penalty. In England, at the time of the Norman Conquest, it was allowed to those in clerical orders accused of any felony, except high treason. Later, the test whether one was exempted became based on the ability to read. This again made the clergy almost the only people to benefit, as they alone were sufficiently educated to be able to read. With the development of printing, however, many laymen were able to read and, accordingly, they, too, claimed the same exemption accorded the clergy. Laymen who claimed benefit of clergy were, however, restricted in some respects. The claimant must not only plead the exemption in open court, but by the provisions of a statute of Henry VII, chapter xiii, if benefit of clergy was allowed to a man who had been convicted and so saved from the gallows, the accused must, as evidence of this, be burnt on his hand with a red-hot iron. Moreover, no layman could claim the exemption a second time after it had once been allowed.

The Alvey case brings up in an interesting way the privilege of benefit of clergy. It appears from the court records that on a previous occasion Alvey had been convicted of " feloniously " killing a maidservant by beating her to death. Alvey had then claimed benefit of clergy. When he showed that he could read the plea was allowed. At the same time the court ordered that he should be burnt " in the brawn of his right hand with a red-hot iron." Alvey, as we have seen, had now been convicted of another felony, grand larceny. When he tried to claim benefit of clergy a second time, the provincial court, in accordance with the usual practice, informed him that he could not claim this exemption again. Just as the attempt of the petty jury to reduce Alvey's offence from grand to petit larceny had failed, so had the prisoner's own attempt to claim benefit of clergy a second time.

His last hope of escaping the gallows gone, Alvey now " threw himself upon the mercy of the court." The presiding justice, however, at once proceeded to give judgment in the following words:

You (Alvey) shall be carried to the place from whence you came, from thence to the place of execution and there to be hanged by the neck till you are dead.

That the noose was never placed around Alvey's neck appears to have been due to the presence in court of some influential friends. As soon as they had heard the judgment, these persons came forward and, "upon their knees," humbly begged the life of Alvey and asked the governor to reverse the judgment. After listening to their plea, Governor Calvert said that he would grant the prisoner "respite of execution and liberty to depart to his own house," but added his Lordship—

The judgment and sentence passed against Alvey must remain in full force and effect on the records during his pleasure and according as the said Alvey shall henceforward behave himself in his remaining course of life.

Almost ten years later Pope Alvey sent a petition to the lower house, humbly imploring the delegates to ask the governor to pardon him. When the delegates acted favorably on this petition, the governor decided to pardon Alvey.[1]

The first law regarding the punishment of criminals was passed a few years after the founding of the colony. This gave the governor, or any member of his council, power to apprehend felons and to keep the peace. These officials could impose or inflict any punishment they saw fit except that in cases "extending to taking away of life or member," then in such cases the offender must be first indicted and afterwards tried by a jury of twelve men. Several years later another statute was passed, the terms of which declared that "all crimes and offences" shall be judged and determined according to the law of the province, or, in the absence of "certain law," then according to the best discretion of the court. No one, however, could be sentenced to death, to lose "a member," or freehold, nor be outlawed, or exiled, or even fined more than one thousand pounds of tobacco, "without law certain of the province." By the provisions of the same act no corporal punishment could be inflicted on a gentleman. During the same year it was enacted that either party in either a civil or criminal suit could ask for a trial by a jury, provided they put up security to pay the cost of the jury. In criminal cases, however, where the loss of "life or member" was involved, the offender could demand a jury trial without having to furnish any security for the jury's expenses.[2]

Service on a jury was compulsory. Any one summoned to serve on a jury in either a civil, or criminal case, must do so unless " by any other office exempted." Failure to serve made the offender liable to pay a fine of five hundred pounds of tobacco. Likewise if any one refused to testify after being summoned as a witness he must pay the same fine. The summons to the witness usually stated that he was required to appear at the next meeting of the court to testify " in a cause depending betwixt, etc., etc., upon peril of forfeiture of five hundred pounds of tobacco to the Lord Proprietor in case he appear not according to summons." There are a number of instances in the early records of where either a juror was fined for not serving or a witness was similarly punished for not coming to testify. Both witnesses and jurors received thirty pounds of tobacco a day for their services. This was about five shillings a day. In Charles county juries would sometimes refuse to render a verdict until each juryman had been paid for his services.[3]

What of the punishment of a witness who was guilty of perjury? In one case John Goneere admitted having perjured himself while testifying under oath. He asked the court to take into consideration the fact that this was the first time he had committed such an offence. The court, however, was not in favour of any leniency, but ordered the sheriff to take Goneere outside and nail him by both ears to the pillory with three nails in each ear. The nails were then " slit out " and Goneere whipped with twenty " good lashes." In another suit " touching a bull," Blanche Howell was said to have committed " a wilful and voluntary perjury therein." In the name of the Lord Proprietor, his Lordship's attorney asked that Blanche should be brought " to condign punishment." Although the woman denied the charge, a jury found her guilty. Thereupon the judges of the provincial court ordered that she should stand nailed " in the pillory and loose both her ears, and this to be executed before any other business in court be proceeded upon." [4]

Charles James, sheriff of Cecil county, was impeached by the lower house of the assembly for perjury. The articles of impeachment stated that when testifying under oath before them, Sheriff James had " wilfully, corruptly and falsely sworn against Abraham Wild," one of the commissioners of Cecil county, in matters relating to his administration of justice in that county. The delegates asked that the governor would no longer allow James to hold any public office in Maryland and that he might be punished " as the nature of so great crimes require." As the charges against the sheriff were proved, he was thereafter forbidden to hold any office in Maryland. No other punishment seems to have been imposed.[5]

During the trial of one case a man by the name of Nathaniel Burrowes, interrupting the proceedings, told the judges of the provincial court that he was of the opinion that some of the witnesses had perjured themselves and that he could prove it. Such a charge was too serious to go unnoticed. Accordingly, the judges ordered Burrowes committed to prison until he gave bond with good security for his appearance at the next meeting of the court. When Burrowes came up for his hearing, he failed to support his charge of perjury and told the judges that he must have been drunk at the time he made such remarks. The court, however, did nothing to punish Burrowes for his slanderous statement; the judges merely fined him one hundred pounds of tobacco for having been drunk in court.[6]

There are several instances of where the veracity of a witness was questioned by the litigants. In one case a man attacked his own witness, whom he had subpoenaed, saying that he was not fit to testify, that he was a rogue, and that he had been found in bed with " that whore, his mistress," in whose behalf he would swear to anything. In another case, a man objected to a woman's testifying, as he claimed she was " a thief and a liar." The court ruled, however, that the woman was qualified to testify, as she had never been convicted of the charge made against her.[7]

There is one case of where a woman was convicted of forgery in the early records. This was when Elizabeth Greene, the wife of William Greene, forced a servant boy in her employ to forge a receipt, or discharge, for a debt. At the trial the youth described how his mistress had " dictated " to him what he should write, and what names he should put down as witnesses, although he and his mistress were the only persons present. When the jury found Elizabeth guilty of forgery, the court ordered her " set on the pillory and loose one of her ears." After this punishment had been inflicted, Elizabeth was to be imprisoned for twelve months and to pay double costs and damages " to the party grieved," if the latter should make this demand.[8]

In order to break up the practice which some colonists had of altering the mark, or quality, of tobacco after it had been sold or attached by a sheriff, a law was passed by the terms of which a person guilty of this offence was not only required to restore fourfold the value of the tobacco " to the party grieved," but also compelled to stand in the pillory for two hours during court time with his offence written on a piece of paper placed upon his back.[9]

It was many years before a prison was built in Maryland. In 1643, when the colony had been settled for almost a decade, Sheriff Edward Packer, of St. Mary's county, said that there was no prison in the county " but his own hands." At about this time, although the proceeds from fines imposed on drunkards could be used " toward the building of a prison," it is doubtful if any such funds were so used. Ten years later, in 1653, there is a reference to a " common gaol " at St. Mary's. It is still questionable whether at this time any building existed which was used exclusively as a prison, since, in the following year, the house of Henry Fox was, by order of the governor, made the prison of St. Mary's county. Fox himself was made " the keeper thereof." In this house was confined Cuthbert Fenwick, a prominent colonist, until he had paid a debt which he owed. Curiously enough, Fenwick was told that if he wished he could leave " the prison " at any time, provided he did not go more than a half a mile away from it.[10]

It was not until the spring of 1662 that the assembly seriously discussed the necessity of having a prison " for the securing of malefactors and other exorbitant persons." It is possible that at this time one of the small buildings on the plantation of Hannah Lee, the widow of Hugh Lee, was used for a prison. The other buildings on her plantation were used for the meetings of the assembly and provincial court. During the next year, however, definite action was taken to erect a prison because " divers inconveniences have happened within this province through want of places for securing offenders." An act was passed which authorized the appropriation of two thousand pounds of tobacco for building, at St. Mary's, a log house, twenty feet square, which was to be used as a prison. Provision was also made for the erection, near the prison, of a pillory, stocks and ducking stool. There was also a gallows which probably stood near this prison.[11]

Only three years after this, or in 1666, the assembly passed another act providing for the construction of still another prison. Apparently the log house prison had not proved adequate. The reason for the erection of the new prison, the delegates said, was that " due respect to the government is not given for want of prisons to restrain offenders, and persons indebted taken care of for discharging their debts." The new prison was built near a spring on the east side of St. Mary's. Provided he constructed a building which was " well and sufficiently built," and furnished it with iron and so forth for " restraining and safe-keeping offenders," the builder, who was Raymond Staplefort, was to receive ten thousand pounds of tobacco. Staplefort was also made keeper of this prison " for and during the term of his natural life." To this prison were sent all criminals who had committed some offense, the punishment for which might be death, or the taking of " member," and also all persons who had failed to pay their debts, unless the county where the debt was incurred already had a place for confining debtors.[12]

This prison had only been standing for eight years when the assembly again began discussing the need of having another prison. By this time, 1674, apparently the building which had been used as a state or court house was also in a state of decay. The delegates thought that the new state house and prison should be constructed of brick. The previous ones had been built of wood. Captain John Quigley was employed to build the new structures at a cost of 330,000 pounds of tobacco. Even if we assume that tobacco was worth as little as a penny a pound at this time, the cost of the new buildings was £1,375, a very tidy sum. The state house, which boasted glass windows, was to be used for the meetings of the assembly and also for the sessions of the provincial court. After the structure was built, provision was made for separating the room in which the council and provincial court held their meetings from the room in which the lower house of the assembly met. At first, it was planned only to have heat in the governor's room, but later a chimney was added to the building with two fireplaces for heating both the first and second stories. " Forms and tables " were placed in the room used by the judges of the provincial court for the convenience of committees and juries. As the room occupied by the members of the provincial court in the old building, used as a court house, had rails, benches and carpets, it is probable that the room in which the court met in the new state house was similarly furnished.[13]

John Broomfield, the crier of the provincial court, was charged with seeing that the new state house was regularly swept and kept clean. When he failed to perform his duties properly, Broomfield was dismissed and John Cullin appointed in his place.[14]

The prison which Captain Quigley was to build was to be constructed of either stone or brick laid in lime and sand. It was to be twenty-four feet long, fifteen feet wide and nine feet high, and partitioned into two parts, with a loft above. The floor of the loft was to be planked, while the floor of the first floor was to be laid with stone or brick. There were to be two windows on the lower floor and one in the loft. The framework of these windows was to be of wood. Each window was to be twenty inches wide and thirty inches high with three upright iron bars and two " a thwart the upright." Each bar was to be an inch and a quarter square. The roof of the prison was covered with tile. In this prison were confined criminals and debtors.[15]

Before 1674 few of the counties had prisons. Even Kent

Island twenty years after its settlement still had no place to confine criminals. The sheriff of the island was compelled to have a guard " stand over " a prisoner. In 1663 the assembly, while giving no directions for the construction of prisons in the different counties, did, in that year, order the county commissioners to set up stocks and pillory near the county court house and a ducking stool in the most convenient place in the county. This was in order that no offender should escape without " due correction." Counties failing to set up these instruments of punishment were subject to a fine of one thousand pounds of tobacco. Baltimore and Talbot counties were not at this time required to comply with this act as both counties were still sparsely settled. All county justices, however, must provide irons for burning " malefactors." One iron was to be marked with the letter " H," probably for hog stealers, and another with the letter "R," possibly for runaway servants. The judges of the provincial court at this time also instructed the sheriffs of every county to provide two irons, one with the letter " M " and the other with the letter " T." Probably these irons were to be used to brand murderers and thieves.[16]

In compliance with the instructions of the assembly, the commissioners of Charles county ordered the sheriff to procure a pair of stocks, pillory and whipping post and place them near the county court house. The sheriff was also to place a ducking stool on Pope's creek. Soon after this William Robisson made an agreement with the county justices by the terms of which he was to receive fifteen hundred pounds of tobacco for the erection of all of these instruments of punishment. It is interesting to note that members of a jury of a court leet of St. Clements' Manor presented Thomas Gerrard, Lord of this Manor, for not having provided stocks, pillory and a ducking stool. The jurors ordered that " these instruments of justice " must be provided before the next court and paid for by " a general contribution throughout the manor." [17]

During the same year, that is, 1674, that the assembly made arrangements for the building of a state house and prison at St. Mary's village, the delegates also took into their consideration the need of having court houses and prisons in all of the counties. A law was passed making it obligatory for all county commissioners to erect a court house and " a good strong house for a prison." The court house must be constructed within two years, the prison within a year. Both were to be built at the expense of the county. Members of the council wished that all sheriffs should be compelled to take their prisoners to the county prisons in order that they might be confined there " in sure and safe custody," and that criminals might receive " condign punishment " and creditors have their debts paid or the bodies of their debtors " duely restrained." Both houses of the assembly thought that gaolers, or keepers, should be placed in charge of the prisons.[18]

A few years after this, or in 1678, the assembly passed another law relating to the county courts. The new act required the county commissioners, for the better administration of justice in their courts, to purchase certain law books, including " the Statute Books of England to these times, named Keeble's Abridgements of the Statutes, and Dalton's Justice of the Peace." These books were to be kept on hand in each county court for reference purposes.[19]

By 1676 prisons had been built in every county for the incarceration of debtors and criminals. The reason that the county commissioners so quickly complied with the law requiring them to build court houses and prisons was that a failure to do so made them liable to pay a fine of ten thousand pounds of tobacco. The first court house and prison to be erected in Baltimore county after the passage of this act was near the head of the Gunpowder river. About ten years later, however, another court house and prison was built on the south side of Winter's Run. This was found to be a more convenient place " for the whole county." [20]

Any sheriff in charge of a prison who allowed any prisoners to escape, either " voluntarily or negligently," was fined twenty thousand pounds of tobacco, half of which was to be used in strengthening the county prisons and paying for the cost of keeping the prisoners. If the escaping prisoner was a debtor, or one against whom a judgment had been entered, the sheriff was liable to pay the amount of the debt, or judgment, to the person to whom it was due. Any prisoner accused of a felony, or other crime " deserving death," who escaped was, upon his recapture, to be placed in irons and then put to death.[21]

Because so many servants ran away from their masters, or criminals escaped into neighbouring colonies, the provincial authorities decided to erect a prison on the plantation of Augustine Herrman, in what is now Cecil county. The prison was to be constructed of logs and to be twenty feet square. Prisoners were required to work in order to defray the cost of their imprisonment and if they would not do so they could be whipped by the keeper of the prison. In this prison were confined fugitives from justice and runaway servants from other colonies, as well as those attempting to escape from Maryland.[22]

Conditions in the prisons must have been very primitive. There was no fireplace and probably straw was all the prisoners had to sleep on. Some of the more serious offenders were kept in irons. One woman in a petition to the governor said that she had been so long in prison that all her clothes were worn out and as a result she was " in great distress." She asked that either she might be brought to " a speedy trial," or else some provision should be made to clothe her. Her husband was compelled by the provincial court to furnish her clothing. In another petition to the governor, a man, suspected of murder, informed his Lordship that he had " long lain in prison in a miserable condition." [23]

Sheriffs in proprietary Maryland were allowed to ask fees for performing different services. He could, for example, ask twenty pounds of tobacco for each day he kept a man in prison, and fifty pounds of tobacco for inflicting corporal punishment. Although the prisoner himself was supposed to pay the sheriff these fees, the counties had been in the habit of paying them to the sheriffs until, in 1662, they objected to doing this any longer. The county commissioners said that criminals should be compelled to pay the cost of their own imprisonment, otherwise it was " an encouragement of offenders." One cannot help but be amazed at the reason given. A man must indeed have been in wretched circumstances if life in one of the jails seemed attractive. Notwithstanding the poor reason given by the county justices, an act was passed by the terms of which no sheriff, or jailer, could thereafter charge their county, " or the public," with any fee involved in keeping a man prisoner, but when the man was finally released from prison then he must pay for the cost of his imprisonment, if he had " sufficient estate," or if he did not, then by " servitude or otherwise." Where a man was executed, or banished from the province, leaving no property, in such cases the sheriff, or jailer, could recover the imprisonment fees either from the county where the criminal had resided, or from " the whole country," as the provincial court should determine.[24]

A difficult question bothered the assembly and that was who should pay for the cost of a servant's imprisonment " the public, or the county, or the master." The servant, of course, had no property of his own with which to pay for the cost of his own imprisonment. The servant, however, could be compelled to serve the county commissioners in some trade or labour after he was released from prison and thus pay for the expenses incidental to his incarceration. But was this fair to the servant's master who was entitled to have his services for the unexpired period of the indenture? This question was finally decided by making the county liable in the first instance for the cost of a servant's imprisonment. After the servant was released from prison he must then return to his master and serve him the rest of the term of his indenture. When these duties were over, then the servant must " by servitude or otherwise " satisfy the county commissioners for whatever expense they had incurred by reason of his imprisonment.[25]

As the punishment of every criminal is mentioned, and often commented upon, in all of the chapters of this book, it would not be advisable to discuss at this point the character, or justification, of the punishment imposed in each particular case. It will suffice to describe some of the more usual forms of punishment and, later, to call attention to some cases where the punishment was of an unusual character. To have a man or woman whipped was one of the most usual forms of punishment. In some instances as many as thirty-nine stripes would be given the offender. Such a case occurred in Cecil county where a

servant because of some " misdemeanor " was ordered to have
this many lashes laid on his back. Edward Inglish, sheriff of
this county, was instructed to administer the punishment. Inglish placed the servant's hands in the whipping post and then
secured some peach tree switches, but finding them too small
went out to cut more " that might give content." While he was
doing this somebody let the prisoner escape. When the servant
ran away, all those standing expectantly around the whipping
post waiting to see a good lashing set up a shout. One of the
Cecil county commissioners seeing what had occurred cried:
" Shame." But the prisoner whose name is not known was indeed lucky. The thirty-nine lashes would have meant a raw and
bleeding back. Apparently no corporal punishment could be
inflicted upon " a gentleman." [26]

Other usual forms of punishment were to set a man in stocks
and, of course, to put him in prison. Some were nailed by their
ears to the pillory. A few were hanged. In one case a negro
was ordered " drawn and hanged." That is to say, he was
dragged on the ground from the prison to the gallows at the
tail of a horse or cart.[27]

Occasionally, a rather unusual punishment was imposed.
Thus one man was not only compelled to acknowledge his
offense in open court, but he was also required to repair and
keep in condition a certain bridge " that it be sufficient to pass
over securely at all tides, and to keep it in the like manner for
one whole year." Another man was forced to stand in open
court " with a paper on his breast declaring his offence." In
Talbot county a man and his wife were ordered by the county
justices to kneel before them and ask their forgiveness. The
commissioners said that they took this way of punishing them
because of their inability " to make any other satisfaction." [28]

While to slander any one is not a crime, John Little, who
was guilty of doing this, was subjected to two rather unusual
punishments. In one case, it appears, he was charged with
having slandered Elizabeth Potts by saying that he saw her
having intercourse with an Indian youth. By way of punishment Little not only had to pay a fine of five hundred pounds
of tobacco, but he was also ordered to stand " for the space of

one hour at the door of the court with a paper in his hat written in capital letters signifying that he hath scandalized the said Mrs. Potts." In another case Little was charged with slandering a man's mother and father. As punishment for this Little was ordered to stand by the whipping post " stripped naked from his waist upwards for the space of one hour with a whip over his head." Later at the request of Little's wife and " divers neighbours," and on account of the prisoner's age and the " unseasonableness " of the weather, the court shortened the time that Little must stand at the whipping post.[29]

One man who was convicted and condemned to death for stealing was pardoned on condition that he thereafter serve as a common hangman. The pardon, however, was to be kept secret until " such time as the prisoner shall be carried to the place for execution and the rope put about his neck, at which time the said pardon was to be produced and not before." [30]

CHAPTER III

HOUSES, CLOTHING AND THEFT

There were comparatively few cases of theft in Maryland during the proprietary period. Several reasons can be given for this, among them that there was comparatively little of any value to be stolen. This is evident when we consider the clothing of the colonists, the small houses in which they lived, and their few personal belongings.

What did the early colonists wear? Among the personal effects of Leonard Calvert, Maryland's first governor, were " a pair of new Holland socks, . . . one pair of shoes, two combs and a hat brush." Calvert also owned a close fitting upper garment called " a jack." This is hardly an elaborate wardrobe. Let us hope that this is not a complete inventory and that he possessed more in the way of clothing to protect him from wintry winds in southern Maryland. The inventory of the personal effects of Captain Robert Wintour, a member of the governor's council, showed that he had worn during his life the following articles of clothing—

A Portugal cap, a sea cap, an old frieze suit, a serge suit and coat, a cloth suit and coat, a hareskin suit, a buff coat, a short coat lined with plush, a pair of frieze breeches and coat, an old gray coat, a Holland jacket and a canvas doublet.

Captain Wintour also owned three pairs of leather breeches, two matchcoats, five pairs of boots, eight pairs of shoes, an old silver belt, two knots of girtweb and three thimbles. Possibly the captain was a bachelor, or, perhaps, Mrs. Wintour may not have come with him to the New World.[1]

When Thomas Adams, a burgess, died he was found to have possessed an old beaver hat, a periwig, an old satin suit, an old satin doublet with silver buttons, a pair of cuffs for the hands, a pair of old stockings, a pair of spur leathers, and—all males take shame—three looking glasses. Zachary Mottershead, another member of the house of burgesses, had among his per-

4

sonal effects two caps, a suit of clothes, two coats, one doublet, a waistcoat, three shirts, four pairs of cuffs, two pairs of stockings, three pairs of boothose, a pair of boots and spurs, sixteen gold buttons and one handkerchief. The handkerchief was greatly prized in early Maryland as is shown by specific mention being made of them in inventories. No colonist had more than one or two of them. Few colonists owned any jewelry, such as rings, precious stones, or articles of silver or gold.[2]

One planter, by the name of Henry Hide, boasted quite an elaborate wardrobe. Among the garments which he wore " with stockings suitable to them all " were:

> One new fashion stuff suit, lined through with Indian silk, and a
>   knot of ribbons of two colours on the shoulder.
> One serge coloured suit, lined with ordinary stuff . . .
> One light coloured suit, lined with ordinary stuff . . .
> One white-flowered satin waistcoat with silver lace.
> One cap with fur, laced over the head.
> One white caster hat and two belts for swords.

From this list we may imagine that Hide was somewhat of a dandy.[3]

From the findings of a coroner's jury we learn much of the way a colonist dressed. The jury, after coming to the conclusion that the man, whose name is not known, had been shot to death, then gave a description of the way in which he was dressed. It appears that the dead man, who was red-bearded, wore a serge doublet, which had " open sleeves " and was faced with half silk damask. His " britches " were also made of serge and he wore a pair of " oldmill stockings." His underclothing consisted of a shirt somewhat worn out and a pair of " canvas drawers." From this description we obtain a fairly good idea of the way a man dressed during the seventeenth century. The doublet was a close-fitting garment for men, with or without sleeves, covering the body from the neck to the waist, or sometimes a little below. The dead man's serge doublet had full sleeves which were open, or slashed, showing through the openings the lining of the sleeves which in this case were made of silk damask. This material was generally woven with a pattern of flowers and the like. The man's breeches, which

came to the knees, were also made of serge. The drawers were
made of canvas; men must have been hardier in those days than
now.[4]

Many came to Maryland during the proprietary period as
indented, or indentured, servants. Any colonist bringing over
a servant to the province was advised to procure the following
articles of clothing for him:

> Two Monmouth caps or hats.
> One suit of canvas.
> One suit of frieze.
> One suit of cloth.
> One coarse cloth, or frieze coat.
> One waistcoat.
> Three shirts.
> Three pairs of stockings.
> Six pairs of shoes.
> Inkle (broad tape) for garters.
> Three falling bands (ruff or band for the neck).
> One dozen of points (laces for fastening the clothing).

This whole outfit cost a little over four pounds sterling. At
the end of his term of servitude a servant was allowed, by " the
custom of the country," a suit of kersey or broad cloth, a shift,
or shirt, of white linen, one new pair of stockings and shoes.[5]

From these descriptions of the planter's and servant's clothing it is evident that in most cases it was only enough to cover
their needs. Perhaps he had an extra suit beside the one which
he wore, several shirts, a few pairs of stockings and possibly
an extra pair of boots or shoes. Clothing was sometimes left
by will. Richard Lee bequeathed his wife's satin petticoat to
Mrs. Lewger, wife of the colony's first secretary. It would be
interesting to know what Mr. Lewger said when he heard of
this bequest.[6]

In what type of house did the seventeenth-century Marylander
live? A suitable house for " a gentleman " only had to be
fifteen feet square. Such were the dimensions of the dwelling
in which Paul Simpson lived. His house, which had " a welch
chimney," was " floored and lofted with deal boards, and lined
with riven boards on the inside." Inside the house there was

" a joined bedstead, one small joined table, six joined stools, and three wainscot chairs." The bed was hung with curtains and a valance which was that part of the drapery around a bedstead from the bed to the floor.[7]

The average size of a dwelling in Maryland during the seventeenth century was, however, larger than Simpson's house. Most of them were about twenty-five feet long and twenty feet wide and had only one story with a loft. Practically all of them were built of wood, either boards or logs, the chimneys alone being constructed of brick. A few houses had brick foundations. Even as prominent a colonist as Captain Thomas Cornwallis lived in a house of " sawn timber," only " a story and a half high," with the chimneys and cellar the only parts made of brick. The captain's house also had locks on the door and glass in the windows. Although those planning to go to Maryland to settle were advised to take with them locks, hinges and bolts for the doors and glass and lead for the windows, very few of the houses in early Maryland had either locks or glass windows. Wooden shutters were used instead of glass.[8]

Most of the houses had one or two chimneys which were built at the ends of the house outside the wooden framework. Bricks were made in the colony. In the early records there is mention of brickmaking, of brickmasons and brickmoulds. Indeed colonists were instructed to take to Maryland servants who were brickmakers and bricklayers. As early as 1634, the year of settlement, it was realized that there was good loam in Maryland which would make as good brick " as any in England." [9]

Two other houses in early Maryland will be described. Richard Wright, a merchant, planned to have a house twenty-five feet square. Unlike most dwellings it was to have two stories. The room on the first floor was to be laid with sawn boards, " the same to be lathed and plaistered," the room on the second floor was to be laid with split boards, " jointed close," with a window out of each side. The stairs were to go up " at the end, the same to be sealed." There was also to be two closets and a fireplace in the room on the first floor and a porch " with windows to jet out of the room." The foundations of this house were of wood.[10]

James Neale, a prominent colonist, lived in a house twenty-five feet long which had two chimneys. These, no doubt, were built outside the wooden framework at each end of the dwelling. This house was to be built by John Tompkinson at a cost of one thousand pounds of tobacco or about eight pounds sterling in English currency. Neale also gave Tompkinson one pair of shoes and stockings for setting up two bedsteads. It was Neale's contention that the house which Tompkinson constructed was not worth a third of what he had paid for it, and that the carpenter wasted a lot of nails in building the house.[11]

From the account which has been given of the houses of the planters in early Maryland it is evident that most of them were rather simple affairs. Indeed, in the village of St. Mary's almost fifty years after its settlement, it was said that the dwellings there, as well as in the rest of the colony, were " very mean and little and generally after the manner of the meanest farm houses in England." [12]

As we would expect to be the case, the houses of the seventeenth-century Marylander were plainly furnished. In them would be found bedsteads, tables, stools and perhaps a chest of drawers. Pillows were in use, as well as blankets and sheets. Some even boasted towels. There were no toilet facilities, however, and few colonists owned a chamberpot. Candlesticks were used to light the house. Articles of tableware included bowls, saucers, platters, plates, saltsellers, flagons and tankards. Although spoons were in use, knives and forks were not. Most of the tableware was made of pewter; woodenware and earthenware was sometimes used. A few colonists owned table-cloths and napkins. Articles in the kitchen, including kettles, ladles, gridirons and pots were made of either iron, copper or brass. Persons going to Maryland to settle during the seventeenth century were advised to take with them the following household articles:

One iron pot, one iron kettle, one large frying pan, one gridiron, two skillets, one spit, platters, dishes, and spoons of wood.

These articles were supposed to supply the needs of six people. They cost about one pound, ten shillings, in England.[13]

When we consider the personal effects of the early Maryland colonists, it is quite evident that there was comparatively little of real value to steal. This does not mean that a thief might not be in need of a suit of clothes, or some household article, or of food. It would have been easy, no doubt, to enter most of the small unprotected houses during the absence of the owner. But would the thief have been justified in taking such a risk? It would seem not for several reasons.

In the first place, what chance would the thief have of escaping to another colony? The twentieth-century thief, possessed of an automobile, can soon escape from the state in which the crime was committed. But in seventeenth-century Maryland, even if the thief secured a horse, where would he go? As there were few paths or roads in early Maryland he could not get very far on horseback. Indeed the colony had been settled for over thirty years before any attempt was made to provide for roads and paths. It was not until the spring of 1666 that the members of the assembly came to the conclusion that it would be " convenient and very much for the benefit of the inhabitants of this province that ways and paths be marked and the heads of the rivers, creeks and branches be made passable." Accordingly, an act was passed directing the commissioners of each county to meet together to discuss what highways and paths should be laid out and marked in each county. The county justices were also instructed to make passable for " horse and foot " the heads of rivers and creeks which the roads, or paths, crossed. In order to carry on this work, the commissioners could appoint overseers and laborers.[14]

How far were these instructions of the assembly carried out? In Kent county, on the Eastern Shore, there were a few paths and roads. A road we learn was about ten feet wide. There were overseers of the highways in each of the hundreds of this county whose duty it was to open new roads as well as to keep the old ones in repair. These overseers were also authorized to build bridges which must be " passable for horse and foot." In building and repairing roads and bridges, the overseer of highways was empowered to employ all " taxables." Persons so employed must provide their own food and tools. Any who, after being summoned to work on the roads or bridges, refused to do so, would be taken before the nearest justice of the peace, who was authorized " to deal with them as he shall think fit." One of the few bridges constructed in Kent county was built so as to be half a foot above " a common high water." It had a railing and was " well-staked and made fast in its place." [15]

In Talbot and Somerset counties, also on the Eastern Shore, " overseers, surveyors and repairers of the highways " were appointed. On the Eastern Shore there was a dirt path or road which ran north and south crossing the heads of rivers and creeks which empty into the Chesapeake. There were also paths along the north and south banks of the Chester river. [16]

Although there must have been several paths or dirt roads in southern Maryland, few references to them can be found in the early records. Because of floods a path, or road, which crossed the Wicomico river in this part of Maryland became so dangerous to travel on that steps were taken to repair this highway. A bridge was also built across one of the tributaries of this river. It was not until near the end of the proprietary period that provision was made for keeping in condition the roads and paths in the vicinity of St. Mary's, the colonial capital and most important settlement. At that time, 1685, the constable of this village was authorized to make an inspection of these roads and to take with him labourers " for the repairing, clearing, amending and making the aforesaid roads passable both for horses, cart and foot." [17]

Two visitors to Maryland near the end of the seventeenth century said that most of the highways in the colony were nothing more than foot, or bridle paths, which were marked by blazing trees, that is, a piece of bark was cut off the tree with an axe at about the height of a man's head. Because of the number of paths so marked, which often ran into or across one another, these blazes, they said, were of little use and often misleading. There were, however, a few roads which were broader and passable for carts. [18]

From this discussion of the paths and roads in early Maryland, it is obvious that their scarcity and condition would impede a thief's escape. What if the criminal wanted to cross one of the

many rivers which empty into the Chesapeake? There was no
ferry crossing the Potomac by which the thief could escape into
Virginia. In Maryland it was thirty years after the founding of
the colony before an attempt was made to establish ferries. By
the terms of a law passed in 1664 small ferry boats fourteen feet
" by the keel " were placed in operation on the St. Mary's and
Wicomico rivers in southern Maryland for the convenience of
those travelling afoot. A somewhat larger boat eighteen feet
long was in use on the Patuxent river for those travelling either
on foot or on horseback. The ferrymen on any of these rivers
must hold themselves in readiness " between sun rising and sun
setting " to accommodate any who applied to him for passage,
provided weather conditions permitted a safe crossing. The
man who ran the ferry over the Wicomico river received two
thousand pounds of tobacco a year for his services. He was
required not only to cross the river " as often every day as passengers shall require," but even if there were no passengers he
was in that case obliged to cross the river at least twice during
the day. Perhaps the exercise of rowing, or sailing the boat,
was to keep him in trim.[19]

The man who ran the ferry over the Patuxent river from
the southern bank to Point Patience was entitled to ask one
shilling for each passenger and horse he transported. The
keeper of this ferry was warned not to carry over the river any
one who was not a freeman. This was in order to prevent the
escape of runaway servants. As servants, as we shall later see.
committed most of the thefts, or other crimes, this regulation
helped to prevent their escape. Near the end of the proprietary
period a ferry was maintained at the Severn river. Here the
ferryman was entitled to ask only six pence for the transportation of each man and horse. There was also a ferry which
carried passengers on foot and on horseback over one of the
Eastern Shore rivers.[20]

The law regarding the necessity of having a pass in order
to leave the province also made it difficult for a thief, or any
criminal, to escape. By the terms of a law passed in 1666 if
any one intended to leave the province, he must give notice of
this intention " by setting up his name " at the office of the

colonial secretary three months before the date of his departure. If within that time no one had any objections to make to his leaving the colony, then the governor, chancellor, or secretary, would issue him a pass for which he must pay two shillings, six pence. While this act was primarily intended to prevent the escape of debtors and runaway servants, it also helped to prevent the escape of thieves or other criminals. At one time it was necessary to have a pass even to leave a county. All colonists were allowed to examine all strangers, or " suspicious persons," who did not have a certificate or pass. If necessary they could take them before the nearest justice of the peace for a further examination.[21]

So far we have considered the difficulty a thief, or any criminal, would have in escaping from the province into another colony. But let us suppose that instead of trying to do this, the thief tried to dispose of the article he had stolen to some inhabitant of the colony. Would he be likely to succeed in doing this? Several reasons can be given to show that he probably would not be successful. In the first place, it should be remembered that during the entire proprietary period Maryland was but sparsely settled. The original pioneers who came over in 1634 numbered about two hundred. By 1660 there were about eight thousand inhabitants, and, by the end of the proprietary period in 1688 this number had increased to twenty-five thousand. With such a comparatively small population it would be difficult for a thief to conceal his identity. Plantations were scattered along the banks of the rivers and streams which empty into the Chesapeake. During this early period it was probably not difficult to know one's neighbours.[22]

The needs of the planters in the way of clothing, tableware, etc., were satisfied by visits of vessels from England bringing these articles. Some of the boats came to St. Mary's, others to the landings of the more affluent planters. In view of this practice, the arrival of an individual at a plantation with something to sell would at once create suspicion. As has already been stated, most of the thefts were committed by members of the servant class. If the man arriving at a plantation happened to be a runaway servant, any one buying from him did so at his own risk. A colonist sheltering even for one night a man, whom he knew to be a runaway servant, might be forced to pay a fine. In other words, "entertaining or harbouring" a runaway servant, as it was known, was forbidden. All colonists ran the risk of paying another fine if they bought goods from a servant without his master's permission.[23]

Having pointed out the reasons which made it unlikely that a thief would escape, or that he could dispose of his stolen goods, we are now in a position to discuss the cases of theft which actually occurred. First, let us consider those cases in which servants were involved. A servant named John Richardson was charged with running away and taking with him some of his master's "goods." He was tried before the members of the assembly who found him guilty and ordered him whipped "three several times." In this case one burgess thought that the servant should be laid in irons and then whipped soundly, while another assemblyman advocated hanging. Thomas Guinn, another servant, was given ten lashes on his bare back "well laid on" for stealing some of the planter's canoes from their landings.[24]

John Smith, Jr., a servant of Francis Bright, was accused of breaking into a chest and taking several things out of it, including a suit of clothes, a shirt and a pair of shoes. Smith also stole some food from his master. When caught with the articles of clothing, the servant said that some Indians had broken open the chest. At his trial Smith was convicted of theft and ordered whipped with twenty-five lashes. The court in imposing this sentence took into consideration the fact that the servant had never taken any of the articles in question out of his master's house, but regretting having stolen them, returned them to his master.[25]

At his trial before the provincial court John Oliver, a servant, was convicted of "the felonious taking of seven shillings and six pence, English money, and one piece of Spanish money called a piece of eight." Upon his conviction Oliver was asked what he had to say for himself and why a sentence of death should not be imposed. The servant at once claimed benefit of clergy. When "the book" was produced, however, it was found that Oliver could not read. Accordingly, the death sentence was imposed. Oliver escaped the gallows, however, when the proprietor pardoned him on two conditions, one of which was that he should act as "general hangman" in the province during the rest of his life, and the other was that he must not only serve his master during the term of his servitude, but at the end of this period he should "make satisfaction" by additional servitude for the cost of his imprisonment.[26]

There is one other case in the early records of where a man was pardoned on similar conditions. This was when James Douglas was convicted and condemned to death for stealing "a horse, bridle and saddle." In a petition to the deputy governors, then in charge of the government, Douglas said that although he had nothing to say in mitigation and extenuation of his crime, he did, however, humbly prostrate himself at their feet, begging for "grace and mercy." Douglas asked the provincial officials to consider "the poor, distressed condition of his poor wife" and his own "tender years." If he was pardoned he promised "amendment for the future." Touched by this appeal, the deputy governors granted Douglas a pardon on condition that he serve as a common hangman "for the future." The pardon, however, was to be kept secret until "such time as the said Douglas shall be carried to the place for execution and the rope put about his neck, at which time the said pardon was to be produced and not before." [27]

During the fall of 1663 a law was passed which attempted to break up the practice which servants had of stealing their masters' or mistresses' goods and then bartering, or selling them, to others. Under the provisions of the new act any colonist buying anything from a servant without the permission of his master, mistress or overseer, was liable to pay a fine of two thousand pounds of tobacco, one half of which went to the Lord Proprietor and the other half to the owner of the goods. If the offender did not have this much tobacco on hand, then he could be given thirty lashes on his back. In one instance, a man bought a hat from a servant with the understanding that the amount agreed upon for the hat would not be paid until the servant was free. This was held a violation of the Act of 1663 and the would-be purchaser had to pay a fine of two thousand pounds of tobacco.[28]

The punishment the servant received under the provisions of the Act of 1663 was indeed severe. If he was caught stealing, or trading with his master's goods, or killing any of his master's livestock, he would, if it was his first offence, receive thirty stripes on his naked back. For the second offence and every offence thereafter, the servant would not only be whipped in this manner, but in addition the letter " R " would be burned on his shoulder with a red-hot iron. One wonders why the letter " R," as this would seem more appropriate for runaway servants. In the case of poultry, however, the servant would only be whipped no matter how often he stole chickens.[29]

While servants seem to have been guilty of most of the thefts, others, too, were accused of this crime. The case of Pope Alvey in the previous chapter will be recalled. Alvey was accused of stealing a cow. In Charles county Robert Wilson was accused of having gone aboard a New England vessel while in port and stolen a skimming dish and some other " small thing." When he came up for trial, Wilson confessed his " petty larceny " and humbly begged pardon " for his erroneous act." The justices, however, ignoring his request for pardon, ordered the sheriff to take Wilson outside the court house and give him " ten sound lashes." In another case, a grand jury indicted John Ellis for petty larceny. Ellis was charged with having stolen a shirt from John Cully. At the jury trial Ellis pleaded not guilty. On the witness stand Cully said that Ellis had taken his shirt and that he had " challenged it upon his back." A woman witness testified that she thought Cully had recovered his shirt. The jury now went out. When they returned their verdict was that Ellis was not guilty. Later he was " cleared by proclamation." [30]

One of the most interesting cases of theft occurred at the house of Mrs. Symon Overzee who died giving birth to a child. This case involved not only the distinction between grand and petty larceny, but also the old English law of theftbote. Taking advantage of the confusion caused by the death of Mrs. Over-zee, John Williams and his wife, Mary, stole some clothing which had belonged to the deceased woman. Considering the plain and simple clothing worn by most colonists, the articles stolen were quite valuable and included lace, gloves, ribbon, neckclothes, coifs, stockings, gorgets, pinners, a pair of bodies (bodice), safeguard (petticoat) and "one bastard Flaunders laced Holland smock." As this clothing was worth about fifty pounds sterling, Mrs. Overzee must have been a well-dressed woman and she and her husband colonists of means.

After considering the evidence in this case, the grand jury returned a billa vera or true bill. As the articles which they had stolen were worth fifty pounds sterling, John and Mary Williams were indicted for grand larceny. Mary, the wife of Daniel Clocker, was also indicted for felony as being accessory before and after the fact, while Thomas Courtney was indicted for being an accessory after the fact. At the trial before a petit jury all the prisoners pleaded, "Not Guilty." The jury, however, found John and Mary Williams guilty as principals "to the value of fifty shillings," but not fifty pounds sterling. Mary Clocker as an accessory was found guilty to the same value, while Thomas Courtney as an accessory was found guilty "to the value of four pence." After the trial or petty jury had rendered its verdict, the sheriff was ordered to keep the prisoners in safe custody awaiting a later summons.

At a subsequent meeting of the provincial court all four prisoners were called to the bar. John and Mary Williams and Mary Clocker were informed that they had been convicted of a felony. When they were asked why judgment of death should not be pronounced against them, they said nothing, only craved mercy. The governor, as the presiding judge, at once passed sentence ordering the sheriff "to return the prisoners from whence they came, and thence to execution, and then to hang by the neck till they be dead." As the articles which Thomas Courtney had received from the Williams were worth less than twelve pence, he was told that the offence of which he was guilty was petit and not grand larceny, as was the case with the others. In view of this he was not sentenced to death. The sheriff, however, was told to give him thirty stripes. Fortunately, for all those involved in this crime, the proclamation of an amnesty soon after their conviction saved three of the accused from the gallows and Courtney from his lashing.[31]

This case, besides bringing out the distinction between petit and grand larceny, involved another old English law, now also obsolete, and that is what was known as theftbote. This was when the owner of goods, which had been stolen, either received them back from the thief, or agreed to accept some compensation from the thief by way of composition. This was considered compounding a felony. This charge was made against Symon Overzee, whose wife's clothing had been stolen. It was made by Daniel Clocker, whose wife, as we have seen, was convicted of being an accessory both before and after the fact. It was Clocker's contention that Overzee had told him that if he returned some of the goods which had been stolen, and also paid him a certain amount of tobacco, he would not prosecute his wife. Overzee denied this and said that Clocker had made the accusation " out of mere malice and spleen." While it was true there was some talk about the payment of tobacco, it was, said Overzee, in satisfaction of a debt which Clocker owed him. When all the evidence in this case had been considered by a grand jury, their verdict was " Ignoramus," that is, " We do not know; we ignore," being the word at this time written on a bill of indictment when the evidence was considered insufficient to warrant the finding of a true bill. In other words, the grand jury did not think the facts justified indicting Symon Overzee for theftbote.[32]

There are several instances in early Maryland where persons were pardoned after they had been convicted of theft. Thomas Gibbons, of Baltimore county, was indicted by a grand jury for stealing " a parcel of black peake " worth forty shillings. When Gibbons pleaded guilty, he was sentenced to be hanged. Before the sentence could be carried out, Philip Calvert, in the name of the Lord Proprietor, pardoned Gibbons " in hopes that for the future he may become a new man." In another case, a woman named Elizabeth Withrington was convicted and sentenced to death by the justices of the provincial court for " the felonious taking of several goods." Elizabeth at once petitioned the proprietor for a pardon. The woman admitted that she had nothing to say in her own defence, nor any reason to hope that her life might be spared, but only depended on his Lordship's " abundant goodness and mercy, which she doth most humbly beg his Lordship to extend to her in the pardoning her crime and granting her life." Elizabeth received a pardon.[33]

By the provisions of a law passed in 1642 larceny was declared a capital offense and any one guilty of this crime, whether principal or accessory, could be put to death, or burned in hand, loose a member, or outlawed, imprisoned for life, etc., etc. About ten years later, or in 1654, an act was passed which declared that any one guilty of stealing must restore fourfold the value of the article stolen. If unable to do this, then the thief must make the fourfold satisfaction by servitude. Should any other action accompany the theft, such as assaulting the person from whom the article was stolen, or breaking open the house, or picking the lock, the thief so offending was, upon his conviction, to be punished at the discretion of the court. A case arose under this statute in the year following its passage. A man convicted of stealing a bodkin was compelled to pay four times as much as it was worth.[34]

In 1681 an important change was made in the law regarding theft. The reason for the change was that the members of the assembly were of the opinion that while " the severity of the laws of England against all thieving, stealing and purloining " was suitable to the mother country and all such thickly populated places, such severity was not " agreeable to the nature and constitution of this province, so meanly and thinly inhabited." The delegates also thought that the provincial authorities were unwise in following the practice of the English courts by which all or most crimes " above petit (or petty) larceny " were punishable by the loss of member, burning in the hand, and even the death penalty. This practice, the delegates maintained, was not only too rigorous a one for the colony to imitate, but also there were practical difficulties in its application. This was due to the necessity of prosecuting all such offenses in the provincial court, the judges of this tribunal alone having " the power of life and member." Many colonists lived a great distance from St. Mary's. In order for them to prosecute an offender before the provincial court, it meant that not only must they come to the court, but also they must bring their witnesses, etc., all of which involved much expense. The result was that many offenders were not punished. As the provincial court only met three or four times a year, this meant that " speedy justice " was impossible and that the accused often had to remain in prison for a long time before being brought to trial.

For the reasons which have been given, the assembly, in 1681, passed a law by the terms of which it was made possible for the justices of the county courts to hear and decide " all thievings or stealings whatsoever "—robbery, burglary and house-breaking excepted—provided the value of the goods, livestock, etc., did not exceed one thousand pounds of tobacco. Any one convicted under this statute could be punished by the county justices by ordering them to be whipped, or placed in the pillory, or both. The county commissioners were not, however, given power to crop the offender's ears or to mutilate any member. The convicted man must also pay the person from whom he had stolen four times the value of the purloined article. The Act of 1681 further provided that if any one after having been twice tried and convicted in the county courts under this statute, was accused a third time of a similar offense, then, in such a case, the justices of the county courts must send the offender to the provincial court for trial, whose judges would try the offender according to the severe English laws.[35]

In the printed records of early Maryland there are few references to burglary or to housebreaking. Burglary, at common law, was the breaking and entering the dwelling house of another, in the nighttime, with the intent of committing a felony therein. Housebreaking, on the other hand, was the act of breaking open and entering, with a felonious intent, the dwelling house of another, by day or night.

A case of burglary occurred in Calvert county. Arthur Nottool, a tailor, was indicted by a grand jury for " burglariously and feloniously " breaking and entering the house of John Hunt during the nighttime and stealing a gun valued at twenty shillings and a shirt worth five shillings. While he was being tried for this crime before the provincial court, the tailor escaped from prison and fled. When later he was captured Nottool had to face an additional charge of breaking prison. At his trial before a petty, or petit jury, evidence was introduced to show that the tailor himself had admitted breaking into Hunt's house and taking a shirt out of a chest. The jury found Nottool guilty on both counts. Asked what he had to say for himself, the tailor craved benefit of clergy, and " the book being given, and demanded whether he read or not, answer, was made that he read." In this way the tailor escaped hanging. The sheriff, however, was ordered to burn Nottool's right hand " in the brawn " with a red-hot iron. This was to show that he had claimed benefit of clergy. If the tailor later committed another crime, he could not claim the same privilege again.[36]

A case of housebreaking can be found in the early records. This was when Robert Dennis was indicted for this offence. The facts were almost identical as those in the Nottool case, except that Dennis did his breaking and entering during the daytime. Just as the tailor had done, Dennis also stole a gun and shirt. Dennis admitted stealing both the shirt and the gun. Dennis, however, never came up for trial as he managed to escape from prison.[37]

In Calvert county Raymond Staplefort was presented by a grand jury for forcible entry and larceny. As stated in the indictment:

Raymond Staplefort, about the 22nd of December, 1664, by force and arms several rooms of John Bayley's did enter, where no entry is given by law, and likewise on the 12th of February, in the year aforesaid, several goods and chattels to the value of twenty thousand pounds of tobacco, then and there found, feloniously did steal, take and carry away, contrary to his Lordship's peace, rule and dignity.

Staplefort pleading not guilty, asked for a trial by a petit jury. At the trial, Bayley himself admitted that the rooms which Staplefort was accused of entering were in a house which they had both purchased, each paying half for it. By an agreement among them the house was to be divided, Bayley having the use of a room and two sheds, and Staplefort the other part of the house. As both men had a right to enter the house, Staplefort could hardly be accused of forcible entry. Apparently Bayley did not succeed in proving to the jury's satisfaction that he had a separate and exclusive interest in the goods which Staplefort was accused of stealing. While the evidence in this case is somewhat conflicting and confusing, the jury was right in not finding Staplefort guilty.[38]

5

CHAPTER IV

LIVESTOCK AND HOG STEALING

Livestock played an important part in the economic life of the early Maryland colonists. Cecil Calvert, for example, recompensed his officials in the province with cattle. To his colonial secretary the proprietor gave, besides twenty barrels of corn a year, two steers and the use of " six milch kine " for his secretary's care and trouble in keeping accounts " and in my other affairs." Governor Leonard Calvert persuaded the soldiers in the colony to take cows and their calves towards the payment of their wages.[1]

Nicholas Cawsine in his will left a steer to a minister " as being a faithful Christian, and desiring the prayers of the Church." Thomas Hebden, a chirurgeon, left a brown heifer to a Catholic priest in order that he might pray for his soul, while William Marshall, of St. Mary's county, by the terms of his will provided that the milk of three heifers and half their increase should be used for " the maintenance of a minister." [2]

Cattle were donated for a boy's " schooling," and livestock was often given to children during the life of the parent in advance of a future distribution. This was known as advancing the child a portion. For example, Thomas Greene, member of the council and later a governor of Maryland, gave one calf to his son, Leonard, " to his own use this day forever to advance him a portion." Kine and swine were also left by will to children or friends of the testator.[3]

Gifts were made of livestock or they were given in payment of services rendered. One man agreed to build a house for another, the consideration being a red cow. Thomas Gerrard, of St. Clement's Manor, gave his servant a cow with a calf " by her side " in order to encourage him in the faithful discharge of his duties. In another case, the cowkeeper for his faithful care in tending livestock was rewarded with a cow calf and her increase.[4]

Kine were bartered or exchanged for beaver skins, salt, corn, swine, land, canoes, and guns. In one instance, six oxen were exchanged for three boy servants, and in another case four " milch cows " were bartered for one manservant. Cattle, of course, were also sold for tobacco which was the usual medium of exchange.[5]

There was much litigation over cattle. One suit was instituted against a planter who had killed one of the oxen belonging to the Lord Proprietor. A jury was impanelled in another case to determine whether a cow and her increase belonged to Captain Thomas Cornwallis or to another man. Cuthbert Fenwick sued because some one had killed his " steer calf." Because the defendant had failed to deliver to him a cow and a calf at his plantation, the plaintiff, in another case, maintained that he had been much " damified in the loss of their increase and the want of the milk for his sustenance." An amusing defence was put up in still another instance where the defendant was accused of taking away a young bull, together with other cattle, whereby the plaintiff sustained loss and damage " both in his breed and milk." In this case, the defendant successfully pleaded that the bull ran after the other cattle and that he could not " force him back." [6]

It is probable that the first cattle in Maryland were brought from Virginia. The King of England asked the Virginia authorities to allow the Marylanders to buy whatever livestock and commodities they could spare. The instructions given those planning to go to Maryland advised them to purchase cattle in Virginia. It was said that four or five pounds sterling would buy a cow in Virginia and about twenty-five shillings a breeding sow. Many of the Virginians, however, were so opposed to the new settlement north of the Potomac that they said that they would rather kill their cattle than sell them to the Marylanders. Whether most of the first cattle came from Virginia or direct from England, it was not long before many Maryland colonists owned livestock. Captain Thomas Cornwallis, a prominent colonial officer, had on his plantation called Cross Manor at least one hundred neat cattle, a great stock of swine and some sheep and horses.[7]

The early colonists soon became concerned about the question

of protecting their cornfields against roaming herds of cattle.
Fearing that the inhabitants would only plant tobacco and not
be self-supporting, laws were passed requiring colonists to plant
at least two acres in corn. Having made it obligatory to plant
corn, other laws were enacted, by the terms of which, all colonists were required to erect fences about their cornfields. The
fence must be four or five feet high and " sufficient and strongly
made." If, after a planter had built such a fence, cattle managed to get in and trample down or eat his corn, then the
owner of the livestock was liable for damages. The owner of
the cornfield was not, however, permitted to kill the cattle that
had caused the damage. Because a few people burnt the fences
erected around cornfields, some advocated that these " ill-dis-
posed and malicious persons " should be liable to the same
punishment as willful burners of houses. Any one guilty of
willfully burning or destroying a man's home might be burnt
in the hand, or even put to death.[8]

The colony was not very old when the livestock of the settlers began to run wild. The inhabitants of Kent Island complained that their tame cattle had been " carried away and
spoiled by wild bulls." The Kent Islanders asked the governor
and council to authorize the rounding up of these wild cattle.
In southern Maryland, too, livestock were running wild, as
several colonists declared that their cattle had been disturbed
by " divers herds of wild cattle." By 1661 there were so many
herds of wild cattle that the proprietary officials allowed any
ranger, or colonist, to kill them provided they " reserved their
tallow and hides to the Lord Proprietor's use." It was said
that some of the rangers would purposely drive the planter's
livestock into the woods in order to collect a reward for rounding them up and bringing them back to their owners. This
abuse was remedied by requiring all rangers to take out a license
and to give good and sufficient security that they would not
injure the cattle of any of the inhabitants, or even presume to
range upon any man's land without the consent of the owner.
Even after this, however, it appears that complaints were still
heard about the injury done to the planter's livestock by rangers. For this reason the rangers were summoned to give an
account of their " actings and doings," and, in the meantime,

forbidden to act any more in that capacity. All colonists were given a chance to air their complaints.[9]

As the cattle of most of the colonists roamed through the woods, it was advisable to have all livestock marked for purposes of identification. Indeed, the law required all inhabitants to enter in the records of either the county court, or the secretary's office at St. Mary's, a description of the markings of their livestock. This regulation applied to swine as well as to cattle. Among owners of livestock who recorded their marks were Josias Fendall, one time governor of Maryland. His mark for his cattle and hogs was " cropped on the left ear and underkeeled on the right ear." John Lewger, first colonial secretary, had as his livestock markings " overkeeled and underkeeled on both ears." Philip Calvert, another secretary of the province, recorded his mark as " the left ear cropped, the right ear overkeeled," while William Calvert had as his mark " the left ear with the fore part taken away, and, on the right ear, the hind part taken away." Other markings were used, including slits in the ear, and cuts made in the form of swallow tails, fleur-de-lis and numbers. Women as well as men recorded their livestock markings.[10]

In the sale or gift of cattle a very careful description was given of the animal. Not only were the ear markings described, but also the body markings, as, for example, one cow " brown of colour . . . white hind legs, white tail and a white patch on her rump." The horns would sometimes be described as well as the udder. Often the name and age of the beast would be given. Heart, Spot, Doe, Nancy, Lady, Boldface and Fortune were typical names. The vendor usually promised that the purchaser would have the cattle " peaceably and without molestation forever," and that he would defend the buyer against any claims that might be made to the livestock.[11]

It was sometimes difficult to locate the bull or cow which had been sold. One man said that he had " sought up and down for the cow, but could not find it," while in another case a bull calf " running in the woods " was sold, provided the purchaser " come to the sight of the said bull." A dispute arose in one instance as to whether a cow was with calf at the time of the sale. When later the cow died, the purchaser cut the animal open and declared that although he " searched the cow to the full," he could not find any calf in her.[12]

The ownership of cattle was often disputed. This would be largely determined by the markings of the livestock in question. Sometimes the charge was made that the marking on an animal had been purposely altered. A court would in some instances appoint " viewers " to look at the beast and report on the markings at the next session of the court. Captain William Evans and Thomas Gerrard, two prominent colonists, went to law about the marking of a cow. Gerrard, in order to prove that the animal in question was his, threw down on a table in the court room the cow's ears and the ears of some of his swine which he had cut off for the members of the court " to judge of the resemblance and nearness of both marks." As the judges were of the opinion that the marks were not the same, the cow was given to Captain Evans. Gerrard, not satisfied with this decision, appealed to the governor and council. He maintained that as Evans had adopted his own markings for livestock he should be compelled to give them up. Gerrard admitted, however, that he had never recorded his mark for cattle and swine as the law required. The governor decided to ask the members of his council for their opinions. These follow:

Mr. Baker Brooke: Captain Evans recording the mark first, to have the mark.

Mr. Nathaniel Utie: According to the law and precedents of the province, the mark to belong to Captain Evans.

Captain John Price: Captain Evans to keep the mark according to the acts of assembly.

Captain William Stone: Mr. Gerrard should keep his mark and stand to the censure of the court for his neglect according to the act.

Philip Calvert, Secretary: The mark to be Captain Evans' as the first mark upon record.

The question was finally settled by Captain Evans, at the request of the governor, agreeing to assign over his mark for livestock to Gerrard. The latter, who was Lord of St. Clement's Manor, lost no time in having this mark recorded for his cattle.[13]

John Harwood faced a serious charge when he was accused

of marking the livestock of others as his own and of boasting that he hoped " within six or seven weeks to be at the marking of a great many more." The court, before whom Harwood was tried, not only fined him one thousand pounds of tobacco but also had him whipped with thirty lashes. In Kent county John Deare was less severely punished for a similar offence. When several persons in this county complained that he had killed their cattle and marked some as his own, Deare was told by the justices that in the future he could not go into the woods to kill any livestock, unless he took with him " two honest neighbours " to see that he did nothing injurious to any planter's livestock.[14]

The proprietary policy regarding hog stealing underwent a number of changes. At first, it was customary to issue licenses to kill swine. As early as 1642, the governor issued a proclamation forbidding any of the inhabitants of St. Mary's to kill any swine, except upon their own land, without a license from him. Before a license would be granted, a man had to put up one thousand pounds of tobacco as security which he would forfeit if he killed any marked swine, except those of his own mark, or any swine at all in his Lordship's forests, and did not bring the ears of the slaughtered beasts, together with the skin between the ears, to the governor or colonial secretary within a month after the killing. A number of colonists took out one of these licenses, including Thomas Gerrard of St. Clement's Manor, and Thomas Hebden. It appears that the latter had to forfeit the security of one thousand pounds when he failed to comply with the terms of the license. Nothing daunted, Hebden took out another license to kill swine putting up an additional one thousand pounds of tobacco.[15]

During the early years of the province, many suits were instituted for injury to swine. Captain Thomas Cornwallis sued two men in an action of trespass " for hunting and killing unlawfully his swine of his known mark . . ." Cyprian Thorowgood and John Hollis, two traders with the Indians, brought similar suits against those who had injured their swine. Hollis asked that the men who had caused the injuries should be bound " in security not to commit the like trespasses again as they have menaced to do." Robert Taylor accused Robert

Brooke, of De la Brooke Manor on the Patuxent river, of killing "divers of his hogs." Brooke, on the other hand, sued
Cuthbert Fenwick for having killed some of his swine, and was
especially grieved by the loss of a large boar, which, he said,
defended the rest of his hogs from wolves. Brooke also alleged
that many of his sows had left his pen "for want of a boar." [16]

During the years 1649-1650, the proprietary officials adopted
a different policy in regard to the killing of swine. As the
licenses to kill hogs had occasioned "some inconvenience and
offence to divers of the inhabitants of the province," the governor decided to revoke all such licenses. At about the same time
a law was enacted which provided that any one who stole,
wrongfully killed, or carried away any marked hogs belonging
to some one else, should pay double the value of such swine
to the owner thereof, and an additional two hundred pounds
of tobacco to the man who had informed on him and three
hundred pounds of tobacco to the proprietor. The same act
made it necessary for any one who knew of any case of hog
stealing to report the same to the constable of the hundred in
which he was residing. Failure to do so made one liable to be
censured and fined as the chief offender. [17]

Many cases arose while this law was in effect. Thomas
Courtney and several other men were indicted "for feloniously
killing hogs." It was charged that Courtney and the men with
him "with dogs and guns" hunted and killed some marked
hogs, and, after killing them, did "unlawfully mismark by
cutting and mangling the ears of them," by which means the
rightful owners were prevented from claiming them as their
own. All this, it was alleged, was contrary to the law of the
colony and "against the peace of the Lord Proprietor, his rule
and government." In accordance with the provisions of the law
then in force, Courtney was compelled to pay double the value
of the hogs to the owner, two hundred pounds of tobacco to
"the informer," and three hundred pounds of tobacco to the
proprietor. [18]

Cuthbert Fenwick sued three men, one of whom was a
servant, for "killing and making away with several of his
marked hogs." After the court had considered all the evidence,
they came to the conclusion that the killing had been done

with " a willful design." The two men who were freemen were compelled to pay the usual damages and fines in tobacco. As the servant did not have any tobacco, or goods, with which to pay, the court ordered that he should " stand three courts in the pillory with his fact (act) written in capital letters over his head, and that he be given twenty lashes on the bare back every court." The sheriff was ordered to keep the servant in his custody and to bring him to the next three meetings of the provincial court until this sentence was carried out. This punishment of a servant was in accordance with the terms of the law which provided that if any one did not have any estate, or goods, he might be censured and punished as the court thought fit.[19]

Many cases arose in the Charles county court in which planters were accused of killing the swine of others. At a winter session of this court several men were indicted " for killing and stealing hogs contrary to the laws of this province and the rule and dignity of the Lord Proprietor." The four men charged with this offence were William Allen, John Muns, John Boyden and John Cabell. James Lee, the owner of the hogs, was the man who not only gave the information leading to the arrest of the four men, but also was the chief witness against them at their trial. Two other men were also called as witnesses by the prosecution. All three men testified at great length which shows the importance attached to a case of this kind. The substance of their testimony was that the four men when caught with the dead swine had admitted having killed them, and that Allen, one of the prisoners at the bar, had even offered to let Lee have some of his own stock in place of the ones which had been slaughtered. Lee, however, was unwilling to agree to this proposition, for, as he explained, he would not take " a pig in a poke," meaning that he would not accept any hogs which he did not know anything about or had not seen.

In defence of their actions, two of the prisoners were able to produce a note signed by James Lee authorizing Allen and Cabell to kill any of his hogs " in Lewis' Neck." The prosecution maintained that the hogs in question were not killed within Lewis' Neck, that the note was over a year old, and that John Muns and John Boyden were never licensed to kill any of Lee's hogs at any time. The final result of this case was that each of the four men were compelled to pay Lee, as the owner of the slaughtered swine, double the value of each hog, an additional two hundred pounds of tobacco to Lee for having " informed " against them and also three hundred pounds of tobacco as a fine to the Lord Proprietor. As tobacco was worth about two pence a pound the total amount which was paid for the killing of a few swine amounted to nearly fifty pounds sterling. In addition to this sum, the four prisoners had to pay " the cost and charge of the suit." This was indeed a severe penalty to pay.[20]

At the same session of the Charles county court, Thomas Standbridge was accused of killing and stealing a hog belonging to Daniel Johnson. At his trial Standbridge attempted to show that he had bought the hog in question of Johnson. The prosecution, however, introduced witnesses who testified that Johnson had declared that Standbridge had not purchased the hog from him, " neither would he smother any of his knavery." The upshot of this case was that Standbridge was forced to pay Johnson double the value of the hog and the usual fines to the informer and proprietor.[21]

During a spring session of the Charles county court, Thomas Baker complained against William Robisson for injuries done to his swine. Baker said that some of his hogs had " come home shot, others had come home torn with a dog or dogs." When evidence was produced to show that the markings of some of the swine had been altered, the court commissioners ordered that Robisson should put up " a sufficient bond " that he would not hunt by himself alone with dogs and gun, but only in company with neighbours until he had cleared himself of this complaint. In answer to this suit against him, Robisson said that no credit should be given to any claim which Baker made about killing his hogs, as himself was known to be a hog stealer. Indeed, it was Robisson's contention, that Baker only made this complaint against him in order to try to make one of his neighbours appear " as ignominious as himself." Baker failed to prove that part of his case against Robisson as regards the charge made of having altered the markings of swine, although it was shown that Robisson had " a pied dog " which ran foul of some of his neighbour's swine, injuring and killing some of them, and that Robisson had refused to kill the dog. The result of this case was that while Robisson was freed of the charge of hog killing, he was, however, compelled to pay the cost of the suit brought by Baker " by reason that he hath partly been the occasion of the suit, or complaint, by not killing his dog upon complaint of his neighbours." [22]

In Kent county two colonists complained against John Salter and William Price on the ground that they suspected them of stealing hogs. The informers, who were William Elliott and John Ringgold, said that several circumstances made them suspect Salter and Price. In the first place, a number of hogs had been lost by their neighbours. Moreover, pork was often seen in the house in which Price and Salter lived, although it was well known that they had no swine of their own. When Price was questioned about the pork in his house and asked where the ears of the hog were, he replied that a dog had eaten them and, added, it was a wild hog which he had a right to kill.

Henry Morgan, a commissioner of Kent county, was one of the witnesses summoned to testify against Price. Morgan told how a boy servant had informed him that when he and Price had been in the woods together Price had shot at and killed a marked hog which belonged to a man named Thomas South, another member of the Kent county court. Morgan added that when he heard of this, he suggested to Price that he had better go to see South and " make his peace with him." According to Morgan, Price had acted on this advice and had admitted to the magistrate that he had " done him an injury."

After hearing all the evidence against Price, the Kent justices came to the conclusion that there was " just cause " of suspecting Price of killing a marked hog in violation of the act of the assembly covering such cases. As Price was unable to pay the fines provided for by this act, the commissioners imposed a rather unusual punishment. First of all, Price must stand in " open court " with a piece of paper on his chest " declaring his offence," and later, had to make a public acknowledgment of his offence. Furthermore, Price was ordered to see that a county bridge was repaired before the next meeting of the court.

John Salter was now put on the witness stand. Salter main-

tained that he had only killed hogs that were wild and unmarked. Mrs. Frances Morgan was then called to testify. She
told how when Jane, the wife of John Salter, had come to see
her she had asked her if she had slaughtered any swine lately
as she would like some pork. When Mrs. Salter replied that
none had been killed, Mrs. Morgan asked her to explain the
singed pork which she knew had been eaten in her house. Mrs.
Salter said that this was a part of a shoulder of pork which had
been given her husband by Matthew Read as a reward for
helping him kill two hogs. Then, according to Mrs. Morgan,
she asked Mrs. Salter whether this was the hog of which the
guts had been thrown into a creek.

Mrs. Salter: Ay, be God was it.
Mrs. Morgan: 'Tis strange that Matthew Read would singe his hogs
and cast the guts into the creek and loose the fat.
Mrs. Salter: Ay, be God did he.

After hearing several other witnesses, the Kent county justices
were of the opinion that there were grounds for suspecting
Salter of hog stealing. Accordingly, they informed him that
in the future he should not kill any swine, unless at least two
of his " honest neighbours " were with him at the time and that
he must produce the ears of the hogs which he did kill.[23]
In Talbot county, Christopher Denny accused Richard Tilghman, a chirurgeon, of killing some of his marked hogs. At
the trial of this case, it appeared that before Denny had brought
suit, Tilghman had " tendered satisfaction " for the hogs which
he had killed, but that Denny had refused to accept it. In
view of this fact, the Talbot commissioners decided that Tilghman would only have to pay for the hogs killed " and no
more," while Denny must pay the cost of the suit. In another
case before the Talbot county court, William Dawson claimed
that he had lost several of his marked hogs and had found one
of them at the house of William Dell whom he suspects had
made away with the rest of his swine. The justices ordered
Dell to give Dawson a hog in place of the one which he had
killed.[24]
In the proceedings of the court leet and court baron of St.
Clement's Manor, of which Thomas Gerrard was Lord, there

are several cases where swine were the subject of the suit. Robert Cole was presented by a jury of this court for killing and marking one of the Lord of the Manor's hogs for which he was fined two thousand pounds of tobacco. Later this amount was " affeered," that is, reduced to half this sum. This case was then appealed to the provincial court where it was shown that Cole had a right to kill the hog in question.[25]

A jury of the court leet and court baron also presented and fined Captain Luke Gardiner two hundred pounds of tobacco for catching two wild hogs and not restoring one of them to the Lord of the Manor as he should have done. John Stanley and Henry Neale were presented at a subsequent session of the same court for killing three marked hogs upon " the Lord's Manor." Edward Connery, who faced a similar charge, was able to show that he had killed the swine " by the Lord's order and license." Several men were fined for keeping hogs on the manor lands.[26]

As the existing law was not considered sufficiently " penal against offender " to prevent hog stealing, the proprietary assembly, in 1662, passed a statute which provided that if any one was convicted of this offence a second time he would have the letter " H " burned on his shoulder with a red-hot iron. Every county court was to be provided with the necessary iron. Four years later, or, in 1666, another change was made in the law relating to hog stealers. Again the reason for the change was that the delegates thought that the punishment for this offence was not sufficiently severe.[27]

The new act classified those who should be considered hog stealers as those who, either as principal or accessory, had killed any hogs in the woods, or upon a plantation, and had then cut off the ears intending to deface or change the markings. A man would also be considered a hog stealer if he killed swine belonging to others and failed to cut off the ears and nail them up on public view for at least a month after he had killed them. By taking this precaution, the colonist would show that he had not tried to conceal his wrongful killing. Also classified as a hog stealer was the man who secretly slaughtered any young pigs, or unmarked shoats, not on his own land, or which were not " sorting or in company " with his own hogs.[28]

The Act of 1666 also provided for the punishment of hog stealers. Any one reputed by "common fame" to be a hog stealer, who had been warned by a planter to keep off his land, and then, notwithstanding the warning, dared to hunt swine with dog or gun on the planter's land, the man offending in this respect would be liable to pay a fine of one thousand pounds of tobacco, half of which went to the owner of the land and the other half to the Lord Proprietor. The Act of 1666 further provided that any person, whether principal or accessory, who was convicted as a hog stealer, should for the first offence of this kind stand in the pillory for four hours and then have his ears cropped. In addition, the offender must pay treble damages to the owner of the stolen hogs. For the second offence, the offender would be "stigmatized in the forehead with the letter H," and, if he offended a third time, the man was to be considered a felon not entitled to claim benefit of clergy.[29]

The new law was indeed very severe on hog stealers. Although in none of the records so far printed of the provincial court and county court proceedings could cases be found where these punishments were inflicted, a number of men must have been convicted under this statute. This is shown by the mention made in the assembly proceedings of 1688 of a negro who had stood in the pillory and had his ears cropped for having killed some hogs. Also in the proceedings of the court of chancery there are several instances of where Charles Calvert pardoned those who had been convicted of hog stealing. The governor, no doubt, felt that the new law was unnecessarily severe. However, the statute probably was effective in reducing the cases of hog stealing, as it is doubtful if many colonists would commit this offence hoping later for a pardon from the governor.[30]

What of the servant who killed the swine or cattle of his master? By the terms of a law passed in 1663 a servant convicted of this offence was to receive thirty stripes on his bare back. If the servant offended a second time, he or she would receive the same number of stripes and in addition would be burned on the shoulder with a red-hot iron with the letter "R." One wonders why the letter "R" and not the letter "H," the mark for hog stealers.[31]

Near the close of the proprietary period, a by-law was adopted by the village authorities of St. Mary's which throws interesting light on the conditions existing in this settlement. It appears that many complaints were made by the inhabitants of St. Mary's of the great damage and injury suffered from hogs kept in the village which uprooted their gardens and fields and even killed their poultry. In order to prevent this, a regulation was passed requiring the people living in St. Mary's to keep their hogs in a pen. In the future, the town constable was authorized to seize any swine running loose and to impound them. Before the owner could recover them, he must pay the constable six pence apiece for each of the hogs which he had impounded.[32]

Although they can hardly be classified as livestock, mention should be made in this chapter of horses and dogs. Like cattle they played an important part in the life of the early colonists. This was especially true of horses. Leonard Calvert, Maryland's first governor, left a mare colt to his godson and directed that Margaret Brent, his executrix, should also give the first mare colt " that should fall this year, and if none fall this year, then the first that shall hereafter fall unto Mr. Temperance Pippett of Virginia." Cuthbert Fenwick in his will, after providing for the distribution of his land, said that:

I give the first mare foal that is foaled unto Teresa, Cuthbert, Jr., and Ignatius Fenwick, provided it live, until it have a mare foal, then the increase of those two to be equally divided amongst the aforesaid children, only the horse colts is to belong to the mother, until the said mare brings a mare foal.

In the inventory of the estate of Thomas Harris there is the following amusing entry: " One mare gone astray about three years ago and never since heard of." [33]

Horses were put to many uses. They were used to run a mill, and draw carts. Rangers, who protected the frontiers against Indian raids, were mounted. Horses were also used on military expeditions against savages. Messengers went on horesback in order to facilitate the more speedy dispatch of official messages. Sheriffs and the doorkeeper of the assembly were authorized to press horse and man for the quicker dispatch of their more important letters. Although a supply of horses was kept at

Mattapany, on the Patuxent river, where messengers could secure a fresh horse, many colonists put in claims for horses " pressed," or taken up, by virtue of some official warrant for the public service. In case of the illness of his patient a chirurgeon would ride a horse in order to reach him more quickly. When Colonel Edward Pye, a member of the council, learnt that " his Lady was on the point of death," this made him " take horse " and ride away during the night to her bedside. Horses were " furnished " with saddles, saddle cloths, girths, cruppers, stirrups and leather, snaffle bridles and curbs.[34]

There is mention of a horse race in Talbot county on the second day of January, 1672. This was probably the first horse race in Maryland. One man bet and won one thousand pounds of tobacco on this race. Although the man who wagered this amount sued and recovered it, in another case, where a similar action was brought for two hundred pounds of tobacco, the defendant successfully pleaded that it was " a wager at a horse race and therefore not actionable." Later, in 1681, a horse race was run near the plantation of James Rumsey.[35]

Many sales of horses were made in early Maryland. In one curious case Mistress Margaret Brent sold to Edward Cotton—

One mare of whitish colour, about four years old, being she that was suckled by Nicholas Harvey's wife. And I, the said Margaret Brent, do warrant the sale of the said mare unto Edward Cotton, or his assigns, from all just claims whatsoever. Witness my hand, January 8th, 1648.
                                                            Margaret Brent.

Horses were usually sold with a guarantee that they were " sound in wind and limb." As the ears of horses were not generally marked as was the case with cattle and swine, the sale would describe the body markings as " one bright bay mare with a star on the forehead," or " one iron gray horse, about two years old, with a gray spot on the forehead." The name and age of the horse were often given. The animals were branded or marked on the buttocks with the owner's mark.[36]

There was sufficient demand for horses in early Maryland to justify the presence of horse traders, or horsecoursers, as they were called. John Bigger was probably the first horse trader in Maryland. Bigger said that " formerly he used to trade much in

horses, in buying and selling them for many years together."
When asked for his opinion of a horse's age, he looked into
its mouth. Denis White, another horse trader who hailed from
the north, also sold horses in the province. While in the colony
he had intercourse with a maidservant of one of the colonists
which resulted in the birth of a bastard. White then decamped.
The master of the servant, in order to provide for the maintenance of the child, succeeded in levying an attachment upon
the property which White had left behind in the province.[37]

There was much litigation about horses. One man was arrested and charged with " unlawfully riding of a mare " belonging to another. John Hammond sought to hold Cuthbert Fenwick responsible for his failure to take care of his horse. He
claimed that after Fenwick had kept his mount for over five
weeks he sent it back to him with the bridle changed, the saddle
torn to pieces, and with no saddle cloth, or girt, and only one
stirrup. Hammond asked that Fenwick should reimburse him
for the abuse and detention of his horse and its " furniture." [38]

Fenwick, himself, sought to hold another man responsible
for injuries to his horses. This was when Thomas Clark, of
Resurrection Manor in Calvert county, was indicted for having
killed two mares belonging to Cuthbert Fenwick, of St. Cuthbert's Manor in the same county, by discharging " a certain gun
laden with gunpowder and swan shot . . . so that of the said
wounds the mares did the same day die . . ." Witnesses were
called who said that they had seen Clark with a gun in his
hands, had heard the report of his gun, and had seen the mares
bleeding and later dead with shots in their bellies. In the trial
before a jury, although it appeared that the horses had leaped
over his fence, Clark was found guilty and compelled to pay
damages. The judges of the provincial court then ordered that
Clark should " find security to be bound with him that he shall
not misbehave himself, and so to continue until the next provincial court." If Clark could not find the necessary security,
then he must remain in the sheriff's custody.[39]

In one case Josias Fendall, one time governor of Maryland,
brought suit alleging that William Kidson, a servant of John
Hatch, did without Fendall's consent, or knowledge, take two
of his young horses " from place to place whereby one of them

6

received his death." Fendall said that although he could not " directly charge Kidson with killing the horse with his own hand, yet by the evidence it may be very probably conjectured." Kidson, in defence of his actions, said that he had not taken the horses from place to place, but that when his master had told him to take them back to Fendall he rode one of his master's mares with the two young horses following along. At the trial, witnesses testified that they had heard Kidson say that he would not be bothered with watching the two horses day and night and that he would send them home " in the devil's name." When leading the horses home, Kidson was seen to have a sharp-pointed stick in his hand. Soon after this one of the horses was seen running about with his " guts hanging out." Although on the basis of this evidence the jury held Kidson responsible for killing the horse, the case was later " compounded " and Fendall had to pay the cost of the litigation.[40]

In Talbot county, Joseph Wickes claimed that William Osborne and others had brought about seventy horses upon his land. For the tresspass and damage done by these animals Wickes asked five pounds of tobacco a head, but failed to recover on his demand. In the same county, Colonel Henry Coursey and Henry Morgan engaged in a controversy over a horse. It was Coursey's contention that Morgan, objecting to one of Coursey's stallions coming near his plantation, had first made the stallion " service " one of his mares and then killed the animal.[41]

In the proceedings of the court leet of Thomas Gerrard, of St. Clement's Manor, John Tenison was presented by a jury of this court " for suffering his horses to destroy John Blakiston's cornfield." Blakiston was at the same time presented for hunting Tenison's horses out of his cornfield, the fence around which was proved " insufficient " by the oaths of two other colonists. At a later meeting of the same court, Henry Poulter was presented for keeping a mare and a foal upon the manor lands. Poulter was ordered to remove the beasts within twelve days or pay a fine. In one instance, the jurors actually presented " a stray horse taken upon this manor and delivered to the Lord." [42]

The proprietary policy in regard to the importation of horses in the province underwent a radical change. Until 1647 every effort was made to maintain and increase the supply of horses in the colony. During the fall of 1647, for " the benefit and preservation of the increase of the stock of horses " then within the province, Governor Thomas Greene forbade the exportation of any horses " upon pain of such severe punishment, as shall be thought fit by the provincial court to be inflicted upon all such persons " as disobeyed his injunction. Greene did this, he said, because horses were " so useful both in peace and war, and the said stock being as yet very small." [43]

By 1671, however, the stock of horses had increased to such an extent that the members of the lower house introduced a bill prohibiting the importation of horses in the province. After discussing the matter, the members of the upper house, or council, said that they might be willing to pass such a law provided that any horse, gelding or mare, fifteen hands high or more, could still be imported into the colony and also provided that any of the inhabitants of a neighbouring province would be allowed to come to Maryland with all their stock of horses. As all the members of the assembly finally agreed that the great number of horses already in the colony were very " destructive to the inhabitants," an act was passed which prohibited the importation of all horses whatever their size. [44]

The Act of 1671 was enforced in Talbot county when three mares which George Robbins attempted to import into the colony were confiscated. In two cases, however, this law was not enforced. In one of them it appears that William Sturdivant, coming from New York on horseback, had his horse seized by a sheriff upon his arrival in Maryland. When Sturdivant said that he was so poor that he would have to sell his clothes in order to purchase a new horse to take him back to New York, the Talbot commissioners wisely decided to let him keep his horse. In the other case, when the horse of John Newnam was seized for violating the Act of 1671, Newnam was also able to show that he needed the animal for his transportation. [45]

The Act of 1671 expired in 1674. The members of the lower house at once asked that it might be revived and further that the governor would not grant any more licenses to any person to import horses. The councillors were willing to agree to this suggestion. They thought, however, that it should still be lawful for the governor to authorize any one who came from Virginia, either on public or private matters, to bring their own mounts with them and to keep them with them in the colony " during their attendance upon public or private business." [46]

By 1680 horses had become so plentiful in the province that they had begun to run wild. John Hawkins, of Talbot county, was granted a license to hunt and round them up. Because of complaints against Hawkins for abusing this license it was later revoked. Several years later Colonels Henry Darnall and William Digges, two members of the council, were appointed keepers of the forest and chief rangers. In this capacity they were authorized to round up all wild horses in the woods and to make whatever disposition of them they thought fit. Darnall and Digges could appoint deputy or subrangers to carry out the powers granted to them. These subrangers were required to take an oath that in the performance of their duties they would not injure the livestock of the colonists. Despite their oath, these rangers were guilty of so many injuries to livestock that they were finally suspended from office. [47]

During the spring of 1682, the members of the upper house sought to limit the number of horses in the colony by proposing that no freeman in the colony could keep a stone horse (stallion), or mare, who did not own at least fifty acres of land, and also that no stallion over a year old and which was not fourteen hands high should be allowed to run in the woods. The delegates of the lower house did not, however, approve of these suggestions for several reasons. First, they said, they did not think it wise until the province was secure from attacks by Indians " to abridge the freemen " in their right to keep horses, secondly, that the proposal would not lessen the number of horses within the province, and, lastly, that the Act for the Encouragement of Tillage, if renewed, would make it necessary to have more not less horses. Regarding the suggestion of the councillors about allowing small stallions run at large in the woods, the delegates, while admitting that some law might be necessary to prevent this, asked, however, that it might be referred to the consideration of the next assembly. In reply to these arguments, the councillors said that as there was no longer any danger of a war with the Indians, this reason which the delegates gave for opposing the limitation of the supply of horses was without foundation.[48]

The members of the lower house now seem to have taken matters into their own hands and drew up a bill of their own for lessening the number and bettering the breed of the horses in the province. Although the records do not disclose what were the terms of this proposed law, we do know that the council objected to the statute on the ground that it would have neither of the effects which it was intended to have and that there would be just as many horses as ever. The members of the council now came forward with another proposal. They suggested that only six plantations in each county should be permitted to keep a stone horse and that these animals, which should be fifteen hands high, must be kept in enclosures. Any man who wanted a foal was to be permitted to bring his mare to one of the six stallions to have it " leap " her. Eighteen pence could be asked as a service charge. No stone horses were to be allowed to range the woods, and if any did so, it would be lawful to kill them. Although the members of the lower house told the council to go ahead and draw up a bill embodying these suggestions, the councillors do not appear to have taken their advice. Until the end of the proprietary period, the limitation and improvement of the breed of horses in the colony remained a matter of controversy.[49]

Many Maryland colonists kept dogs. Indeed, they were advised to take with them to the province a good mastiff as a protection for their homes and a water spaniel for hunting. During this early period with the plantations often separated by great distances, the possession of a dog must have given the planter a certain sense of security. When the colony was sparsely settled, visits from neighbours were probably infrequent. Unused to strangers, dogs would sometimes bite them. Occasionally, they were actually used to attack trespassers.[50]

In some instances the owner of a dog might be dragged into court, as when Francis Armstrong was warned to appear to

answer to a charge of breaking his Lordship's peace " by assaulting with his dog " another man. In another case, Thomas
Courtney sued Francis Anketill for keeping " an unlawful dog "
which had bitten and injured his wife. Anketill defended himself by saying that the biting was " accidental " and that after
it had occurred he had killed the dog. He added that neither
he nor any one else had " set the dog " on Courtney's wife and
that certainly he would not have done so as Courtney's wife was
his own sister. The provincial court were not long in deciding
that Courtney had not shown a cause of action.[51]

In Charles county James Fox alleged that when he went to
the house of Arthur Turner, a bitch which the latter owned
had attacked and bitten him in the leg. As a result, Fox said,
he had been so lame that he doubted whether he would be able
" to make a crop " that year. Although he had asked Turner
to pay him for the cure of his leg and for the time lost from
his work, Fox maintained that he had refused to do so " notwithstanding the bitch was known to have bit divers others
before." To prove the vicious character of the dog, Fox had a
woman, whose name was Mary Tarline, testify. She said that
the bitch had once bitten her but had drawn no blood.

Turner, defending this case, claimed that the only reason
the dog had bitten Fox was that he had carelessly trod on it.
To show his regret for the mishap, Turner said that he had
killed the animal.

After hearing all the evidence in this case, the county justices
decided that Fox had not shown any cause of action and his
case was dismissed. The justness of this decision is questionable. It would seem from the testimony of Mary Tarline that
Turner must have known of the vicious character of his dog and
therefore kept the bitch at his own risk. The dog had had its
" first bite." After that the owner should be held responsible
for further injury by the same animal.[52]

Dogs were sometimes used to hunt swine that were running
wild. In doing this, however, the owner of the dog might get
himself into trouble with the owner of the hogs. Richard Wollman, a commissioner of Talbot county, accused another man of
using his dog " to hunt, foul and mangle divers of his hogs."

During a session of the court baron and court leet of Thomas Gerrard, of St. Clement's Manor, Richard Foster was presented by a jury for allowing his dogs " to worry and lug " some hogs on the manor lands. Foster was fined twenty pounds of tobacco. At another session of the same court, a jury presented Joshua Lee for injuring hogs belonging to another man " by setting his dogs on them and tearing their ears and other hurts." For this Lee was fined one hundred pounds of tobacco.[53]

CHAPTER V

SERVANT DISCIPLINE AND PUNISHMENT

The importance of servants in early Maryland is shown by the many tasks which they were called upon to do. Indeed a colonist going to Maryland was advised " to furnish himself with as many as he can," in order that they could perform the various duties assigned to them. Of course, the number of servants which any one took to the province depended upon the colonist's means. Six pounds sterling must be paid for the servant's passage and to feed a servant and keep him in clothing, etc., for one year cost about thirteen or fourteen pounds more. Assuming that a colonist had the means he would take several servants with him. Prospective settlers in Maryland were told to take a servant who was a carpenter by trade. His services would be invaluable in helping the colonist to erect his house after his arrival in the province. Others whose services, it was said, would be found useful included boatwrights, millwrights, brickmakers, coopers, smiths and leather dressers. Servants were also useful in clearing the ground, gardening, fencing and taking care of livestock.[1]

Any healthy young man who was ready and willing to work, even if he knew no trade, would be in demand in the province. If the servant did nothing more than plant vegetables and tobacco during the summer and made pipe-staves for tobacco hogsheads during the winter, it was estimated that such labour would result in a yearly profit of about fifty pounds sterling. Even after allowing six pounds sterling for the cost of transporting the servant and thirteen or fourteen more pounds for feeding and clothing during the year, the colonist who had a servant in his employ would still make a nice profit.[2]

All colonists, who were able to do so, carried out the suggestion about taking servants to Maryland with them. Captain Thomas Cornwallis, appropriately dubbed by one historian the " Miles Standish " of early Maryland, had about twenty servants. William Mitchell, a member of the governour's council, had about the same number, some of whom were artificers, workmen, and "divers other very useful persons." Robert Brooke, another councillor, brought twenty-eight servants to the province, while Robert Wintour, also a councillor, had but seven. It was considered necessary for any "gentleman" to have at least one servant " to get him wood to burn in his chamber and do him such service as he shall command." Every colonist eventually hoped to be able to have " a servant, or two, maybe three." [3]

What of the character and type of persons who came to the colony as indentured, or indented, servants? Captain Thomas Cornwallis said that some of those whom he had brought over were of " very good rank and quality." On the other hand, Dr. Luke Barber, a chirurgeon, said that he brought with him from England only a very low class of men and women, rogues and whores, some from Newgate, others from Bridewell. Newgate was an old London prison, Bridewell, a house of correction in the same town. That many of this class were brought to Maryland during the seventeenth century is shown by the passage of a law intended to break up the practice that ship captains, merchants and sailors had of importing " notorious felons and malefactors," who had been convicted of crimes in England, and selling them as servants in Maryland. Some of them were taken out of jail for the purpose of sending them to the colony.[4]

Robert Wintour, a member of Governor Leonard Calvert's council, said that it was probably true that the servants in the different colonies were " for the most part the scum of the people taken up promiscuously as vagrants and runaways from their English masters, debauched, idle, lazy, squanderers, jailbirds and the like." Captain Wintour maintained, however, that the class of servants who came to Maryland was better than the average. Almost all of the servants in the colony came from England or Ireland. Few came from any other place.[5]

While most of the servants in early Maryland were indentured, or indented, many, too, worked for wages. The indenture was an agreement entered into between the master and the servant. According to the provisions of most indentures, the

servant agreed to serve his master " in such service and employment," as his master should employ him.   In return the master
usually agreed to feed and clothe the servant during the period
of years stated in the indenture.   An indenture of this kind was
executed in duplicate, the two parts being indented by a notched
cut, or line, so that the two pieces of papers, or parchment,
fitted together when laid along side of one another.   The master retained one copy of the indenture, the servant the other.

Any number of such indentures can be found in the proceedings of the provincial and county courts.   There was, of course,
nothing to prevent the master and servant inserting special provisions in the indenture, such as a statement of the character of
the employment.   For example, one master agreed that during
the three years that the indenture was in force he would only
employ the servant as a cordwainer.   In another case, the master
promised that while the servant was in his employ he would
" learn him his books and to write," while by the terms of another
indenture the master agreed that if during the period of the
agreement he did not succeed in teaching the servant to read,
then he would teach him the trade of cooper or carpenter.   As
beating at the mortar was very hard work, one indenture stated
that the servant should not be required to do this kind of work.[6]

As has already been stated, many servants worked for wages.
Some were paid for their services in tobacco, some in money, or
perhaps in cattle, corn, clothing, or land.   One servant sued no
less a person than Governor Leonard Calvert for wages in
tobacco and clothing due for his services.   He begged the members of the provincial court to consider his condition which was
" naked and remediless."   His claim was allowed.[7]

Cecil Calvert, the Maryland proprietor, was not in favour of
having indentured, or apprentice servants, as he called them,
working on his plantations in Maryland.   It would be cheaper,
he thought, to hire servants in Maryland by the year which could
be done for a yearly wage of fifteen hundred pounds of tobacco.
This was better, he thought, than sending indentured servants
to Maryland for whom he would have to send supplies every
year from England.   Calvert wanted only such servants hired by
the year in Maryland that were necessary to look after his cattle

and his farm land. Their wages could be paid from such reve-
nues and profits as accrued to him in the colony.[8]

Before returning to a further consideration of indentured
servants, it is interesting to note some of the duties which both
wage and indentured servants performed. As it is sometimes
difficult to determine from the records whether an indentured or
wage servant is being discussed, this is a good reason for discussing all duties undertaken by those of the servant class. One
servant we find employed as an interpreter of the aboriginal
language. Others as coopers, carpenters, sawyers, joiners,
tailors, brickmakers and smiths. One man was employed by a
smith to keep his forge supplied with coal, beat his bread and
help him gather his tobacco.[9]

Most of the duties were performed out-of-doors. One servant was supposed to hunt and fish for his master, or to be " a
bayly for husbandry, or do any other work in the country." One
man worked " in a crop of corn and tobacco; " another acted as
a cowkeeper. Women as well as men worked in the field planting and raising tobacco and corn. The more fortunate maidservants acted as housekeepers, ladies' maids, or as gover-
nesses of children. In one instance, the mistress was to teach
the maid to read and sew.[10]

According to some Dutch visitors to Maryland during the
latter part of the proprietary period, most servants were employed in raising tobacco. Then, after working all day in the
tobacco fields, the servants in the evening were often compelled
to grind and pound grain, generally maize, for their master's
family and their own food. The Dutchmen cited one case as
an illustration of the importance planters placed on tobacco
raising. One master, they said, when a servant of his was dying
made him dig his own grave. If the master had made any of
his well servants do this, it would have taken them from the
tobacco fields.[11]

The same Dutchmen said that most servants were given
nothing but corn to eat and water to drink and yet with this
sparse diet they were compelled to work very hard. A servant
was required to work on Saturday afternoons. Members of the
Maryland assembly thought that no servant should be allowed

to rest on that afternoon. One wage servant was discharged by his master because he refused to work on Saturday afternoon.[12]

Some men acted as overseers of servants. One overseer's duties were to see that the servants under his charge "made good and sufficient crops of both tobacco and corn." He was also to look after all the livestock owned by his master. Another man acted as overseer " for and until the perfecting and finishing of one crop of tobacco and doing and performing of other services and labours." [13] ·

Let us return again to a consideration of indentured servants. What was the length of time a servant was required to serve under the terms of an indenture? The agreement, or indenture, might state the number of years in which case this provision would govern. An unskilled labourer was likely to have to serve longer than a workman who knew some trade, such as a carpenter, boatwright, cooper, etc. As the prospective settler often would need the services of such men, he would, in order to induce them to go with him to Maryland, offer them shorter terms of service. Some indentures were printed. The number of years which a particular servant agreed to serve would be inserted in the space provided for this at the time the agreement was made. The average period was five years.[14]

One of the reasons for exacting a certain number of years of service from the servant was that the master had paid the cost of the passage of the servant to the colony. This, as we have seen, usually amounted to about six pounds sterling. Therefore, it was considered reasonable that having obtained a free passage to the colony, the servant both on the voyage over and after his arrival in Maryland should work for his master a certain number of years in order to pay for his passage.[15]

Many of the indentures were executed in England before the colonist and his servant went to the New World. Some such agreements were entered into voluntarily by the servant. Other servants, however, especially those under age, were taken against their will, or kidnapped, by some enterprising sea captain who later sold their services to some one in Maryland. This was called " spiriting." It was said that some of the servants of Captain Thomas Cornwallis, distinguished Marylander, were very young children stolen from their parents. It was in order to break up this practice that the English authorities required certain formalities to be complied with before a servant could be taken to the colony. If over twenty-one, the indenture must be signed in the presence of some justice or magistrate, who must be satisfied that the servant made the agreement of his own free will. If, however, the servant was under twenty-one, then every precaution was taken to see that every boy or girl had obtained the consent of his or her parents, or of the master, if they were apprentices.[16]

The above regulations were not in force until 1682, or until near the end of the proprietary period. This was the reason perhaps that before this so many servants arrived in Maryland without any indentures at all. As the colonists were anxious to have the benefit of their labour, they, as well as the provincial authorities, were not particular in finding out the why and wherefore of their coming to the colony. The only question that seems to have concerned the proprietary officials was how long should the boy or girl serve whose passage had been paid to the province.

Beginning in 1654, laws were enacted governing this question. While it is impossible to discuss in detail all these statutes, the Act of 1666 can be taken as typical. By the provisions of this act, if any one paid for the transportation into the province of a servant, who had no indenture and was over twenty-two, then the servant must serve five years after his or her arrival in the colony. If the boy or girl was between the ages of eighteen and twenty-two, then he or she must serve six years, if between fifteen and eighteen, seven years, and if under fifteen, then until he or she reached the age of twenty-two. Obviously, this act was unfair in some respects. For example, any young person of twenty-three would only have to serve five years or until he or she was twenty-eight, while a boy or girl who was only twenty-two would have to serve six years. A child of fourteen had to serve eight years, and if only ten, twelve years. The only defence which can be offered for this difference in the period of servitude is that no persons twenty-two or younger were in a position to take care of themselves. For this reason, the provincial authorities wanted all boys and girls supported by the master or mistress until they reached years of discretion. It will be recalled that by the terms of indentures the master or mistress must feed and clothe the servant during the period of the agreement.

Still another injustice can be pointed out. Was it fair to compel a young person over twenty-two to serve five years? The cost of the servant's passage was six pounds sterling, the cost of feeding and clothing the servant for one year was about fourteen pounds sterling, or seventy pounds sterling for five years. Including the passage money, this meant a total outlay of about seventy-five pounds sterling. As we have seen, even the labour of a field hand would result in a yearly profit to the master of about fifty pounds sterling or two hundred and fifty pounds profit for the five-year period. If we deduct the seventy-five pounds sterling from this amount, the master made a total profit of one hundred and seventy-five pounds sterling. The only defense that can be offered for allowing a master to make such a large profit is that young persons over twenty-two arriving in the colony were hardly in a better position to support themselves than those under that age. During the five-year period, however, servants would not only become acclimated and hardened physically, but also, if they had a good master, they might be taught some trade. Most indentures had a provision stating that at the expiration of the agreement, the master must, at a slight cost to himself, furnish the servant with a few articles of clothing, corn, farming implements, etc. This was to help the servant to become self-supporting.

All who owned or kept servants were obliged to bring them within six months after they had acquired them before the justices of the county in which they lived in order to have their ages determined and entered in the county records.

It was necessary to do this no matter how the servant had been acquired whether by paying for the transportation, purchase or otherwise. Failure to do this made the master or mistress liable to pay a fine of one thousand pounds of tobacco. The only case in which it was unnecessary to do this was where the servant was over twenty-two years of age, as in this case the longest period of servitude that could be demanded was five years. If any owner of a servant, or any servant, objected to the age which the county justices had stated the servant to be, then in such an event the master, or the servant, was at liberty at any time during the period of servitude to produce a certificate showing the servant's correct age. No servant, once the time to serve had been decided upon by the provisions of the indenture, or, if there was none, by the determination of the court, was free to enter into another indenture which would oblige him to serve a longer period.[17]

One can easily imagine the feelings of a child, young girl or youth, as he or she stood before the justices of a court in order to have their ages determined. Newly arrived in a strange country, probably with no relatives or friends, it was a momentous time in their lives. Would the master or mistress whom they must serve prove kind or cruel? This and other thoughts must have crossed their minds.

Many prominent Marylanders brought their servants before judges in order to have their ages determined, including Colonel Henry Coursey and Thomas Gerrard, of St. Clement's Manor. In Charles county, there were many cases of where servants coming to the colony without any indentures were haled before the county justices to have them judge their ages. Henry Adams, himself a commissioner of the county court, brought servants before his fellow commissioners to have them pass on their ages. While some youths twenty-two or older came before the county court, most of them were younger, boys and girls ranging in age from eleven to twenty.[18]

In Kent county, too, there were many instances of where boys and girls were brought before the county commissioners to have their ages determined. Seth Foster, a justice of this county, brought a lad who was judged to be only eleven. He was told that he must serve Foster for ten years. In one case, the Kent county commissioners ordered a girl to serve her master for six years, but the justices added that if the original indenture was found, then the terms of that agreement would govern. Later an indenture was discovered by the provisions of which the girl only had to serve four years. A young boy, who rejoiced in the name of James Bringgergrass, appeared before the Kent justices who decided that he was fourteen years old. This lad was told that he must serve eight years unless within that time he could produce an indenture, or some record, which would show that he was older than fourteen.[19]

The case of Ricckett Mecane shows how a young boy might be compelled to sign an indenture, which, from his point of view was unjust. In a petition to the governor and council, young Mecane told how he had been taken by force, or spirited away from England. Upon his arrival in Maryland, he had been sold to Thomas Gerrard with the understanding that he would serve him for fifteen years. Ricckett said that as he had already been a servant of Gerrard's for six and a half years, he should be allowed his freedom. To require him to serve eight and a half years more, said the nineteen-year-old youth, was " contrary to the laws of God and man that a Christian subject should be made a slave."

It appears that young Mecane was one of eight Irish boys who had arrived in the colony at the same time. Most of the lads were so young that some one who saw them remarked that they should have cradles to rock them. Captain Robert Hinfield was the name of the sea captain who brought these children to the province. One witness, who was present at the sale of the boys, said that the captain had threatened them all with a whipping unless they signed agreements to serve for fifteen years. Another witness, however, said that the lads had signed " without force or constraint." Whether they had signed voluntarily, or because threatened with a birching, the boys, including young Mecane, were at such a tender age that they were incapable of knowing or understanding what they had signed. The provincial court declared that when he was twenty-one Mecane should be freed.[20]

There are several cases in early Maryland where a husband died leaving his wife destitute and faced with the problem of taking care of her children. Take, for example, the case of Andrew Hanson, a Swede who had come to Maryland from the Delaware. When he died he left four small children and a wife " big with child with the fifth." As Mrs. Hanson had no

way of supporting all these children, there was nothing for
her to do but "to dispose of some of them" to her friends.
The oldest of her children was Hance, a lad of nine. He was
bound over to Captain Joseph Wickes who was to have the
benefit of his services until he was twenty-one. Wickes, on his
part, agreed to feed and clothe the boy and also "to bestow
education and learning upon him, as reading in the English
tongue, writing of a good legible hand, and casting of accounts
in four several rules of arithmetic." [21]

When Mrs. Anne Gess found herself facing a similar situation, she bound over her three-year-old son to a justice of the
peace for a period of eighteen years. The lad was to serve
Henry Adams, the commissioner in question, "in all such services and employments," as his master should think fit. Adams,
on his part, agreed to furnish the boy with "sufficient meat,
drink, washing and lodging fitting for a servant." Mrs. Gess
also bound over her daughter-in-law to Thomas Baker, another
Charles county commissioner, to serve him as a maidservant.[22]

Two rather unusual agreements are to be found in the early
Maryland records. In one of them, Hannah Mathewes agreed
to serve Thomas Greene, then governor of Maryland, for a
period of four years if she failed to pay a certain amount of
tobacco and corn to his Lordship by a specified date. Walter
Gess, the husband of Anne Gess, made a similar agreement with
Richard Trew, a boatwright.[23]

There are many instances in early Maryland where a servant
was sold, or assigned, to another master or mistress for the
unexpired period of the indenture. Usually the agreement to
sell the servant specified the age of the servant and the time for
which the servant was to serve. Sometimes the sale described the
servant as being "a sufficient servant," or "an able manservant," or "a likely boy." In one case a jury was called upon
to decide whether an agreement to sell "an able servant" had
been carried out. The jury's verdict was that the man tendered
by the vendor was "a likely manservant to outward appearance." [24]

Of course, women servants were sold as well as menservants.
One man agreed to deliver a maidservant who was "sound

7

and in perfect health." An amusing case involving the sale of a maidservant occurred in Charles county. James Neale had a woman servant whom William Marshall wished to buy, provided Neale was willing to sell her to him. Neale said that he could dispose of her, as he had another girl " to serve his turn and that he could make a shift with her." Neale added that while the girl servant Marshall wished to buy had her good qualities she also had her faults. She was a good cook, Neale declared, and could make good butter, but she was also both " a whore and a thief." If Marshall could break her of these habits, Neale was sure that she would make an excellent servant. The records do not disclose whether Marshall finally decided to buy the servant. If he did, Neale told Marshall that he could also have the clothes which he had bought for her at the same price which he had paid for them. The girl's clothing included three pairs of shoes, two pairs of worsted stockings, " a pair of bodies (bodice)," and a petticoat made of broadcloth.[25]

Whenever a servant was sold or assigned to another master during the period of the indenture, the new master must assume all the duties required of him by the terms of the agreement. The same was true of the servant. Indeed the law so required. The new master, for example, must at the end of the indenture furnish the servant with corn and clothing as this provision was in almost all indentures. It did not make any difference how often a servant changed masters. Whoever was the master at the time was bound by all the conditions incorporated in the original indenture.[26]

Not all servants were sold for so many pounds of tobacco. There were instances where they were exchanged for livestock, for " a house and plantation," and for a boat. Sometimes a manservant would be exchanged for a maidservant.[27]

There were several ways in which a servant could obtain his or her freedom. End of the term of the indenture was the most usual case. If the master was very brutal in his treatment of the servant, the latter might be freed, or at least sold or assigned, to another master. They might be freed by the provisions of their master's will. Robert Mackey obtained his freedom when it was shown that his master, Bassill Little, had

declared before his death that Mackey had been " so faithful a
servant to him in his sickness that if it pleased God to let him
live, or take him out of this world, he would make Mackey a
freeman, or if please God he should live, he would make
Mackey as good as a freeman." In view of this statement the
judges of the provincial court ordered that Mackey should be
" released, and set free from all claims of servitude." Before
coming to Maryland, it appears that Mackey had been " a groom
to some great persons both in Scotland and England for a
space of six years." [28]

During the term of the indenture, the master might sign a
statement, or make an oral declaration, freeing the servant, or
he might in " open court " set his servant free and " publish
him as free a man as himself." One master said that if he died
before the servant's term of servitude had ended, then the servant should be free. The servant himself, or a friend or relative, might buy the servant's freedom. One servant obtained
his freedom on condition that for the next three years he would
give his former master each year " fifty dressed skins." One
master agreed to free his servant " in consideration of his
curing his leg." [29]

Bridgett Nelson, a maidservant, succeeded in obtaining her
freedom when she informed the members of the provincial
court that her master, Quintin Counyer, had before his death
offered to pay for her freedom out of his estate and that the
same intention was shown by a letter which he had written. It
seems that on his deathbed Counyer had also declared that he
was married to Bridgett and to prove it they had broken a
piece of silver which was then divided between them.[30]

It was the policy of the provincial authorities to provide servants at the end of their period of indenture with enough in the
way of food, clothing, etc., in order to help them to become
self-supporting. The ambitious servant hoped to become a
planter himself. By the terms of a law passed in 1640, every
manservant must be given at the end of his servitude one
good suit of kersey or broadcloth, a shift, or shirt, of white
linen, one new pair of shoes and stockings, two hoes, one axe,
three barrels of corn and fifty acres of land, five of which should

be plantable.   A woman servant was to receive a year's provision of corn and the same proportion of clothes and land.[31]

About fifteen years later, or in 1654, a slight change was made
in the character of the clothing to. be given each manservant.
Thereafter each man was to have, besides the clothing he wore,
one cloth suit, one pair of canvas drawers, one pair of shoes and
stockings, a new hat or cap, and two shirts instead of one.   By
the terms of a particular indenture a servant might receive even
more than the law required.   Take the case of Cornelius Cannaday, a brickmaker.   Not only did Thomas Gerrard, of St.
Clement's Manor, who was Cannaday's master, promise to
clothe and feed him, but at the end of his period of indenture
Cannaday was to receive two hundred acres of land, with " a
sufficient house upon the same twenty-five feet long and sixteen feet wide," a bed filled with " feathers or flock," a pillow,
one rug, two dishes, one pot and six spoons.   In addition,
Gerrard promised to give Cannaday two cows or heifers with
calf, two sows with pig, two goats with kid, and five barrels
of corn.   Probably the reason that Cannaday was promised so
much was that as a brickmaker his services were prized and he
could therefore demand very favorable terms.   Such special
agreements were upheld by the local courts.[32]

Although it was only twenty years after the founding of the
colony, under the provisions of the Act of 1654, mentioned
above, the servant no longer was allowed to receive fifty acres of
land at the expiration of his period of servitude.   The reason
for this change, provincial authorities explained, was that the
master himself now received only fifty acres of land for having
paid for the transportation of the servant to the colony, whereas
formerly a master was entitled to receive one hundred acres.[33]

There were many suits instituted by servants to recover their
allowance of clothing, corn, etc., according to " the custom of
the country."   Most of them were successful.   Hercules Hayes,
for example, asked for and obtained his " corn and clothes."
Hercules maintained that he should not be compelled to serve
his master any longer, as whatever " deficiencies of service " he
had been guilty of during the past years, he had been corporally
punished for them.   And the court agreed with him.   In one

case, however, the master in answer to the servant's suit for his allowance of corn, clothing, etc., said the latter had left his service before the time stated in his indenture had expired and that he had never made good the time lost by additional service.[34]

Some of the following cases showing the treatment of servants at the hands of their masters will probably explain why all the suicides were among members of the servant class. In the next chapter dealing with homicide two masters are hanged for beating their servant boys to death. In one instance, we find a servant ran away from his master to live with Indians. This man said that he would rather remain with the savages than return to his master " to be starved for want of food, clothing, and have his brains beaten out." [35]

Servants must be respectful in their relations with their masters. When David Stevens admitted that he had " scandalously " abused Richard Preston, his master, the members of the provincial court ordered the sheriff to give him ten lashes on his bare back. Even more severe was the punishment administered to another servant named Owen Morgan who was employed by William Hopkins, and who complained to the provincial court that when he was giving Morgan some "correction," the latter struck him with a club and threatened him "with many uncivil and opprobrious words, with cursing, swearing, blaspheming, etc." For such actions Morgan was given thirty lashes.[36]

Servants were frequently disciplined by their master or mistress. Although it was reported that Sarah Goulson had beaten her maidservant " two hours by the clock," this charge was never proved. When James Woosey had an argument with his mistress, she boxed him on the ears and struck him " divers times." In another case, where the maidservant of Mary Taylor took some sugar which did not belong to her, her mistress " beat her for it." Mary Bradnox, the mistress of Anne Stanley, threatened to have her whipped if she said anything about having seen her mistress rem in rem, that is, copulating with John Salter. Mrs. Bradnox, being the wife of one of the Kent county justices, naturally wished to be discreet. Anne said that the threatened beating was " all the prompting " her mistress used. What more could Mrs. Bradnox have used? [37]

If one occupied the status of an indentured servant, it was not always wise to complain of the character of the food which a master provided. Four servants employed by Richard Preston found this to be the case. These four men complained to the provincial court that their master did not allow them " sufficient provisions for the enablement to our work, but straightens us so far that we are brought so weak, we are not able to perform the employments he puts us upon." The petitioners added that although they wanted no more than what was necessary for their sustenance, yet Preston allowed them nothing but bread and beans.

To this complaint Richard Preston replied that his servants had refused to work unless he supplied them with meat. Preston said that as he had been unable to get any meat for them, he had supplied his servants with fish, sugar, oil and vinegar. In view of the fact that it was not his fault that he had been unable to get any meat, Preston thought that his servants should receive some punishment for being so perverse lest " a worse evil by their example should ensue by encouraging other servants to do the like."

The judges of the provincial court agreed with Preston and ordered that the petitioning servants should be " forthwith whipped with thirty lashes each." The court also ordered that " two of the mildest servants," who had not been so refractory as the others, should be pardoned and the two so pardoned should whip their companions. The threat of a lashing made all the servants quickly change their attitude. Kneeling before the judges of the provincial court, they begged their master, Mr. Preston, and the court to forgive them " for their former misdemeanors " and promised obedience in the future. In view of their humble attitude, the members of the court declared that they would suspend the sentence of the whipping for the present, but at the same time warned the four servants that they must be on their good behaviour towards their master " ever hereafter." [38]

That the master was not able to prove insubordination in every case is shown by the suit which Robert Kedger (or Cager) brought against William Black for neglecting his employment as overseer of his servants. It was Kedger's contention that Black tried to persuade his servants to be disobedient and " to villify and abuse him and his wife." When the case came up for trial, several witnesses testified how Black had told the servants, of whom he was in charge, to cut clubs to knock their master on the head, and that it would be " a good deed " to poison their master. In order to carry out the latter plan, it was said that Black had two pones (corn pone) made with holes in them for the insertion of poison. The testimony of another witness was quite different. He said that when Kedger had accused Black of stirring up the servants under his charge against their master, the latter had denied it. Although the weight of evidence supported the charge which Kedger had made against Black, the jury, before whom the case was tried, decided that Kedger had " no cause of action " and dismissed his suit. The members of the provincial court confirmed this verdict.[39]

As we shall later see, there were a number of cases where servants were maltreated by their masters and the latter escaped without any punishment, or, at least, without being adequately punished for their brutality. There were, however, other cases in which the servant received justice at the hands of law. In such cases the court usually either ordered the servant freed, or sold or assigned, to another master. This was especially true where a maidservant had been ill treated. Thus when Susan Frizell complained of " extreme usage " at her master's hands, the provincial court decided to set the servant free. So, too, when Mary Jones convinced the members of the same court that she had been treated too harshly, the judges ordered that neither her master or mistress should any more " meddle with her for matter of correction," but should sell or exchange her " with all convenient speed that may be." [40]

Sarah Hall, a maidservant to Thomas Wynne, in a petition to the members of the provincial court, complained that she had been ill used by her master and mistress " in beating and abusing her." At the trial of this case Sarah Evans, also a servant in the employ of the Wynnes, said that when her mistress asked Sarah Hall a question to which she made no reply, Mrs. Wynne boxed her on the ears. Nicholas Rawlings was then called to the witness stand, who testified that he saw—

Mr. Wynne give his servant, Sarah, a kick on the breech and a box on the ear, and threatened to knock her down with a chair, but did not do it, but that for what occasion he knows not, it being just as he came in the doors and did not see the beginning of the difference.

In view of the evidence which had been produced as to the ill treatment received by Sarah Hall, the judges of the provincial court asked two sworn appraisers to report what in their judgment Sarah's services were worth, she having still two more years to serve. The appraisers estimated Sarah's worth to be one thousand pounds of tobacco " at the most." After receiving this report, the provincial court ordered that Sarah should be sold to some other master or mistress.

The decision of the provincial court in this case seems fair and reasonable. While Mrs. Wynne might have been justified in boxing Sarah Hall on the ear when she refused to answer, her husband was hardly warranted in kicking the girl " on the breech " and threatening to hit her with a chair no matter what " the difference " between them.[41]

Among the most brutal masters and mistresses of servants were Captain and Mrs. Thomas Bradnox. The captain was one of the justices of the peace of Kent county. Indeed, on one occasion, it was reported that when he was drunk, Bradnox swearing " like a madman, . . . took tobacco stalks and beat his servants." The treatment of a servant girl by the name of Sarah Taylor by Captain and Mrs. Bradnox was especially cruel. It is little wonder that Sarah complained to the county authorities. At the hearing witnesses said that they had seen black spots on her back and arms. A rope had been used to beat the girl. It seems that Sarah ran away from her master and mistress and took refuge with some other planters. Although one colonist who harbored Sarah was not punished, the other who had played the part of a good Samaritan was compelled by the Kent county commissioners " to ask Captain Bradnox's forgiveness in open court and promise never to commit the like again." As for Sarah, one of the county justices thought that she should be whipped for running away. The rest of the court, however, thought otherwise. They told Sarah, who was present in court, that they thought the stripes already given her by her master were sufficient corporal punishment. She must, however, on her knees ask Captain and Mrs. Bradnox's forgiveness for having run away " and promise amendment for the future." Sarah did as she was told, though we imagine rather reluctantly.[42]

Back again with the Bradnoxes, Sarah was subjected to more brutal treatment. No doubt, her master and mistress resented her having complained to the county commissioners about their previous maltreatment of her. And probably, Captain Bradnox, himself a justice of the county, did not relish the reflection on his conduct. At any rate, during the next few years, we find Sarah again complaining of ill treatment, of " abuses and stripes " given her without cause. The poor girl ran away again and not knowing to whom to turn lived in the woods and almost starved " with eating trash." Captain and Mrs. Bradnox, on the other hand, maintained that they were justified in chastising Sarah for " her neglect in several things." They produced in the Kent county court the stick with which they had beaten her. The Bradnoxes even accused Sarah of having stolen some clothing belonging to them, but a jury acquitted the girl of this charge.[43]

Unable to endure the treatment which she was receiving at the hands of her master and mistress any longer, Sarah appeared in person before Captain Robert Vaughan, James Ringgold and Nicholas Pickard, who were commissioners of the Kent county court. Sarah told these justices how, on one occasion, when she was working in the kitchen her master and mistress entered and " suddenly fell upon her." Mrs. Bradnox, Sarah continued, held her down while her husband beat her " with a great rope's end . . . and so unreasonably that there is twenty-one impressions of blows, small and great, upon her back and arms." After they had finished whipping her, Captain Bradnox remarked, "Now, spoil me a batch of bread again."

When she had concluded her account, Sarah undressed and showed the three commissioners the marks on her back and arms which her master had inflicted with a rope's end. Upon the conclusion of the interview with Sarah, Justices Vaughan, Ringgold and Pickard drew up a report incorporating what Sarah had told them and also an account of the condition in

which they had found her body and the fact that Captain Bradnox admitted beating her and would answer for it. This report
was considered by all the Kent county commissioners at the next
meeting of that court. At the same session, a man by the name
of Joseph Newman, who had been living with the Captain and
Mrs. Bradnox, informed the county justices how, on one Sunday
morning, the Captain grabbing " a three-footed stool " hit Sarah
Taylor on the head with it. According to Newman, Bradnox's
only reason for doing this was that Sarah had picked up a book
to read. " You dissembling jade," Bradnox had exclaimed,
" what do you with a book in your hand?"

After Newman had finished testifying, the Kent county court
was not long in deciding " to discharge Sarah Taylor of her
apprenticeship" with Captain and Mrs. Thomas Bradnox.
They did so, they declared, because of " the imminent danger
likely to ensue by the inveterate malice of her master and mistress toward her." This " malice " was shown by the way in
which Sarah had been punished.[44]

Captain Bradnox at once tried, but without success, to upset
this decision of the court by claiming that Sarah Taylor had
made false statements. Mrs. Bradnox then had the audacity to
claim that the county commissioners had no right to set her
maidservant free. The governor, to whom the case was
appealed, asked Henry Coursey and Edward Lloyd to decide
this question. After hearing both sides present their views,
Lloyd and Coursey compelled each of the Kent county justices
to pay Mrs. Bradnox two hundred and twenty pounds of tobacco
for having freed her maidservant. It is hard to understand or
to justify this decision. Having so brutally treated Sarah Taylor, it would seem that not only should she have been freed
without compensation to either her master and mistress, but
that both should have been fined or punished in some way.[45]

In the cases just discussed, it was always a maidservant, who,
because of the bad treatment she had received, was ordered
either freed or sold to another master or mistress. The same
decision was sometimes reached in the case of a boy servant
who had been maltreated. Take, for example, the case of John
Ward. The Charles county court commissioners summoned

Arthur Turner, his master, to explain why he so ill treated this youth " that the voice of the people crieth shame thereat." Turner was ordered to bring the lad before the justices of the court together with a copy of his indenture. Turner did as he was instructed. By the terms of the indenture, which was made when Ward was a small boy of nine or ten, it appears that John had agreed to serve Turner until he was twenty-one. Turner, on his part, agreed not only to clothe and feed the lad, but also to teach him the trade of a cooper, or carpenter, in case he could not bring your Ward " to reading." When Ward reached his majority, his master agreed to give the youth " double apparel, three barrels of corn, a cow and sow, and fifty acres of land."

As he stood before the county commissioners, young Ward must have been a pitiful sight. The lad was almost naked. What little clothing he had on was ragged and torn. On one of his bare legs could be seen " a most rotten, filthy, stinking ulcer " that was most loathsome to " the beholders thereof." The youth's hair seemed " to be rotted of ashes." The county justices lost no time in freeing young Ward who was now almost twenty-one. Let us hope that the county commissioners also compelled his negligent master to give the boy the clothing, corn, livestock and land, promised him by the provisions of the indenture.[46]

There are several cases of maltreated servants, who, although not ordered freed or sold, were, however, afforded some relief. Thomas Markeen, a servant of William Chaplin, complained that though he was physically unable to perform his duties as well as he " formerly hath done," yet his master tried to compel him to do so by threatening and beating him. The provincial court ordered Chaplin to stop punishing Markeen until the latter's complaint was " heard and determined " at the next general court.[47]

Another servant by the name of William Ireland complained to the provincial court about the treatment he received at the hands of his master, Captain Philip Morgan. He said that Morgan " did unhumanly beat him," that his master made him and other servants in his employ " beat their victuals " during the night, and that they often lacked sufficient food. After taking these complaints under consideration, the provincial court judges ordered Captain Morgan to stop beating or striking Ireland " unlawfully," that his servants should be compelled to work only at " seasonable times," and that he supply them all with sufficient food.[48]

Mr. and Mrs. Thomas Ward, it would seem, escaped with rather light punishment for the beating they administered to Alse Lutt, a maidservant in their employ. For having run away, Mrs. Ward whipped Alse's bare back with a peach tree rod. After this chastisement was over, Mrs. Ward put salt on the wounds made by the rod. When she was doing this, the girl cried out and asked her mistress to use her like a Christian. To this Mrs. Ward replied, " Do you liken yourself like a Christian? " Not long after this the girl died. The jury, before whom this case was tried, found that while the punishment given by Mrs. Ward was not the cause of the servant's death, it was, however, unreasonable considering " her weak state of body." The county justices then imposed a fine of three hundred pounds of tobacco for the " unreasonable and unchristian-like punishment." [49]

While the imposition of a tobacco fine was slight punishment indeed, there are cases in the early records where the owner of a servant escaped scot free. One of such cases was when Anne Nevill, the wife of John Nevill, was tried before a jury for having feloniously killed her maidservant, Margaret Redfearne, by beating her to death. Among the witnesses who testified in this case was Michael Farmer who said that he had seen Mrs. Nevill strike Margaret with her shoe and then had ordered the servant to go inside the house and strip herself naked. Mrs. Nevill then went in and proceeded to give Margaret a sound beating with a peach tree rod. The marks of this beating were later seen on the servant's thighs by another witness.

Susan Barbary, when called to the stand, testified that upon her examination of Margaret's body she had noticed black and blue marks on her throat and breast and also across her back. According to Margaret, these had been caused by her mistress whipping her after she had been thrown across a log. Several witnesses testified that Margaret had told them that if she died it would be because of the beatings she had receivd from her mistress. Indeed the girl herself on her deathbed said that her condition was due to the " bad usage " which she had received from Mrs. Nevill. After Margaret's death, a coroner's jury was impanelled to examine her body. They reported that the corpse showed signs of having been beaten. Despite all this incriminating evidence, the jury before whom Anne Nevill was tried brought in verdict of " Not Guilty." The case is indeed a travesty on justice.[50]

A curious example of servant maltreatment, arose in connection with a suit for defamation. William Parrott claimed that his maidservant, Alice Brasse, had injured his reputation by spreading abroad a report that while she was being whipped her master had broken two of her ribs and that if she died within a year she " would lay her death to him." Witnesses called in this case seemed to have differed as to the severity of the beating which Parrott had administered to Alice. One man said that although the girl, immediately after her whipping, lay down as if she was about to die, the next morning she went about her business " in dressing victuals and righting up the house." A woman who examined Alice about two weeks after her beating, testified that her body showed black and blue marks and she was "swelled down the back." In other words, there seems to have been sufficient evidence to show that Alice Brasse was not slandering her master, but merely telling the truth. Accordingly, the judgment of the provincial court was that there was no cause of action and dismissed the case. But what of poor Alice? Was she freed because of her maltreatment, or was her master warned to be more considerate? The records do not disclose the answers.[51]

There are other cases in the early records where the master or mistress after maltreating a servant escaped with no punishment, or if some was imposed, it was not adequate. It will be recalled how Sarah Taylor, a servant in the employ of Captain and Mrs. Thomas Bradnox, was finally freed because of the bad treatment she received at the hands of her master and mistress. Sarah was not the only servant whom the Bradnoxes maltreated.

Thomas Watson, another servant, died under circumstances that made it look as if the captain and his wife had beaten him to death. Some of the evidence produced before a grand jury was both convincing and revolting. Sarah Taylor, at that time still in the employ of the Bradnoxes, testified how for six days Watson had been confined in a building near the main house without food or water. Sarah said that she had seen Watson drink his own urine. She and the other servants were afraid to take the poor man any food or water as they had been warned by the captain and his wife that they would be whipped if they attempted to do this. When Watson finally managed to escape, he was so weak that he had to crawl on his hands and knees to the main house. On one occasion, Sarah described how when Watson was cooking and having difficulty in turning the spit, his master, angered by the delay, struck him so hard on the face that blood issued from his mouth and nose. According to Sarah Taylor, Watson had declared two or three days before his death that his master and mistress were the cause of his death.

Thomas Southern, another servant in Captain Bradnox's employ, told what happened when he and Watson were sent out to cut some wood. While they were resting for a few minutes from their labour and having a smoke, or " pipeing," as Southern expressed it, Bradnox came out and asked them if that was the way they worked. With that the captain cut a hickory stick and proceeded to give young Watson " many a stripe," in fact about " fifty odd." On another occasion when they went out to a thicket to work, Southern said that Bradnox beat Watson with a stick until he broke it and then he cut another and told John White, a fellow servant, " to drive him along."

White himself testified that Watson had " very bad usage, not fit for a Christian," and how Captain Bradnox had followed the youth all one morning with a stick in his hand to make him fetch wood and had whipped him " more like a dog than a Christian." Both White and Sarah Taylor said that before his death Watson had complained of an abscess on his back which was due, he said, to a blow which Mrs. Bradnox had given him with " a cowl staff." White further testified that he had examined Thomas Watson's members and that the skin " did drap away . . . in his sickness " because of his bad usage and want of proper care.

In order to refute the testimony of these witnesses, Captain and Mrs. Bradnox attempted to show that Watson had died not as a result of their maltreatment but from disease. One man, who examined the youth before he died, thought he was suffering from dropsy, while another, who made a similar examination, declared it was scurvy, and well it might have been as this disease is due to the lack of proper nourishment.

John White, who had given such incriminating testimony against Bradnox, was again recalled to the witness stand. White now admitted that he had heard Watson say that his disease would eventually kill him. White also made a statement about Sarah Taylor which tended to lessen the importance of her testimony. This was by saying that he had heard Sarah remark that if she could not get the upper hand of her mistress she would run a knife into her bowels. In view of the treatment which the maidservant received, we cannot blame her for expressing this wish.

After White had finished testifying, the Bradnoxes asked a youth by the name of Charles Hollingsworth to take the witness stand. The testimony of this lad tended to lessen the credit which could be given to any of John White's statements. Hollingsworth said that White was " an idle runaway and of no credit . . . that he was forsworn, or perjured, and that White had broken open a storehouse as the general report went . . . "

Finally, the defence produced evidence to show the result of a blood test made on the dead body of Watson. For this reason they had a witness testify that when Captain Bradnox touched the corpse no blood came from it. This was supposed to prove the innocence of the captain, as had he been guilty of beating the youth to death, many at this time thought the corpse would bleed.

During the trial of this case, Captain Bradnox died. But Mrs. Bradnox, as we have seen, had also been accused of beating young Watson. Despite the evidence produced, the grand jury decided not to indict her. Proclamation was then made—

That if any man have anything to say on the behalf of the Lord Proprietor against Mary Bradnox, the relict of Thomas Bradnox, concerning the death of Thomas Watson, they shall be heard.

When no one came forward, Mrs. Bradnox was "cleared by proclamation." [52]

Thomas Watson and Sarah Taylor were not the only servants whom Captain and Mrs. Bradnox abused. James Wilson, another man in their employ, received whippings. When this man also died, suspicion again rested on the Bradnoxes. A coroner's inquest was ordered who found that—

The material cause of his (Wilson's) death was an intermitting fever joined with dropsy, or scurvy, as commonly understood, and further that the stripes given him by his master not long before his death were not material.

In view of all these instances where Captain and Mrs. Bradnox were accused of beating their servants, one cannot help but feel that neither of them were properly punished for their actions. Perhaps the fact that Captain Bradnox was an influential man, being a member of the Kent county court, had something to do with his escaping punishment.[53]

Whatever his position, the captain could not have been a man of good character. Although he called himself a gentleman, he could not sign his own name. Not only was Bradnox faced with charges of beating his servants, but on one occasion he was arrested for drunkenness and disturbing the peace. Although not punished for this offence, he was at the same session of the court fined for his profanity. At one time, the captain was charged with having "wickedly and feloniously killed and eaten . . . a two-year-old steer." After his death, it appeared that he had "converted to his own use" about six or seven thousand pounds of tobacco with which he was supposed to have bought "drums and colours" for the use of the company of soldiers that he commanded.[54]

Captain Thomas Bradnox sided with Richard Ingle in his attempt to overthrow the proprietary government in Maryland. At this time, Bradnox, in command of "a crew of rebels," seized the house of Margaret Brent on Kent Island making it

" his garrison." He also killed some of her cattle and " spent and wasted corn and other provisions of hers." He and his men also made an attack on the house in which Captain Robert Vaughan, a fellow commissioner, lived. During the attack, two of Vaughan's servants were killed and the justice himself was taken prisoner by Bradnox, who confined him in his own house " for the space of three weeks, or thereabouts." The governor's prosecuting attorney was right when he said that Captain Bradnox had, at various times, been guilty of " the crimes of rebellion, sedition, rapines, thefts, robberies, and other such-like felonious practices." That a man of this character would maltreat his servants there seems little doubt. What is more remarkable is that a man of Thomas Bradnox's type should be appointed to office. County commissioner, sheriff and captain of militia, these were the positions which Bradnox held. Surely the standards of seventeenth-century Maryland must have been low indeed to allow such an ignorant and vicious man to hold these offices.[55]

John Grammer was another man who escaped being punished for maltreating a servant named Thomas Simmons. After the latter's death, a coroner's inquest was held. Stripes were found on the servant's body which had evidently been inflicted with a whip. These stripes, members of the inquest thought, might have been " a furtherance to his death."

The verdict of the coroner's jury was, however, contradicted by the findings of two chirurgeons who made a similar examination. These two men reported that they could find no contusion upon the fleshy parts of Simmons' body and that the corpse seemed clear of " inward bruises." The two chirurgeons were of the opinion that the lungs, liver, and heart of the deceased servant were in such a diseased condition that they doubted if the servant could have lived much longer under any conditions.

Witnesses of the beatings which Simmons had received were examined. Christopher Anderson gave one of the most incriminating accounts. Apparently, he acted in the capacity of overseer of the other servants in Grammer's servitude. On one occasion, it seems, Anderson was told by Grammer to give

8

Simmons a whipping for having run away. The overseer did as he was instructed, tying the youth's hands and stripping him naked before he lashed his back with a small whip cord or cat-o'-nine-tails. After this punishment young Simmons went out in the fields to do some weeding. When Grammer, the boy's master, decided that he was not doing the job very well, he came out from his house and whipped the lad with a knotted rope until he fell to the ground. Grammer kept on beating Simmons until he made him get up again. At this moment Anderson intervened and told Grammer that he thought the boy was almost dead. This remark only angered Grammer who said that the youth was " a dissembling rogue " and proceeded to give him another whipping. After this last beating, Anderson said that he was now sure Grammer had killed the young man. At this Grammer exclaimed, " Lord have mercy upon us," and added that he would clear himself of the boy's death. After this they carried Simmons back to his master's house. When asked how he felt, the boy replied that he was very ill and that his master had killed him. Soon after this he died.

Sarah Hunt, a maidservant in Grammer's employ, testified that Christopher Anderson had told her that he had heard Grammer threaten to kill Simmons. When Sarah asked Anderson if he was not sorry that he had given young Simmons a whipping, Sarah said that he retorted, " No, I could have given him ten times more." According to Sarah, after the youth was punished his back was washed with salt and water.

Other servants in Grammer's employ testified as to the brutal treatment which Simmons had received. One of them said that Grammer had made Anderson beat young Simmons and that the overseer had given the youth almost one hundred blows on his bare back with a cat-o'-nine tails. These blows, this witness thought, were " the occasion of Simmons' death." Other servants said that they had seen Grammer himself give Simmons a severe whipping with a rope while the boy was working in the fields.

A grand jury considered the above evidence in connection with the following indictment which had been drawn by the attorney general:

Let it be inquired for the Right Honourable, the Lord Proprietor, whether John Grammer, upon the ninth day of July, 1664, commanded Christopher Anderson the body of Thomas Simmons to whip, and whether the said Anderson the said Simmons did with a whip, commonly called a cat-o'-nine-tails, with one hundred stripes whip and beat, and whether John Grammer on the day aforesaid the said Thomas Simmons with a rope with a knot at the end did afterwards beat with twenty stripes in such unlawful and unreasonable manner that of these blows the said Thomas Simmons on the 11th day of July aforesaid did die and so whether the said John Grammer and Christopher Anderson the said Thomas Simmons feloniously did kill, contrary to the peace of his said Lordship's rule and dignity.

When the grand jurors refused to indict Grammer with the evidence they had before them, the judges of the provincial court asked them to state in writing their reasons for having reached such a decision. The grand jurors did as they were directed. Among the reasons which they gave for their opinion were that although it might have been true that Simmons had been whipped as often as the indictment alleged, yet there was no evidence to show that any of the blows administered by Grammer, or Christopher Anderson, his overseer, caused Simmons' death. The grand jurors also relied on the statement of the two chirurgeons, who, they said, had reported that none of the stripes given Simmons had touched " any principal part " of his body which might have hastened or caused his death. As the judges of the provincial court were satisfied with these reasons, they discharged the grand jury. Grammer thus escaped punishment for his brutality.

It is hard to agree with the final determination of this case. As against the opinion of the two chirurgeons which was favorable to Grammer, we have the finding of the coroner's inquest and the testimony of many witnesses, all of which went to show that the beatings that Simmons received must have at least hastened if they did not cause his death. If the youth was sick, this should have made his master reluctant to whip him at that time.[56]

That Grammer must have been a brutal master is borne out by the fact that not long after the death of Simmons another of his servants died. A coroner's inquest was held. Their finding was—

That having viewed the corpse we do find no impression of any stripes upon his body, but unanimously concur in our judgment that want of good diet and lodging has been the chief furtherance and cause of his death. This is our verdict to the best of our judgments, skill and knowledge in this business.

Although he was confined in prison for almost a year awaiting trial, Grammer appears to have escaped without being punished for failing to furnish this servant proper " diet and lodging." [57]

One more case of extreme brutality to servants remains to be considered and that was when Pope Alvey, a cooper by trade, beat his servant, Alice Sandford, to death. Members of a coroner's inquest held after her death reported:

Having viewed the body of the aforesaid servant, we found no mortal wound, but the body beaten to a jelly, the entrails being clear from any inward disease, to the best of our judgments and the doctors that were with us. But, if it were possible that any Christian could be beaten to death with stripes, we think the aforesaid servant was. And this is our joint verdict.

From the testimony of two witnesses we are able to get an idea of the kind of treatment which Alice received at the hands of her master. John Bessick, one of these witnesses, spoke of finding Alvey and Alice in the woods one day. When he approached them, the girl exclaimed: " Take notice that my master has killed me." Angered by this remark, Alvey explained that he could not get his servant to walk, or, as he said: " This damned whore, I cannot get her along no further than I baste (thrash) her." Thereupon Alvey cut a stick and swearing that he would make her walk lifted up the skirt of her waistcoat and beat her upon her naked back. In order to protect her back, Alice put her hand back of her. Alvey thereupon put his foot on her hand and whipped her again. This beating continued until he had broken three sticks upon her. Still Alice refused to walk and again Alvey beat her. Finally, with the help of another man named Charles Alexander they succeeded in carrying the poor girl to a nearby plantation. When Alice started to cry, Alvey grabbed her by the nose and tried to stop her breathing. Shortly after this she died. Lifting up the girl's head Alvey remarked:" I think really that she is dead." Charles Alexander told how just before Alice died Alvey took a porringer of cold hominy, and, after opening the girl's mouth with a pair of tobacco tongs, poured the contents of the porringer down her throat.

After considering the verdict of the coroner's inquest and the testimony of the two witnesses, a grand jury indicted Alvey for feloniously killing his maidservant " contrary to the peace of his Lordship, his rule and dignity." In his subsequent trial before a petty jury, he was found guilty. When Alvey was asked what he had to say for himself, he craved benefit of clergy, which was granted him, and " the book being given and demanded whether he read or not, answer was made that he read." Thereupon, the provincial court ordered that Alvey should be forthwith burnt in the brawn of his right hand with a red-hot iron. If any one deserved to be hanged, it was the brutal master of Alice Sandford. By pleading benefit of clergy, however, Alvey was able to save himself from the gallows.[58]

A petition presented to the provincial court by Paul Marsh throws additional light on the character of Pope Alvey. It seems that Marsh had hired out one of his servants to Alvey in consideration of a stated amount of tobacco. Not only had Marsh not received the tobacco, but he was also afraid that his servant would never be returned to him. The grounds for his fear, Marsh said, were that Alvey had already been on trial for having caused the death of one servant by his maltreatment and that recently another servant had died in his employ who had " laid his death " to Alvey. Marsh said that it was reported that the last servant had died as a result of being kicked by Alvey. In view of these facts, March not only wanted Alvey to pay him the tobacco he owed him, but he also wanted some " security " that his servant would be returned to him at the end of the period for which he was hired. This case appears to have been settled out of court.[59]

It was this same Pope Alvey, who, it will be recalled, was

accused of stealing. As was the case with all brutal masters or
mistresses of servants, they were a pretty low order of human
beings.[60]

In view of the treatment they sometimes received, it is not
surprising that servants often ran away from their master or
mistress. It was not likely, however, that the servant would
escape. As soon as his departure was discovered, " a hue and cry "
would be raised and a party would start out in pursuit of him.
Constables of each hundred in every county were instructed to
see that this was done. A reward of two hundred pounds of
tobacco was promised any one who seized a runaway servant.
This amount had to be paid by the master of the servant. As
soon as a runaway was caught, he was turned over to the nearest justice of the peace. Notice of the servant's capture was then
posted in the county court near where the servant was seized,
as well as in the other county courts and at the provincial court.
This was in order that all persons might " view the same and
see where their servants are and in whose custody." [61]

In Talbot county, a number of persons claimed the reward of
two hundred pounds of tobacco for having captured, or " taken
up," a runaway servant. In order to recover three servants who
had run away from his service, John Kinemant had to pay a
total of six hundred pounds of tobacco. Kinemant must have
been a severe master to have so many servants run away. On
one occasion, it was said, that he thrashed one youth until he
himself was " aweary." [62]

The help of the neighbouring Indian tribes was solicited in
capturing runaways. Any savage who captured a runaway
servant and brought him before some magistrate would receive
a matchcoat as a reward. Indeed, the Indians were told that
they must apprehend all runaway servants and return them and
that they would be held responsible if they helped, or assisted,
a runaway to escape.[63]

Steps were taken by the proprietary officials to recover servants who had escaped to other colonies. A reward of four
hundred pounds of tobacco was offered to any person in Virginia, or in the other colonies, who captured a runaway servant
and returned him to Maryland. A letter was written to the

governor of New Netherlands asking him to return all servants who had fled to his colony. The Dutch governor was promised that the Maryland authorities would give him " the like help and concurrence." Peter Stuyvesant, in turn, asked the proprietary officials to return all runaways from the Dutch settlements. If they failed to do this, Stuyvesant threatened to retaliate by freeing all servants and slaves who came to New Netherlands from Maryland. Stuyvesant had ample grounds for making this threat, as several of the inhabitants of New Amstel on the Delaware had gone to Maryland where they were not only protected, but also naturalized and made " free dennizens " of the colony.[64]

During the spring of 1669, a special arrangement was made to prevent the escape of runaway servants or criminals. This was done because so many persons of this class were escaping to northern colonies, or entering Maryland from the provinces to the north. In order to break up this practice, if possible, a prison made of logs was erected on the land of Augustine Herrman in what is now a part of Cecil county, although at that time it was still a part of Baltimore county. All runaways that were caught were to be imprisoned in this log cabin. Herrman was to receive four hundred pounds of tobacco for each servant or criminal he captured.[65]

The proprietary officials appear to have made every effort to capture and return to Virginia all servants who ran away from that colony. Sheriffs were authorized " to press boats and hands " to transport the runaways across the Potomac and deliver them to the proper authorities. Virginians would sometimes bring suit in the Maryland courts in order to recover servants who had come to that colony. If they proved their claims, the servants would be restored to them.[66]

Before discussing the punishment imposed on a runaway servant who was captured, mention should be made of the penalty, or the punishment, imposed on a man, or woman, who persuaded, or helped, a servant to run away, or who harbored, or protected, a servant after he had escaped.

As early as 1641, it was made a felony to accompany a runaway servant, and again, in 1649, it was made a felony if any one was accessory to an apprentice, or indentured servant, who ran away. By the terms of a law passed in 1654, any person who was an accessory to any servant's running away " either by enticement, persuasion, promise, contract or approvement " was to be censured by the court. Nor could any one transport out of the province any servant who belonged to another. The person guilty of doing this had to pay the servant's master all damages which he had sustained by the loss of the servant's services.[67]

A number of cases can be found in the early records where a man was accused of persuading, or helping, a servant to run away. John White was one man against whom this accusation was made. It appears that White asked a maidservant of Jane Cockshott whether she would like to go with him to Virginia, and that if she would do this she might be freed from " the service wherein now she lived." At first White asked for a jury trial, but " anon repented himself and put himself for trial upon the court." Members of the provincial court found him guilty of a misdemeanor and ordered him whipped with thirty stripes.[68]

Robert Clarke asked one thousand pounds of tobacco damages from a man, who, he claimed, had taken his servant out of the colony without his consent. Nathaniel Pope demanded four thousand pounds damages of another man for " concurring, aiding, and assisting to the running away " of two of his servants. When Captain Joseph Wickes alleged that Thomas Hailings had enticed away two of his servants and then had helped them escape, the county commissioners, before whom this case was tried, ordered Hailings to give security for his good behaviour " towards all his Lordship's inhabitants, especially towards Mr. Joseph Wickes and all his family." [69]

James Jolly, innkeeper, and William Howes, both of whom lived on the Patuxent river, near Mattapany, were accused of helping two runaway servants to escape. It seems that Jolly and Howes had stopped the two servants, taken their guns from them, and then let them go free. The servants belonged to Robert Slye, who claimed this action showed that Howes and Jolly were " accessory to or acquainted with the design of the runaways."

In defending their actions, Howes declared that the two runaways had first said that they had come from the Severn river, but then had changed their story and said that they had come from Virginia and had fled from there because they were in debt. They planned, they added, to go and live with the Indians. In order to prevent their firearms falling into the hands of the savages, Howes and Jolly took their guns away from them. After this, the two runaways had presented an Indian with a white blanket in order to induce this savage to take them to see his king whose aid and assistance they wished to solicit. James Jolly gave the same account of what had happened as had Howes. In this case, as the judges of the provincial court were of the opinion that there was no evidence that Jolly and Howes had any hand in helping the two servants to escape, they both were acquitted.[70]

In Talbot county, there were several instances of where a colonist was accused of enticing away a servant from his master's service. In one case, Jonathan Sibery, a justice of the peace in that county, claimed that John Davis took away one of his servants without his consent, made him drunk and as a result he was drowned. Sibery, however, failed to recover the three thousand pounds of tobacco damages which he demanded of Davis.[71]

Next to consider those cases in which a colonist was accused of harboring, protecting or "entertaining," a runaway servant. By the provisions of the Act of 1649, reenacted in 1654, any one who "knowingly harbored, or entertained," any runaway servant during his absence from his master, or mistress, could be "fined and censured," as the court before whom such a case was tried, thought fit. Still later, in 1662, it was made the law that any person who "wittingly or willingly" entertained any runaway servant, even for one night, who did not have a pass or certificate from his master, mistress or overseer, would be liable to pay all such damages as the owner of the servant should sustain by the servant's "unlawful departure." Again, in 1666, another law was passed which fixed the penalty that would be imposed on those harboring, or entertaining servants, without the permission of their owners. If such a servant was entertained for one night, a fine of five hundred pounds of tobacco was imposed, and for the second night one thousand pounds, and for every other night after that fifteen hundred pounds of tobacco.[72]

There are many cases in the early records of where the charge was made that some colonist had been guilty of detaining or harboring a servant. John Price, muster master general of the colony, complained to the provincial court that John Norman was " unlawfully harbouring " his servant. In another case, John King, of Calvert county, was fined ten thousand, five hundred pounds of tobacco for entertaining for twenty-one days a runaway servant of Gerard Slye. Half of this extremely large fine went to the proprietary authorities, the other half to Slye. In a petition to the governor and council, John King asked that the half which was to go to the colonial officials should be remitted. Unless this was done, King said it would mean " the total ruin of him and his six small children." The governor and council granted his request.[73]

Father Philip Fisher, a Jesuit priest who went under the assumed name of Thomas Copley, was accused by Richard Blunt of harboring and detaining one of his servants whose name was Nicholas White. Blunt, a Virginian, said that he had been put to great expense by coming to Maryland in order to claim his servant. In defending this suit, Father Fisher maintained that he took White with the understanding that he would only keep him until Blunt sent for him and that when Blunt had sent for his servant he had been willing to let him go. In the meantime, however, it appears that White had run away possibly with the connivance of Father Fisher.

This case was tried before a jury. After all the evidence had been heard, the court instructed the jury that it was for them to decide whether Father Fisher had " injuriously " detained the servant when Blunt had sent for him, and also whether Blunt, or the priest, was the cause of the servant's running away. After some deliberation, the jury brought in a verdict which found Father Fisher guilty of both detaining Blunt's servant and of being the cause of the servant now being absent.

In view of this verdict, the court rendered a judgment which required Fisher to surrender Nicholas White, the servant, to Blunt, and, at the same time, pay the latter one thousand pounds of tobacco in satisfaction of the damages suffered by Blunt through the loss of White's services. If the priest failed to deliver the servant, then he must pay Blunt three thousand pounds of tobacco " in full satisfaction for the said servant and all damages sustained by the complainant therein." [74]

Of course, when a master, or mistress, had the reputation of being cruel to their servants, it was quite natural that when one of them ran away and came to some other colonist's home, the latter would be likely to offer the maltreated servant food and water. It will be recalled how Sarah Taylor was brutally treated by Captain and Mrs. Bradnox, her master and mistress. On one occasion, when she had run away, John Smith finding her lost in the woods brought her to his own house. Captain Bradnox at once brought suit against Smith for detaining and concealing the girl. The captain claimed that the girl spent the night in Smith's home. In his own defence, Smith said that he was too tired that evening to take the servant anywhere, but that the very next morning he planned to take her back either to her master, or to some constable or magistrate. After hearing all the testimony, the Kent county commissioners decided that Bradnox had not proved his case and therefore dismissed his suit.[75]

John Deare was another man whom Captain Bradnox sued for entertaining Sarah Taylor, his maltreated servant, without his consent. In Deare's case the Kent justices ordered him to ask the captain's forgiveness in open court and promise never to commit the same offence again.[76]

In another case in Kent county, Ellis Humphrey had to pay Christopher Andrews fifteen hundred pounds of tobacco damages for entertaining and taking into his service one of Andrews' servants. There were several actions brought in Charles county against some colonist for harboring or entertaining someone else's servant. In one instance, Joseph Edmonds brought a suit against Richard Pinner for having " received and harbored " for over three months Patrick Humes, a runaway servant. The jury, before whom the case was tried, found there was no cause of action. The county commissioners thereupon dismissed Edmonds' suit.[77]

The last question to be considered in this chapter is the punishment of the runaway servant who was overtaken and captured. As early as 1641, it was declared a felony for any servant to leave his master's service with the intention of leaving the province. The servant, if caught, might be put to death, although the governor might " exchange such pains of death into servitude." This meant that the runaway must first return to his master and serve two days for every day that he was absent, and, after he had completed his entire period of servitude with his master, then the runaway must serve an additional seven years. Eight years later, or in 1649, the law was changed. Nothing was said about the death penalty in the new statute, but merely that a servant whether indentured, or hired, must double the time of his unauthorized absence from the service of his master or mistress. By the provisions of an act passed in 1666, a runaway servant was required to serve ten instead of two days for each day he was absent.[78]

During the spring of 1662, a great many complaints seem to have arisen by reason of the increasing number of runaway servants. Accordingly, a law was passed prohibiting any servant travelling more than two miles, later increased to ten, from his master's or mistress' house without a pass or certificate from them. All colonists were authorized to examine all strangers and other suspicious-looking persons, whether servants or freemen, and if they did not have such a pass, or were not known " by integrity or common fame," to take them before the nearest justice of the peace.[79]

In one case, in order to make good their escape, two servants forged a pass, or certificate, in their master's handwriting. Both men were ordered whipped with twenty lashes on their bare backs, but one of them later escaped this punishment upon his promise of " future good behaviour " and also that he whip the other servant who had helped him forge the pass.[80]

When caught, runaway servants would sometimes offer various excuses for their unauthorized absence. Two servants alleged that they were induced to run away by " the persua-

sion " of another servant. A maidservant said that her reason
for running away was that she had been beaten and abused by
some of her fellow servants. The court held this was not a
sufficient excuse and compelled the girl to serve additional time
for having been absent from her master's service. Two Irish
servants ran away because, they said, their master had abused
them " in giving them correction." As the judges of the provincial court thought that their master had " just cause " to
discipline them, the two servants had to serve additional time
for their running away.[81]

A number of cases can be found in the early Talbot county
court records where a servant was compelled to serve ten additional days for each day he was absent without the permission
of his master or mistress. Thus, for example, if a servant was
absent for twenty days he must serve two hundred extra days
after the period provided for by his indenture had expired. In
some cases, the master would not exact the full penalty. Thomas
Vaughan had a maidservant who left his service for seventy-one
days. Instead of requiring this girl to serve seven hundred and
ten days, Vaughan said that he would be satified with only a
year's additional service provided she behaved herself. Francis
Armstrong was even more lenient with a servant who had run
away. Armstrong said that if the man did not leave without his
permission again, he would not require any additional service of
him.[82]

Not all servants who ran away were punished by being required to serve additional time. Sometimes they were whipped.
William Marshall, a justice of Charles county, presented to
his fellow commissioners three of his servants for punishment
all of whom had run away. A servant named Mathew Brown
was apparently the ringleader. The county commissioners at
once ordered that young Brown should be taken to the whipping post, stripped to his waist, and given twenty-seven lashes.
The other two servants, one a boy and the other a girl, who had
run away with Brown " for company," the county justices
decided should " for company's sake at the whipping post in
the public view of the people receive their reward." The youth
was given nine lashes on his bare back; the girl, too, was strip-

ped to receive " a reward " of seven stripes on her bare back.[83]

At another session of the Charles county court, Humphrey Warren had a list of grievances to offer against his manservant, Richard Lamb. First, he said that Lamb sold his clothes when he had no right to do so, secondly, that he had on several occasions left his service without his permission, and lastly, that when it came to swearing he was one of the worst offenders in this respect. The county commissioners ordered the sheriff to take Lamb into his custody and " cause him forthwith to receive twenty lashes upon his bare back in the public view . . ." [84]

In Kent county, Matthew Read, one of the justices of the local court, brought his servant, William Mouse, before his fellow commissioners for being " a constant runaway " from his service. When Mouse had nothing to say in his defence, the court ordered him taken outside and given twenty-five " good and sound " lashes. The youth was also informed that if he ever ran away again, any persons who found him would have the right " to whip him home again to his master . . ." [85]

When a runaway was caught, the master would sometimes whip the servant himself. This occurred in Talbot county when John Kinemant chastised Anthony Pecheco for having run away. Later Kinemant appealed to the Talbot county commissioners stating in his petition that as Pecheco had been absent for twenty days he wanted " satisfaction according to law." Witnesses were called who testified as to the severity of the beating Pecheco had received. One said that Kinemant had whipped the lad until he himself was exhausted. Another witness declared that Kinemant beat the youth with a rod, then rested and gave him another thrashing. In view of this testimony, the Talbot justices were of the opinion that Pecheco had had sufficient punishment for his running away.[86]

CHAPTER VI

HOMICIDE, ASSAULT AND SUICIDE

In early English common law homicide was divided into felonious, justifiable and excusable homicide. Felonious homicide is either manslaughter or murder. Justifiable homicide is where a person kills another in the performance of some legal duty, as in the execution of death sentence, or arresting an outlaw, or to prevent the commission of some atrocious crime. Excusable homicide is where the killing is done without any criminal intent and by accident, or misadventure, or in self-defence. Although at the present time neither justifiable nor excusable homicide involve any legal guilt or punishment, at the early common law the slayer in the case of excusable homicide was liable at least to imprisonment, from which he was usually later released by a pardon.

Under the provisions of an early Maryland statute any one convicted of murder could be put to death and all his property confiscated. There were, however, comparatively few indictments for murder in proprietary Maryland and even fewer executions for this crime. Jacob, a negro slave owned by Colonel Nathaniel Utie, was one of the few persons to be hanged for murder. This negro was suspected of murdering the colonel's wife, Mary. A coroner's jury, after examining Mary's corpse, declared that it was their opinion that the wounds which she had received in her arm were the cause of her death. A grand jury investigating the facts of the case called before them two witnesses. One of them was a youth by the name of Anthony Brispo. He testified that he had seen the negro stab Mrs. Utie about ten o'clock one night, that Jacob had given her two wounds in the right arm, and that Mrs. Utie died the following night after bleeding "a day and a night." As she seemed to be in perfect health before this attack was made upon her, Anthony was certain that Mrs. Utie had died as a result of the wounds which had been inflicted. Francis Stockett, who had dressed the

119

injuries, was then called to the witness stand. He also testified that he thought Mrs. Utie had died of these wounds.

After hearing this testimony, Jacob was presented by the grand jury in an indictment which stated that " the negro slave *and servant*" of the Uties had—

By force and arms, to wit, with a drawn knife of two pence value, which the said Jacob then and there in his hand did hold, upon the aforenamed Mary, then his mistress, then and there in the peace of God and his Lordship's, being voluntarily and of his malice forethought an assault did make, and the same Mary then and there with the said knife feloniously and traiterously upon her right arm strongly did strike a stab, giving her a mortal wound four fingers broad, in the upper part of her right arm, of which mortal wound she, the said Mary upon the fourth day of October did die, and so the aforenamed Jacob, at Spesutie, of his malice aforethought the same Mary Utie, his mistress, in manner and form aforesaid, willing, wittingly, feloniously and traiterously did kill against the peace of his Lordship, his rule and dignity.

Jacob, the negro slave, was now brought before the judges of the provincial court. After the indictment was read to him, he was asked whether he pleaded guilty or not guilty. The negro, refusing to answer, stood "in a manner mute." Apparently, Jacob did not even ask for a trial by a petit jury. When the members of the court declared that it was their opinion that the negro was guilty of petty treason, the governor immediately passed sentence of death using the following words:

You, Jacob, shall be drawn to the gallows at St. Mary's and there hanged by the neck till you are dead.

This case brings up the question of what constituted petty treason. Treason was of two kinds, high treason, when committed against the king, and petit, or petty treason, when committed by a vassal against his lord or superior. The Treason Act restricted petty treason to the killing of a husband by his wife, of a master by his servant, and of prelate by an ecclesiastic owing obedience to him. In the Jacob case, it will be noted, in order to bring his offence within the provisions of the Treason Act, the indictment alleged that the colored man was " the negro slave *and servant*" of the Uties. Although petty treason is now punished as murder, during seventeenth-century Maryland it was considered a much more serious offence for which Jacob, the negro, was " drawn and hanged." That is to say, he was dragged on the ground from the prison at the tail of a horse, or cart, to the gallows where he was hanged.[1]

Before considering the other executions for murder, the case of Antonio, another negro, should be mentioned. Antonio died under circumstances which made it appear that his master, Symon Overzee, might have beaten him to death. The negro, ugly in appearance, was lazy and had been caught stealing food and trying to run away. On one occasion, when Overzee ordered him to go to work he refused to do so, laid down and would not move. Overzee at once gave the negro a beating with some " peach tree wands or twigs." When Antonio continued to act stubbornly, his master, after having his doublet removed, gave him another sound whipping on his bare back. But still the negro refused to work " and feigned himself in fits." Overzee then resorted to pouring hot melted lard on Antonio's back. Still adamant, Overzee commanded an Indian slave to seize the negro and tie his hands above his head to a ladder outside of the house. There seems to have been some doubt as to whether Antonio was suspended from the ladder by his wrists or was able to stand upon the ground. At any rate, while thus bound half-naked a cold wind blew up. Soon afterwards the negro died.

The circumstances under which Antonio, or Tony, as he was sometimes called, died, were referred to a grand jury to determine whether his master, Symon Overzee, " an assault did make . . . and did feloniously kill " the negro. This jury was of the opinion that the evidence did not warrant the finding of a true bill, that is to say, bringing an indictment against Overzee. Proclamation was then made by the sheriff that Overzee " stood upon his justification, and that any one that could give further evidence, should come and give evidence for the Lord Proprietor." When no one appeared, Overzee was acquitted " by proclamation." It is impossible to agree with the findings of the grand jury in this case. Surely the beatings, pouring hot lard on

9

his back and tying the negro naked out-of-doors, was sufficient evidence to warrant an indictment. Whether Overzee would have been found guilty of murder in a trial before a petit jury is, of course, another matter.[2]

Besides the Jacob case discussed at the beginning of this chapter, there are only three other cases in the printed records of the proprietary period where persons were hanged for murder. In two of them masters were accused of beating their servants to death, in the other a woman was accused of killing her bastard child. Indeed, it might be said that there were only three executions for murder as Jacob, the negro, was really hanged on a charge of petty treason.

First, to discuss the cases where masters were hanged for beating their servants to death. When Jeffery Haggman, man-servant of Joseph Fincher, was found dead his master was suspected of having whipped him to death. Some of the testimony which was produced was favorable to Fincher. Edward Ladd and Thomas Whyniard were two witnesses who said that Fincher had not given Haggman a very severe beating. Ladd did not think that the whipping could have hurt Haggman, while Whyniard said that in disciplining Haggman Fincher had only used a small stick. The finding of a coroner's inquest held after the death of Haggman was also favorable to Fincher. The members of the coroner's jury reported that they could find " no mortal wound " which might have caused Haggman's death, but that the young man was " a diseased person " who had died of an abscess and scurvy.

The most incriminating testimony was given by William Gunnell. This youth recounted how on the day before Haggman died Fincher gave his servant a great number of tobacco plants to carry. When Haggman complained that he could not carry such a heavy load, Fincher warned the boy, " Sirrah, go or else I will beat you as never was dog so beaten." To this Haggman replied, " Master, I cannot carry them although you knock me in the head." Seeing the servant staggering under his burden, Fincher came over, kicked him and beat him with his fist, saying, " I'll either knock thee in the head, or starve thee, rather than I'll lead this life with thee." On the next day, which was the day Haggman died, Fincher again kicked and beat the boy.

Gunnell then described how Fincher's wife had also maltreated young Haggman. When the boy did not go to a spring to fetch some water quickly enough, she overtook him and struck him with her hands. While returning from the spring Haggman fell down. As Mrs. Fincher could not get the servant to his feet again, she called on her husband for help. Fincher coming out, took the youth to a nearby tobacco house where he stripped him and beat him until he cried for mercy. Not long after this young Haggman died. Gunnell testified how, on another occasion, Fincher had boasted that he had beaten his servant until he had broken two tobacco sticks " about the sides of him."

Susanna Leeth and Thomas Miles, when called to the witness stand, gave testimony which was similar to Gunnell's. Miles said that when he came to the tobacco house where Haggman had been whipped, the boy was dead.   Miles added that the lad's nose was bloody and he saw black spots and bruises upon the youth's " body and hinder parts." According to Miles, Fincher seemed very anxious to bury the boy before any one else saw the body.

Robert Lloyd, a chirurgeon, was the last witness asked to take the stand. His testimony tended to contradict the finding of the coroner's inquest to which reference has been made. Curiously enough, Lloyd himself had been one of the members of that inquest. The chirurgeon now testified that when he had " viewed " the naked body of Haggman after his death he had found blue spots on the " forepart and hinder parts," and also " two strokes " and a sore on the youth's side. This sore, Lloyd said, he had once attempted to heal.

After hearing all this testimony, the grand jury decided to indict Fincher for murder. The bill stated:

The jury for the Right Honourable, the Lord Proprietor, do present Joseph Fincher, of Rhode river in the county of Anne Arundel, for that the said Joseph Fincher, on the 27th day of August, 1664, upon Jeffrey Haggman, servant to the said Fincher, by force and arms, an assault did make and with certain sticks of no value which he the said Fincher in his right hand then and there did hold, divers blows on the body of the said Jeffrey Haggman did strike so that of the said blows the said Haggman the day aforesaid did die—

And so the said Joseph Fincher the said Jeffrey Haggman then and there feloniously and of malice forethought did kill and murder, contrary to the peace of his said Lordship, his rule and dignity.

Asked how he pleaded to this indictment, Fincher said :" Not Guilty," and asked for a trial before a petit jury. The sheriff was then instructed to impanel " twelve men, good and able." After these were chosen, the indictment was read to them and the same witnesses, who had testified before the grand jury, were again " called, examined, and sworn as before set down."

After hearing all the evidence, the petit jury, of which Samuel Chew was the foreman, retired to consider their verdict. It was not long before they returned to the room in which the provincial court was holding its sessions. After each of the jurors had answered to his name, the foreman handed over the bill of indictment, on the back side of which was written the word: " Guilty."

Then, by way of confirming this verdict, the presiding judge questioned the members of the jury as follows:

Judge: Are you agreed of your verdict?
Jurors: Yes.
Judge: Who shall say for you?
Jurors: The foreman, Samuel Chew.
Judge: Gentlemen of the jury, you say Joseph Fincher is guilty of the murder whereof he stands indicted?
Foreman: Yes.
Judge: You all say so?
Jurors: Yes.

After this verdict was delivered, Joseph Fincher was ordered to hold up his hand. The presiding judge now informed him that although he had pleaded, " Not Guilty," to the charge of murdering his manservant, he had been found " Guilty " by the jury before whom he had been tried. In view of this finding, the judge asked the prisoner what he could say for himself, and why, according to the law, the death penalty should not be imposed. " What sayest thou, Joseph Fincher? " queried the

judge. The prisoner answered: " If I deserve it, I must die."
When asked if that was all he had to say for himself, Fincher
replied in the affirmative.

Sentence of death was now imposed and Fincher was ordered
hanged by the neck. Captain William Burgess, sheriff of Anne
Arundel county, was directed to see that this sentence was carried out within three days. Fincher probably deserved to be
hanged. While it is true that the finding of the coroner's jury
tended to exonerate him, the testimony of several witnesses is
rather convincing. From what they said it seems likely that
Haggman died as the result of the brutal treatment which he
received at the hands of his master. It is interesting to note that
Edward Ladd and Thomas Whyniard, the two young men
whose testimony was favorable to Fincher, were later arrested
" for suspicion of felony by them as is said committed." Although we are not informed of the nature of the charge against
them, we do know that later, when called to the bar of the
provincial court, no witnesses appeared against them. Proclamation was then made that the two prisoners " stand upon their
delivery." After this proclamation was made three successive
times and when no one appeared, the provincial court ordered
the two young men " cleared by proclamation." [3]

John Dandy, a smith, was the other master accused of beating a servant to death. Many witnesses were called in this case.
From their testimony we are able to piece together an account
of circumstances surrounding the death of the servant boy
whose name was Henry Gouge. The naked body of this youth,
it appears, was found floating down a small creek near a mill
at Newtown, where Dandy lived. Although a search was made
for the dead boy's clothes, they were never found. The corpse
itself was black and blue showing that the lad had been whipped
with a switch or small rod. When the body was dragged ashore,
those lifting the corpse out of the creek were surprised that no
water ran out of the mouth, but only blood from the nose and
from a deep gash on the forehead, which, it seems, had been
inflicted by Dandy some time before the youth's death. Dandy
was present at the time helping to bring the body ashore. It
was noticed that when he touched the corpse the wounds started
bleeding again.

On the day before the body of Gouge was found floating in the creek, Dandy had sent the boy to a coal kiln to fetch some live coals for his forge. When the youth failed to return as soon as Dandy thought he should, the smith went to the coal kiln to see what was the matter. While there several persons heard the lad cry out as though he was receiving a severe whipping. Dandy later returned to his forge. He said that the boy had run away. It appears that Dandy had been so much in the habit of beating young Gouge that the smith was heard to remark on one occasion that some day he would probably be hanged for killing the youth.

After considering all the evidence, a grand jury indicted Dandy for the murder of his servant boy. These jurors were then dismissed and a petit jury chosen that found Dandy guilty of " maliciously and feloniously " murdering Gouge. The provincial judges at once passed a sentence of death. Dandy was hanged on an island in the Patuxent river. Did the facts warrant the verdict of the jury and the judgment of the court? Clearly Dandy deserved some punishment for his maltreatment of young Gouge, but is it likely from the facts that he whipped the boy to death? More probably, the lad to escape further chastisement went and drowned himself. As against this theory, it does seem strange that the youth should have taken the trouble to undress before plunging into the water. And if he did undress why weren't his clothes found somewhere along the bank of the creek? Possibly Dandy had removed the lad's clothing before administering the final and fatal beating and had then concealed them.[4]

The fact that the wounds on the dead body of the boy bled afresh when Dandy touched the corpse may have influenced the jury in reaching their verdict. Although an absurd belief, many at that time thought that if a man touched the body of some one whom he had murdered the corpse would bleed. This test was actually applied in two other Maryland cases. In one of them, a girl named Catherine Lake died under circumstances which made it appear that Thomas Mertine might be responsible for her death. Mertine had shoved Catherine with his hand and also given her " a kick upon the britch." Shortly after this treatment the girl had a fit " and departed this world within one hour." A coroner's jury was impanelled to examine the corpse in order to determine, if possible, the cause of her death. This jury required Mertine to lay his hands on the dead body. When at his touch no blood came from the corpse, nor did the body show signs of having been beaten, the jury decided that her death was due to " the Providence of the Almighty." [5]

The other case in which the blood test was applied was when Thomas Watson, a servant of Captain Thomas Bradnox, died under circumstances which made it appear likely that his death might have been the result of abuse received at the hands of his master. In order to try to prove his innocence, Captain Bradnox shogged (jostled) the corpse and thrust his thumb into it to show " how the flesh did dent," but that no blood came from the dead body.[6]

The remaining case to consider where the death penalty was imposed for murder was that of Elizabeth Greene, a spinster, who was accused of " being brought to bed of a bastard," and then " feloniously " making away with it. It was alleged that the woman, after giving birth to the baby, had thrown it into a fire. Elizabeth admitted that she had burnt " something that came from her," but insisted that it was not a baby. Later, however, she confessed having a bastard child, but said she had not murdered it, nor could she say whether when it was born " it was a child formed or not," as she was " gone but four months " when she delivered it, " being put into a fright."

Among the witnesses summoned to testify was a Mrs. Grace Parker who told how Elizabeth had admitted having a child and then burning it. Mrs. Parker was of the opinion that Elizabeth had " gone near her full time " when the baby was born. While Mrs. Parker had only seen Elizabeth once during the time she was pregnant, she had, however, on that occasion examined the woman's breasts and found the milk in them " hard and curdled." A man living in the same house with Elizabeth testified that although he had not seen any child he had heard something cry " like a pig or a child." Another witness said that while he did not know whether Elizabeth had been pregnant she did appear to be very large.

In view of the evidence produced, Elizabeth Greene was presented by a grand jury for the murder of her bastard child. The indictment stated that—

Elizabeth Greene, being big with child by God's Providence, was delivered of a certain living man child which said living man child she the said Elizabeth Greene did throw into the fire, and so the said Elizabeth Greene the living man child by throwing into the fire in manner and form aforesaid, then and there feloniously and of malice forethought did kill and murder contrary to the peace of his said Lordship, his rule and dignity—

The grand jurors were now dismissed and a petit, or trial jury, summoned. These twelve men found Elizabeth guilty of murdering her infant. The judges of the provincial court then asked the woman what she had to say for herself. Elizabeth replied that she " threw herself on the mercy of the court." Again being demanded if that was all she had to say in her defence, the woman answered, " Yes." The members of the provincial court now passed sentence of death in the following words:

Elizabeth Greene you shall be carried to the place from whence you came, from thence to the place of execution, and there hanged by the neck till you are dead, and so God have mercy upon your soul.

Thomas Dent, sheriff of St. Mary's county, was instructed to carry out the judgment of the court within two days after it was rendered, " betwixt eight and nine of the clock in the morning." The sheriff did as he was ordered and later returned the writ to the provincial court with the notation: " Executed by me, Thomas Dent." Elizabeth went to her death without mentioning the name of her seducer.[7]

There were other cases in which women were accused of murdering their infants, although in none of them was the woman convicted. Elizabeth Harris, for example, a spinster and maidservant, was presented by a grand jury for murdering a baby boy to which she had given birth. The evidence on which the grand jury based their indictment was as follows. Robert Joyner, one witness, said that he had encountered Elizabeth Harris carrying a bundle out of which something was hanging which looked like fish guts. When Joyner asked the woman what was in the bundle she replied that it was fish guts. She would not, however, permit Joyner to see for himself, but flung the package into the river. Joyner, his suspicions aroused, pulled the bundle from the water and was surprised to find a dead infant inside. John Gee, another witness, confirmed what Joyner had said.

Margaret Marshguy, on the other hand, gave testimony which tended to exonerate Elizabeth. Margaret, who was a servant in the same house with Elizabeth, said that although they had slept together she had never noticed that Elizabeth was pregnant, nor had any one else ever noticed it. Moreover, added Margaret, no one had ever told her that Elizabeth had given birth to a child. In her trial before a petit jury, Elizabeth was acquitted on the charge of murdering her infant. The testimony of Margaret Marshguy probably influenced the jury in reaching their verdict.[8]

Judith Catchpole, another maidservant, faced a number of charges. First, she was accused of murdering a child she had borne. Secondly, she was accused of cutting the throat of a woman on board the ship which was bringing them both to Maryland. It was said that Judith had done this while the woman was asleep and then had sewed up the wound with a needle and thread. Lastly, Judith was accused of stabbing a seaman in the back. All these accusations were based on statements made by a man who had since died.

In order to determine whether Judith had murdered her infant, the provincial court impanelled a jury of women to examine her body to see if she had ever borne a child. After making an examination, these women came to the conclusion that Judith had not given birth to a baby " within the time charged." Witnesses appearing before the provincial court testified that the man who had made all the accusations against Judith was not " in sound mind." This testimony and the decision of the jury of women brought about Judith's acquittal on all counts.[9]

In another similar case, when Hannah Jenkins was suspected of murdering her infant, another jury of women was ordered "to search the body of said Hannah" to determine, if possible, whether or not she was "delivered of a child." This jury's verdict was that Hannah was "clear from child-bearing and never had a child to the best of their knowledge." Thereupon the court ordered the sheriff to clear Hannah by proclamation.[10]

Francis Brooke was brought before the provincial court where he was accused of having so maltreated and beaten his pregnant wife that he had caused the death of the infant in her womb. From the evidence, it appears that Brooke had been in the habit of administering many beatings to his wife, striking her with a cane, a board, and even with "the great end" of a pair of tongs. Brooke declared that his wife was a whore and that he did not care if she did have a miscarriage as the child within her was "none of his." When Mrs. Brooke had labour pains, her husband gave her wormwood to drink, remarking at the time that his wife must either have "the pox or the devil." A midwife was later sent for and when the child was delivered it was still born. Its body, covered with bruises, was "blood black." Mrs. Brooke gave two contradictory accounts, one that the child was bruised because of the maltreatment she had received at the hands of her husband, and the other that the bruises may have been caused by her falling out of a tree. Although the evidence produced seems convincing, Brooke does not appear to have been convicted.[11]

There are several other cases of murder in proprietary Maryland, but in none of them was there a conviction. When Thomas Allen was found murdered not far from St. Mary's, a coroner's jury was impanelled to investigate the cause of his death. The jurors found that Allen had been shot under the right shoulder, but whether by a musket or a bow and arrow they could not determine as his body was so eaten by worms. Allen had also been scalped and his skull broken. Although two Irishmen were suspected of having committed this murder, there is no record of their having been brought to trial. In another case, Hannah Rogers, a servant of Colonel Samuel Chew, of Anne Arundel county, was presented by a grand jury for having attacked and killed Richard Stevens, another servant in Chew's employ. Hannah, it appears, had hit Stevens over the head with a hoe. In the indictment Hannah was charged with feloniously killing and murdering Stevens " contrary to the peace of his Lordship, his rule and dominion." The petty jury, however, before whom Hannah was tried, brought in a verdict of not guilty.[12]

Before considering the cases of manslaughter in early Maryland, it is well to remember that this offense was the unlawful killing of a human being without malice express or implied. In the murder cases, which we have considered, it will be recalled, that the indictments would state that the person was killed with malice aforethought, which could be express or implied. By the terms of an early proprietary statute, passed in 1642, the punishment for manslaughter, or " homicide," as the act defined it, was stated. Any convicted of this offense, whether principal or accessory before the fact, might be sentenced to be punished in any one of the following ways. He might be put to death, or burnt in the hand, loose " a member," or " corporally corrected, or put to shame," as the members of a court should think " the crime to deserve." The offender might also be outlawed, exiled, or imprisoned for life, and, except in the case of " a gentleman," might be ordered to serve the Lord Proprietor for a term of seven " or less years." The man or woman guilty of manslaughter might also have to forfeit all their property and personal belongings.[13]

There are as few cases of manslaughter to be found in the printed court records as there were of murder. In one of them the accused, although indicted for murder was finally convicted of manslaughter. This was when a man named Patrick Due, of Calvert county, was presented by a grand jury for having assaulted and " feloniously and of malice forethought " killed and murdered Richard Morton. The circumstances surrounding the death of Morton were related by a number of witnesses. Morton, it appears, had come to Maryland from England as a member of the crew of a ship then visiting the province. One day, while on shore leave, Morton and other members of the crew approached a plantation of which Patrick Due was the overseer. Seeing a canoe full of oysters near the shore, Morton and his companions started to eat some of them. When Due

saw what was happening he sent a young boy by the name of
Robert Hobbs " to forewarn the seamen from eating the
oysters." Due instructed the sixteen-year-old lad to take the
dogs, including " Towser," and " set them " on the intruders.
The youth did as he was directed. When Robert asked the
seamen what right they had to eat the oysters, they replied that
they did not see why they should not eat them since they had
cost nothing. But, retorted the boy, they had " cost him his
labour, for that he had been all the day in getting of them."

Soon after this, Due, the overseer, himself came down to the
shore and shouted at Morton and the other seamen: " Damn
me, you dogs, I will kill you, if there be no more sea dogs in
the world." One of the sailors told Due that if they had done
him any wrong in eating the oysters they would pay for them,
and to back up his remark, " heaved a quarter of a piece of
eight " in Due's direction. But the latter said that he would
have none of their money. Due then opened fire on the seamen,
wounding two of them, one of whom was Richard Morton.
This sailor later died of his wounds.

When brought to trial for the murder of Morton, Due
pleaded not guilty. The petit or petty jury before whom he was
tried brought in a verdict of manslaughter but not murder.
After committing the overseer to prison to await the next session, the court adjourned. When, at this meeting, Due was
called to the bar and asked what he had to say for himself, he
replied that he craved benefit of clergy. The request was
granted, " the book " was produced and Due showed that he
could read. The provincial court judges thereupon ordered that
" the said Patrick Due be forthwith burned in the brawn of the
hand with a red-hot iron." [14]

John Dandy, the smith, who was accused of murdering his
boy servant, Henry Gouge, was involved in other difficulties
before he was hanged for that crime. He was also accused of
killing an Indian youth, and, as we shall later see, of assaulting
a young colonist. In the manslaughter case Dandy was indicted
by a grand jury for having made an attack on the Indian boy,
whose Christian name was Edward, and continued the indictment:

One gun charged with bullets against the said Edward did discharge, and therewith did wound the said Edward in the right side of his belly near the navel, so that it pierced his guts, of which said wound the said Edward afterward within the space of three days died, feloniously and contrary to the peace of our Sovereign Lord, the King, and contrary to the peace of the Lord Proprietor, etc.

Although Dandy pleaded not guilty, the trial jury found the smith guilty of " felony and murder." The inconsistency between the indictment and the verdict of this jury should be noted. There is nothing said in the indictment about Dandy killing the Indian lad with malice aforethought which must be express or implied in the case of murder. Nevertheless, the jury found Dandy guilty of murder. Dandy was sentenced to death, but probably because the court felt that the smith should have been found guilty of manslaughter and not murder, the death sentence was commuted on condition that Dandy serve the Lord Proprietor for seven years and also act as public executioner in the colony. Later, because of his good behaviour, the smith was freed from his servitude and no longer required to act as public executioner. In requiring Dandy to serve the proprietor for seven years the court was within the terms of the Act of 1642 which stated that this sentence could be imposed on those guilty of manslaughter.[15]

So far all the cases which had been considered are those of felonious homicide which includes murder and manslaughter. No cases of justifiable homicide could be discovered in the early Maryland court records so far printed. Justifiable homicide, it will be recalled, is where a person kills another in the performance of some legal duty as in arresting an outlaw, or to prevent the commission of some atrocious crime.

There are, however, in the printed court records several cases of excusable homicide. This is where the killing is done without any criminal intent and by accident, or misadventure, or in self-defense. At the early common law, the slayer in case of excusable homicide was liable at least to imprisonment from which he was usually later released by a pardon. Let us investigate the cases of excusable homicide. At a meeting of the provincial court held at St. Mary's during the winter of 1668, the sixteen members of a grand jury were asked to consider—

If Thomas Cocher of Charles county in Port Tobacco creek, planter, on the 24th day of October in the said year at the house of Clement Theobalds, in Port Tobacco creek aforesaid, upon Richard Turner by force and arms an assault did make with a certain gun of the value of ten shillings, which he the said Thomas Cocher in his right hand then and there did hold, divers wounds in the body of him the said Richard Turner did make so that of the said wounds the said Richard Turner immediately did die and so the said Thomas Cocher the said Richard Turner then and there feloniously did kill contrary to the peace of his said Lordship, his rule and dignity.

After hearing the testimony of several witnesses, the grand jury returned a true bill, or "billa vera," as they called it, meaning that they thought the evidence warranted the presentment or indictment of Cocher. The latter pleaded not guilty to the indictment and asked for a trial by jury. After hearing all the evidence, these jurors returned the following verdict—

That Richard Turner was killed with a gun by Thomas Cocher and they pray the advisement of the Court. If the court do find it manslaughter, we find it manslaughter, otherwise we find it manslaughter (or homicide) by misadventure.

The judges of the provincial court then asked the jurymen where Turner was killed. To this they replied: "Without doors upon some logs asleep." The members of the court also inquired of the jurors if Cocher had tried to escape after the shooting. "No, not to their knowledge," they answered.

In view of the jury's verdict and their answers to these questions, the provincial court declared that in their opinion it was a case of manslaughter, or homicide, "by misadventure." Cocher was then committed to prison, but soon afterwards received a pardon from Cecil Calvert, the proprietor, who said that his reason for granting it was that Cocher had killed Turner "by chance and not by felony or of malice forethought." [16]

John Richardson, who lived on Tred Avon creek in Talbot county, received a similar pardon. Richardson, it appears, had been presented by a grand jury for assaulting his wife with a tobacco stick, and, as the indictment read, " feloniously did strike " his wife of which blow she died so that Richardson " of his malice before thought the said Mary Richardson, his then wife, did voluntarily, feloniously and wickedly kill and murder." To this indictment Richardson pleaded, " Not Guilty." A jury was impanelled to hear the testimony of witnesses. After hearing all the evidence, the jurors' verdict was that Richardson was guilty of causing his wife's death "by misadventure." It is a pity that the testimony of the witnesses does not appear in the court records, as it is difficult to understand how beating a woman to death with a tobacco stick can be called homicide " by misadventure." Whatever the extenuating facts, the proprietor pardoned Richardson and released him from prison. His Lordship was of the opinion that Richardson had not killed his wife with malice aforethought, but, as the jury had found, by misadventure.[17]

There are two other instances of where the proprietor used his pardoning power in cases of excusable homicide. This was when Thomas Floyd was indicted and convicted for killing another colonist " by misadventure and against his will," and also when Thomas Curre, a youth under " the age of discretion," was presented by a grand jury for killing a woman under similar circumstances.[18]

A case of accidental killing occurred in Newtown, where James Langworth and John Greenway lived together in a small house. At the time of the fatal accident there were several persons in the dwelling, including Mary, Greenway's wife, Robert Clark and Phillip Anther. It appears that Clark had left his pistol on a table which went off in Mary's hands " unawares " as she was handling it. Clark then reloaded the pistol and put it back on the table. A few minutes later James Langworth came into the room. Although he knew that the pistol had been discharged, he did not, however, know that it had been reloaded. While examining the weapon it went off again, this time with a fatal result, killing Phillip Anther who was sitting near the table. The shot from the pistol penetrated his brain, " whereupon he suddenly died."

When Clark and Langworth were brought before the provincial court, the judges charged the jury to determine how and by whom Anther had been killed, and whether it was done " maliciously, or willfully, or unwittingly and unfortunately."

The jury brought in a verdict that Anther had been killed
" accidentally and unwittingly " by James Langworth and that
they found no negligence or carelessness on the part of either
Clark or Langworth. After hearing this verdict, although the
court acquitted Clark " in every respect touching Phillip Anther's death," they fined Langworth five hundred pounds of
tobacco and ordered him to pay all the court costs. The fine
was later remitted by the governor in compliance with special
directions received from the proprietor. An obvious injustice
was thus righted.[19]

One of the first to die a violent death in the colony was John
Bryant, a planter. His death was caused not by another man
but by a falling tree. Bryant's death under these circumstances
resulted in the application of the old English law of deodand. This word came from the Latin deodandum, to be given
unto God, or a thing given or forfeited to God. In English
law, the word applied to a thing, which, because it had been
the immediate cause of the death of a person, was given to
God, that is, forfeited to the crown to be applied to some pious
or charitable use. If, for example, a wagon ran over a man and
killed him, the vehicle was forfeited as a deodand. After
Bryant's death, a coroner's inquest decided that as the falling
tree had " moved to the death " of John Bryant it should be
forfeited to the Lord Proprietor as a deodand.[20]

In Charles county, about thirty years later, a boy by the name
of Thomas Greenhill was killed under somewhat similar circumstances. The youth, it seems, was at the time of the accident engaged in felling trees just as Bryant had been when he
met his death. When one of the trees fell on the lad, a friend
ran to his assistance. " For Christ sake, Thomas Greenhill,
speak," cried the man. There was no reply as the boy was dead.
After investigating the facts in this case, a coroner's jury gave
as their finding that the youth met his death " accidentally and
for want of care the tree fell on him and killed him." But the
jurors did not ask that the tree should be forfeited as a deodand.[21]

In the Talbot county records there are two cases of persons
being killed by a falling tree. In one of them, a servant of Rich-

ard Tilghman, a chirurgeon, met his death in this way. In order to prove that this man died as a result of an accident, Tilghman produced two witnesses. One witness said that:

Two of Mr. Tilghman's servants were felling a tree which stood near the fence side, and the tree falling, it hit an old tree and broke it into two pieces, and one of the pieces hit the man on the head and beat out his brains and broke his thigh and his leg.

The second witness testified that he saw the man lying on the ground besmeared with blood with part of the tree over his body.

In the second case, a child was killed by a falling tree. A coroner's jury investigating the case declared that they had found the tree was " the only cause of the child's death." They added that they could not find from the evidence that any one was responsible for its death. In neither of these two cases, however, was the tree forfeited as a deodand.[22]

A planter by the name of Thomas Lisle while climbing a tree fell from it to his death. Neighbouring planters described how they found Lisle lying under the tree, " having his breeches much rent." Lisle died soon afterwards. Of course, there was no reason to apply the law of deodand in this case any more than there was in a similar case where a man who was riding " the draught tree " of a mill fell off and killed himself. In neither of these cases did the tree fall on the man.[23]

There were several instances in early Maryland where persons met death by drowning. Usually, it was the result of an accident. In Charles county, for example, where a maidservant was drowned, a coroner's jury was of the opinion that the young woman had come "accidentally by her death." In the same county the body of a man by the name of Stephen Wood was discovered one day in a canoe full of water. One of the legs of the deceased was hanging over the side of the canoe and even in death his hands clutched a short board which he had probably used as a paddle. A coroner's jury was instructed to make the usual investigation. It was the opinion of these jurors that Wood had been taken sick a short time before his death. Accordingly, they thought he might have died while he was in his canoe and that later the wind had blown the canoe

10

near the shore filling it with water. The jurors had been unable to find on Wood's naked body any wound " that could be mortal unto him." [24]

In Kent county, a youth was told to walk out on the ice in order to recover a goose which had been shot. While doing this, the ice cracked under him and he was drowned. After an investigation a coroner's jury gave the following verdict:

That the said boy came by his death, as far as they can discern, not by his own will, intent, or purpose, nor by the intent of any other, but according to the evidence, fell through the ice and perished before any help could come to save his life.[25]

The three cases so far considered, with the exception of the Wood case, have been those where the person was accidentally drowned. What if a man intentionally drowned himself? In such a case he would be considered a suicide and his property, if he had any, would be forfeited to the Lord Proprietor. Nor would he be entitled to a Christian burial. A case of intentional drowning occurred in Talbot county. This was when John Short, a servant of Thomas South, a commissioner of that county, was found drowned. From the testimony of several witnesses it appears that while Short and his master were working together on South's plantation, an argument arose between them. Resenting some act of insubordination on the part of his servant, South decided to give the lad a whipping. Grabbing a tobacco stick South started toward young Short. The youth in order to avoid a beating ran to water side, plunged in, waded out over his depth and was drowned. South, his master, made no effort to save the boy. With this evidence before them, a coroner's jury declared that the servant was a felo-de-se, that is, a felon of one's self, and therefore not entitled to a Christian burial. It is hard to defend this finding of the jurors. It would appear that the youth had no intention of taking his own life, but jumped into the water solely for the purpose of avoiding a whipping.[26]

There are several other instances of suicide in early Maryland, all of them among the servant class. As in the case of John Short, they probably killed themselves in order to escape the brutal treatment they sometimes received at the hands of their masters. Two servants of Anthony Salway (or Salloway), of Anne Arundel county, took their own lives. Anne Vaughan, a maidservant, cut her throat with a pair of scissors and ran a knife " into her belly." Thomas Teedsteed, a manservant, also cut his own throat. Both were found by a coroner's inquest to be guilty of willful self-destruction, or, as the jurors described it in the Teedsteed case, of " feloniously and willfully murthering of himself." [27]

In order to escape further maltreatment, Ann Beetle, maidservant of William Hunt, took her own life. Witnesses said that they had heard many angry words pass between Mrs. Hunt and Ann and that they had seen Mrs. Hunt shove Ann from her bed. As a result of her brutal treatment Ann had a cut over her eyebrow and her face and clothes were covered with blood. Shortly after this Ann was found drowned. A coroner's inquest was ordered. After examining the body of Ann, the members of this inquest rendered the following report:

> We the jurors do find a wound upon her left eyebrow and having had it searched by a chirurgeon we do find it not mortal, but do according to the best of our knowledge judge that she drowned herself wherefore we the jurors of inquest do indict the said Ann Beetle, she not having the fear of God before her eyes, of wilfully murdering herself and so give up our verdict with one consent by our foreman—
> Samuel Chew, coroner.   Francis Holland.

Thus, even after death had mercifully intervened, man-made justice attempted to further punish a brutally treated servant.[28]

In some cases the master would try to show that the servant had not taken his life because of any maltreatment which he had received at his hands. Such a case occurred in Charles county when the body of Roger Evans, a servant of Thomas Baker's, was found along the bank of the Potomac river. To prove that he was in no way responsible for the death of this man, Baker, who was a commissioner of the county court, introduced several witnesses. One man testified that Evans had been treated by his master " as if he had been his own child." Another witness, who had been a guest of Baker's for several weeks before Evans' death, declared that he had never heard the servant complain about his master, but, on the contrary, had always praised his master as a good man.

In view of this testimony the county commissioners exonerated Baker. Whether he had drowned himself, or a return of his old sickness had caused his death, it was apparent in any case, the justices said, that his death was due to his own " idleness and roquish absentment." [29]

There was one other case in Charles county where the master of a servant was freed of all blame in causing a servant's death. This was when John Constable, a servant of William Heard, was found drowned. After a coroner's inquest had concluded their investigation, the jurors declared:

The verdict is that having viewed the dead body of the abovesaid Constable, we find it clear and without stripes and to the best of our judgments was the cause of his own death by willfully drowning of himself.[30]

In a case which arose in Anne Arundel county, an attempt was made not to hold the master responsible for the death of his servant, who had been drowned, but for failing to take the proper steps after the servant's death. To quote the grand jury's indictment:

The jurors for the Right Honourable, the Lord Proprietor, doth present John Holmwood, of Anne Arundel county, for that he having a servant man drowned in a creek adjoining near his house . . . did not cause a jury of inquest to inquire concerning his death, and, being summoned by the commissioners of the said county to appear before them to answer the same, he, in contempt of his Lordship's authority, refused to appear.

From the testimony of witnesses, it appears that neither John Holmwood, nor his wife, were at their home when Charles Hodges, their servant, was drowned. Both were at a meeting house which was some distance from their plantation. Other servants, however, also in the employment of the Holmwoods, had tried in vain to save Hodges from drowning. The body did not come to the surface until the following day. Although the corpse had been in the water only a short time, it was already much disfigured, having been eaten by crabs, and it "stunk extraordinarily." Immediately upon learning that the body of his servant had been found, Holmwood sent one of his servants to give this information to the nearest magistrate. When this official sent back word that he would not order a coroner's inquest, or take any action in the matter, Holmwood went ahead and buried the corpse.

All this testimony helped Holmwood to clear himself of the charge made in the indictment that he had not taken steps to have a coroner's jury summoned. As to his having failed to appear on a certain day before the commissioners of Anne Arundel county, Holmwood said that on that very same day he had also been summoned by a writ from the Lord Proprietor to appear as a member of the assembly. Obviously, he had to obey his Lordship's orders. When no one appeared before the provincial court to contradict what Holmwood had said, the judges ordered that he should be " cleared sine die." The decision of the court seems fair.[31]

There are a number of cases of assault and battery in the printed records of the provincial and county courts. In none of them, however, did the man committing the assault go so far as to cut or " pluck out " another's eyes or tongue. Had he done so, a statute provided that the offender might be put to death, or burnt in hand, etc., etc.[32]

First, as to the cases of assault and battery which came before the provincial court. Edmund Hudson sued Thomas Munday " in an action of battery." Hudson claimed that Munday ran at him with " a naked sword " in an attempt to drive him out of his own home. At the trial, when Hudson was unable to prove this charge, Munday was dismissed " without day," that is, without the court naming a day for a further hearing of the case. In another case, Thomas Allanson, overseer of the governor's servants, brought suit against William Brookes for having assaulted him. Allanson maintained that one night as he was leaving the governor's house on his way back to the servants' quarters, Brookes, armed with a cudgel, set upon him " in cold blood and in the dark " and with " several cruel blows " knocked him down. Allanson said that as he was unarmed he was unable to make any defence. When two women on the witness stand confirmed what Allanson had said about the attack on him, a jury awarded Allanson three hundred pounds of tobacco damages.[33]

John Wright sued Captain Thomas Smith in an action of assault and battery. Wright claimed that while he was attempting to load some hogsheads of tobacco on board of Captain Smith's boat, the latter not only refused to let him have bills of lading for the tobacco, but calling him a cheating knave and rogue, struck him with a large piece of rope. Wright asked that Smith should be compelled " to repair his reputation in open court " and also make reparation for " the unjustifiable battery." [34]

Elionar Martin, a widow, brought suit against George Willson for assaulting her in her own house. While they were drinking together, an argument arose during which Willson knocked the widow's head against a wall. It seems that Willson accused Mrs. Martin of cheating him, while the widow, in turn, said that Willson had been " the ruin of her husband." [35]

Thomas Maidwell sued John Dandy in an action of assault and battery. This is the same man, who, as we have already seen, was accused of killing an Indian youth and later of murdering a boy servant in his employ. In the present instance, Maidwell accused Dandy and his wife of assaulting him and knocking him to the ground. Maidwell, it appears, was working for the smith at his forge near St. Mary's. The trouble started when young Maidwell began to pay too much attention to a girl who lived in the same house with the Dandys and was probably their servant. One day this girl gave Maidwell some peaches. Mrs. Dandy at once remonstrated with the youth for having accepted the fruit, " giving him very ill language." So did her husband who started walking towards Maidwell with a hammer in his hand. In the meantime, Mrs. Dandy, armed with " a smith's cindar," advanced to the attack. Both struck the young man, knocking him down. Maidwell struggled to his feet and ran out of the forge calling for help. Although it would seem from these facts that Maidwell had a very good case against the Dandys for some reason he did not prosecute. Not long after this the youth died, probably from the injuries inflicted by the smith and his wife.[36]

We have last to consider the cases of assault and battery which came before the county courts. In Charles county Richard Roe and his wife, Mary, brought an action of assault and bat-

tery against John Nevill and his wife, Joan. The Roes alleged
that the Nevills—

In or about the month of June last passed vi et armis in and upon
the said Mary made an assault and her the said Mary did violently
beat, bruise and prejudice to the great prejudice, hurt, injury and
detriment of her the said Mary . . . whereof the plaintiffs sayeth in
fact they are damnified to the value of three thousand pounds of
tobacco . . . .

In order to substantiate these allegations, the Roes called
three witnesses to testify in their behalf. The first was Richard
Dodd who said that he heard Joan Nevill and Mary Roe quar-
relling with each other near the log blockhouse in which the
Roes lived. Apparently Joan tried to kindle a fire near the
wall of the blockhouse. When Mary tried to prevent her doing
this, a free-for-all fight ensued during which John Nevill came
to the assistance of his wife. Nevill beat Mary with a stick,
knocking her to the ground. While on the ground, Nevill and
his wife kept " hauling and pulling " her until she cried " murder." As a result of the attack upon her, Mary Roe, according
to Dodd, was " most sadly abused and torn about the face."
Thomas Baker, a Charles county justice, was next summoned
by the Roes. He said that he saw part of the blockhouse in
which the Roes lived on fire and smoking, that Mary Roe and
the Nevills were having an argument and that Mary's face was
scratched and bleeding. The third witness, Robert Cockerill,
testified that he, too, had seen Mary Roe " all scratched and
bloody."
After hearing all the evidence in this case, the jury found
the Nevills guilty of assault and battery, ordered them to pay
ten groats (forty pence) damages and the cost of the suit. It
is hard to justify this award of such a small amount of damages
which were only nominal. The Roes had asked for three thousand pounds of tobacco damages, or about twenty-five pounds
sterling. While this was an exorbitant demand, at the same
time forty pence seems too little.[37]
There are two cases of assault and battery in the Talbot
county records. In the first of these, Robert Knapp declared that
Seth Foster and his wife had " violently set upon him and beat

him so much that he was forced to keep to his bed." Knapp asked for " satisfaction for the abuse." When, however, he failed to prove his declaration by " letting fall his action," the county court ordered Knapp to pay the cost of the suit.[38]

In the other Talbot county case of assault and battery, William Ladds sued Francis Finch. According to Ladds, Finch came to his plantation while he was away and on one occasion grabbed his wife by the throat and at another time hit her on the head with a stick. As this was done without provocation, Ladds asked two thousand pounds of tobacco damages. Witnesses who testified in this case gave an entirely different version of the matter. According to them, Mrs. Ladds had made the first assault and had sworn that she would kill Finch. The Talbot justices relying on their testimony, ordered that Mrs. Ladds should be bound " to her good behaviour " and that the sheriff should take her into his custody until she had " entered into bond with sufficient security." [39]

In Calvert county, Thomas Manning was accused by William Dorrington of assaulting his daughter, Sarah, aged twelve, or, as Dorrington alleged, Manning " did by force of arms assault, wound, beat and evil intreat Sarah," and does still threaten to do so. As a result his child was " afraid of her life or loss of limbs." The sheriff of Calvert county was ordered to require Manning to give security for his appearance to answer this suit and to hold him in prison until he gave the necessary security.[40]

Actions of assault and battery can be found in the records of the meetings of a court leet and court baron of St. Clement's Manor, in St. Mary's county. At one session of this court, a body of jurors presented Samuel Harris for breaking the peace " with a stick and that there was bloodshed committed by Samuel Harris on the body of John Mansell. . . . " Although Harris was fined forty pounds of tobacco, this sum was later remitted. At a subsequent session of the same court, Derby Dollovan was presented by a jury for " committing affray and shedding blood " in the house of another man. Dollovan was ordered to give " sureties for the peace." It is interesting to note that Thomas Gerrard himself, Lord of this Manor, once told an acquaintance to kick another man " on the breech." [41]

CHAPTER VII

DRUNKENNESS, PROFANITY AND WITCHCRAFT

No less a person than Thomas Gerrard, Lord of St. Clement's Manor, and a member of the Governor's Council, was accused of drunkenness. The charge was made by Richard Smith, his Lordship's Attorney General, who alleged that Gerrard " to the great offence of Almighty God, dishonour of his Lordship and the whole Council, hath divers times misbehaved himself and offended in drunkenness and lewd behaviour." On one occasion, in particular, the Attorney General said that Gerrard was intoxicated while on board of a vessel anchored in the St. George's river. Smith asked that the Lord of St. Clement's Manor might be brought " to condigne and exemplary punishment," and gave as his reason for making this request a letter which Cecil Calvert, the Lord Proprietor, had written asking that any councillor guilty of excessive drunkenness should be suspended from office.

In order to support the serious charge which had been made, Captain Nicholas Guyther and Colonel Henry Coursey were summoned as witnesses. Guyther said that he was on board of the vessel in the St. George's river when Gerrard came on board. It was obvious, he declared, that Gerrard had been drinking. Coursey, who was present on the same occasion, testified that Gerrard had " drunk something extraordinary, but he was not so much in drink but he could get out of a cart's way." At the conclusion of this testimony and after the depositions of several others were offered in evidence, Gerrard asked and received permission to answer the charges made against him at the next meeting of the provincial court.

At this session, Gerrard submitted " a long declaration of his former merits and sufferings " which he asked the members of the court to consider. The judges did as they were requested and then arranged for the examination of witnesses " in the matters informed against Mr. Gerrard by the Attorney General." Now, for some unknown reason, no further action was taken against Gerrard, who resumed his seat on the council. But the Lord of St. Clement's Manor was not satisfied with this settlement of the matter. He wrote Cecil Calvert that it was "small satisfaction" to him to have the charges against him dropped when he was ready and willing to defend himself. Gerrard asked that he should not be required to sit as a member of the council while his reputation "lies under any blemish," and until he had "some reparation in his honour." As the Lord Proprietor considered this a reasonable request, he gave directions that the complaints against Gerrard should be carefully investigated and his councillor either acquitted or "censured."

The question of his drunkenness was never finally determined, however, as Gerrard was soon faced with a far more serious charge and that was of being implicated in a rebellion fomented by Josias Fendall. Fendall, it seems, had tried to change the government of Maryland "into the form of a Commonwealth," of which he aspired to be head, to "the great prejudice of his Lordship's rights." The Lord Proprietor, while not going to the extent of asking that Gerrard should be put to death for his association with Fendall, did want Gerrard banished from the province and his estate confiscated. When he learnt of this decision, Gerrard submitted a petition asking for "mercy and favour." The result was that the Lord of St. Clement's Manor was allowed to remain in the colony with the understanding, however, that he should never thereafter hold any office in the province, nor have the right to vote. Gerrard also had to give "a recognizance for his good behaviour" towards the proprietary government.[1]

Gerrard, as we have seen, was a member of the governor's council. There were other cases of drunkenness among proprietary officials, especially in the case of some of the county commissioners. One of the heaviest drinkers in Kent county was Captain Thomas Bradnox, a justice of the peace there. On one occasion, he was so drunk that he was scarcely able to move. When he attempted to rise from his chair he fell to the floor.

It was said that when Bradnox was drunk he swore " like a madman." The captain was presented for being drunk and disturbing the peace. One of the persons who gave information about Bradnox's drunkenness was a woman by the name of Deliverance Lovely, an appropriate name for an informer. Perhaps because he was a commissioner Bradnox escaped punishment for his excessive drinking. Other inhabitants of Kent county, including Matthew Read, another justice of the peace, were summoned to answer for their misconduct and " deboist (debauched) drinking." When these men promised " amendment," the Kent county commissioners, taking into consideration that this was their first offence, reprimanded them and let them go after they had paid the court costs.[2]

Some men in the province entertained very decided ideas on the question of drunkenness. It was Cecil Calvert's wish that any one holding an office such as councillor, or justice of the peace, if twice convicted of being " an usual drunkard, swearer, or curser," should be suspended from office. William Joseph, one of the proprietary governors of Maryland, had even more pronounced views regarding drunkenness. He thought that severe laws should be enacted to punish drunkards. When intoxicated Joseph thought that men were more likely to be profane and " to huff and hector in fudling shops and such like places of sottish behaviour." As for drunkenness, Joseph said:

It is a sin of all sins the most dangerous for not content in itself, it strangely leads us into all manner of sin and vice, depriving us not only of God, but even of ourselves, for it basely unmans us to that degree that it makes us become greater beasts than beasts themselves, so shameful it is to see dogs have more sense than their masters and horses more understanding than their riders.[3]

There seems to have been quite a lot of drinking done in early Maryland. Two Dutchmen visiting the province said that the Maryland planters upon the arrival of any vessel in port would spend so much in buying wine and brandy that they would often neglect to buy enough clothing for their children. Mention is made in early Maryland records of " merry drinking and dancing " and of fiddlers. One drinking party resulted in the death of one of the participants. This was when Daniel

Clocker went to visit John Furnifield. While with Furnifield, Clocker asked him " to give him a cast over the river " in his canoe. Furnifield not only agreed to do this, but " withal fetched out about a pint of drams " which he and Clocker consumed. Before embarking they secured an earthen pot containing about a quart of drams and drank most of this, too. When they tried to paddle the canoe Furnifield fell out and was drowned.[4]

There are two cases in the early records of men drinking themselves to death. William Styles was one of them. When he was found dead in bed one morning an investigation was made by a coroner's jury. It was the juror's opinion that Styles while drunk had over gorged himself with food and as a result had choked to death. David Anderson, of Talbot county, was the other man who died from the effects of excessive drinking. One evening after imbibing very freely he " laid himself down to sleep." In the morning he was dead. A coroner's jury was summoned to investigate. After hearing the testimony of several witnesses and also examining the body of Anderson, the jurors decided that as he " came to his death by being surfeited with drink," liquor was " accidentally " the cause of his death and therefore he should have " a Christian burial." In other words, this verdict meant that the jurymen did not think Anderson had intended to commit suicide by drinking too much.[5]

Several laws were enacted in early Maryland for the punishment of drunkards. By the provisions of these early statutes any one who should " abuse himself by frequent drunkenness," or should be convicted of being drunk by the testimony of two witnesses, would be liable to pay a fine of one hundred pounds of tobacco. If the offender were a servant, without property, he could be imprisoned or set in stocks or bilbos, without food or water, for twenty-four hours. A bilbo was a long bar or bolt of iron with sliding shackles, and a lock at the end, to confine the feet of prisoners. All magistrates and other officers were instructed to make every effort to bring drunkards to trial and punishment. Any one else who saw another drunk and did not within three days report it to the nearest magistrate was liable to pay a fine of one hundred pounds of tobacco. Storekeepers, shipmasters, or heads of families, who allowed drunkenness in their store, ship or house, as the case might be, were also liable to fine of one hundred pounds of tobacco.[6]

By the terms of a later statute the punishment for drunkards was made more severe. Any person who was thereafter convicted of drunkenness could, for the first offence, either be set in stocks for six hours, or pay the customary fine of one hundred pounds of tobacco, half of which went to the man or woman who " informed " on him. If the same person was convicted a second time of drunkenness, he would either be " publicly whipped," or have to pay a fine of three hundred pounds of tobacco. Should the same man be convicted a third time for this offence, then he could be adjudged " a person infamous," and declared incapable of voting or holding any office in the province for three years after his conviction.[7]

A number of colonists were brought before the provincial and county courts charged with drunkenness. As Roger Scott had been drunk for three successive days, he was convicted as " a common drunkard." The provincial court compelled him to pay one hundred pounds of tobacco for each day on which he had been intoxicated. Both in Charles and Kent counties the commissioners fined planters for being drunk. In one case, when James Gaylourd was charged with misbehaving himself and drinking too much, the court ordered Gaylourd committed to the sheriff's custody until he was again " called for." For allowing his prisoner later to escape, the sheriff had to pay a fine of one hundred pounds of tobacco.[8]

What did the seventeenth-century Marylander drink? Those planning to go to Maryland to settle were advised to take with them for use on shipboard, and for some time after their arrival in the colony, burnt claret wine and canary sack. They were also advised to take with them kernels of pears and apples for " the making hereafter of cider and perry," and malt for beer. It was considered fitting that " a gentleman " should always have on hand a sufficient supply of wine and liquor. What was a proper yearly allowance? The provincial court judges answered this question by declaring the amount which a gentleman could consume in one year to be " one anker of drams, a tierce of sack, and a case of English spirits." As an anker contained about ten gallons, a tierce about forty-two gallons, this was a total of fifty-two gallons or over two hundred quarts of drams and sack. Even without allowing for the case of English spirits, a gentleman could, if he wished, drink more than a pint a day.[9]

Some of the early settlers were of the opinion that the grapes of Maryland might produce a good wine. Captain Robert Wintour, member of the governor's council, thought that wine might be made out of the wild grapes in Maryland which would " plentifully supply the inhabitants with good drink." He did not, however, think it would pay to export the home-grown product. Father Andrew White, a Jesuit priest, wrote to Cecil Calvert that—

It would be very expedient to try what wine this land will yield: I have a strong presumption that it will prove well for this autumn (1638). I have drank wine made of the wild grapes not inferiour in its age to any wine of Spain. It had much of muscadine grape, but was dark red inclining to brown. I have not seen as yet any white grape, excepting the fox-grape, which hath some stain of white, but of red grape I have seen much diversity; some less, some greater, some stain, some do not, some are aromatical, some not.

Now, if your Lordship would cause some to plant vineyards, why may not your Lordship monopolize the wine for some years, to your Lordship's great profit, especially if all sorts of vines be gotten out of Spain and France. True it is you must have patience for two or three years before they yield wine, but afterward it is a constant commodity and that a very great one, too.[10]

The priest's suggestion does not seem to have been carried out. During the early period, the Maryland colonists confined most of their efforts to making cider, perry and beer. According to some visitors to Maryland near the end of the seventeenth century, most of the planters drank cider. Cider time was about the last of July or the beginning of August. Thomas Gerrard, on his plantation at Mattapany, had in the upper yard house one hundred and eighty gallons of liquor, in the pear orchard house two hundred and forty gallons and in the lower cider house four hundred gallons. This was a total of over eight hundred gallons of cider and perry on one plantation. It is little wonder that Gerrard was prone to drink too much.[11]

Brandy was made in Maryland of peaches or apples. The planters were said to be very generous with their brandy and that when one of them had a gallon of this liquor all his neighbours would come to help him consume it. Often a colonist would go ten miles to get a drink of brandy. Some planters made it practice to have a morning draft of brandy. The delegates of the assembly enjoyed drinking it. Mention is made in the early records of other kinds of alcoholic beverages, of metheglin, an old-fashioned beverage usually fermented, made of honey and water, and of a liquor called " white coin." Possibly the latter was a relative of our modern " white mule." Sack, a strong white wine, seems to have been much in use. In order to establish friendly relations with the neighbouring colony of Virginia, the first Maryland colonists were instructed to present the governor of that colony with " a butt of sack." Rum was also consumed and there is frequent mention of drams. This may have meant a small drink, or draft, of distilled alcoholic liquor of any kind, such as a dram of brandy, although in most instances it would seem that the colonists had in mind some particular kind of liquor which they called " drams." For example, in one instance " four gallons of drams " were furnished the members of a jury, and, on another occasion, " a bottle of drams " was given to a sick man. Punch, made of brandy and rum, was also consumed in early Maryland, as were cordials.[12]

During the seventeenth century, inns or taverns in Maryland were known as ordinaries. These hostels played an important part in the life of the early colonists. Distances between plantations were great and the hardships involved in travel even greater. Some travelled on foot, some on horseback, and many by boat. Pinnaces, sloops, shallops and even canoes were used on the Chesapeake Bay and its tributary rivers and creeks. When any one arrived at their destination, after enduring the hardships of land or water travel, even the primitive hospitality which the ordinaries offered must have been most welcome. When the general assembly was in session at St. Mary's, the ordinaries would be a convenient place for the burgesses, or delegates, to stay. Also when a meeting of the provincial court

was held, the judges, jurors, witnesses and litigants would so-
journ at one of the ordinaries. So, too, when the county courts
convened, an inn at the county seat was a convenient place to
lodge. Strangers in the province would naturally want to put
up at a tavern or ordinary.[13]

All colonists agreed that ordinaries were necessary. The ones
that were in existence were sometimes so crowded that accommodations were not available, or the provisions were all consumed. Not all the colonists, however, were in agreement as
to the best places to establish ordinaries. Some thought that in
the counties, inns should only be maintained near where the
county courts met and that all other ordinaries should be suppressed. Others, however, were of the opinion that these little
taverns were equally necessary at other places, as, for example,
near a ferry across a river, or creek, or where " many ships do
ride." As the province grew in size, it was not long before
many of the counties had more than one ordinary. In Charles
county, besides the inn at the court house there was one " at
the mill and one upon the race." In Calvert county there were
four " at the town " and also one " at John Grigg's," while in
Anne Arundel county there was one at the court house, one at
Richard Hill's and another " at the Red Lion." [14]

It was necessary to obtain a license before any one could
maintain an ordinary. At first the governor granted the licenses,
but with the growth of the colony the commissioners in each
county were authorized to do this, and, in St. Mary's, the village
officials enjoyed a similar power. In issuing licenses, " the ease
and conveniency of inhabitants, travellers and strangers," was
to be considered. The personal qualifications of the applicant
for a license were also taken into consideration, and whether
the people who lived in his neighbourhood regarded him as " a
man meet to keep an inn or ordinary." Licenses to keep an
ordinary generally lasted for one year and could be renewed.
Keeping an ordinary in most cases merely meant that the owner
of a private house was authorized to furnish board, liquor and
lodging to the public. An ordinary keeper was obliged to
maintain order on his premises. He must not " suffer any
evil rule, or order, to be kept in his house," nor could he allow

any servants, or apprentices, "to remain tipling or drinking in his house without their master's privity." The innkeeper must not allow his other guests to drink too much, or fight and quarrel. If the man who maintained an ordinary failed to keep order, then his license would not only be revoked, but he would also have to forfeit the deposit he had made to obtain his license.[15]

In order to prevent ordinaries, or inns, charging too much for the alcoholic beverages which they dispensed, laws were passed which regulated the prices that could be asked for beer, wines and liquors. The price list in force during the year 1671 is given here:

WINES, LIQUORS AND BEER	SHILLINGS PER GALLON
Brandy	ten shillings
Dutch drams	six shillings
English drams	ten shillings
Rum	six shillings
Canary wine	twelve shillings
Malaga wine	ten shillings
French wine	six shillings
Rhenish wines	six shillings
Madeira, Fayal, Porto Port and other Portugal wines	six shillings
Sherry	ten shillings
Strong beer or ale (made in colony)	two shillings
Beer (imported)	one shilling, six pence
Cider, perry and quince (domestic)	one shilling, six pence
Cider, perry and quince (foreign)	one shilling
Mumm (a type of strong ale or beer)	three shillings

It is interesting to note that less could be charged for imported beer, cider, perry and quince than if the same beverages had been brewed in the colony. One would think that in order to encourage domestic production less could have been asked for the colonial product. Of course, a man did not have to buy a whole gallon. If he purchased more or less than a gallon the same rates were in force, that is, a proportionate part of them. If more than the fixed prices were asked, the ordinary

11

keeper would not only be unable to recover the amount which he had demanded, but he would also have to pay a heavy fine.[16]

The ordinary keeper was entitled to ask one shilling for a meal and six pence for a night's lodging. This latter rate entitled the guest to a bed but not to a separate room. The members of the assembly once had a dispute as to the number of beds which an ordinary keeper should have for his guests. Some thought that in addition to his own, he should have either four good feather beds or the same number of flock beds. The price at which corn, oats, hay or straw and pasturage could be furnished the horses of guests was also fixed. An ordinary keeper must provide stable room for horses.[17]

Several years after the price list of 1671 for wines and liquors was adopted, a change was made in the proprietary policy regarding their sale. The change was due to the fact that it was deemed unfair to compel the ordinary keeper to sell his alcoholic beverages at a fixed price when it was not known how much he had had to pay some merchant for them. In other words, while the innkeeper had to sell at a fixed price, the merchant who imported liquors and wine did not. Accordingly, every person must now bargain with the ordinary keeper as to what price he would ask for liquors. A fixed rate, however, was still asked for beer, meals and lodging, which were about the same as before, although they were now stated in pounds of tobacco instead of English currency. The new law provided that the guest at an ordinary might either pay in tobacco the sum required or give an equivalent amount in pence.[18]

Still another change was made in the proprietary policy regarding the sale of stimulants. By the terms of an act passed near the end of the proprietary period, the commissioner of each county and the village officials of St. Mary's were authorized, during the month of January of each year, to fix the prices at which wines and liquors could be sold for the ensuing year. Ordinary keepers were allowed to be present when the prices were discussed. No doubt they would inform the county, or St. Mary's officials, what they had paid the merchants from whom they had bought their wines and liquors. After the prices to be asked had been agreed upon, the lists were then posted in some

public place in each county and at St. Mary's. All ordinary
keepers who asked more than the posted rates ran the risk of
paying a fine. The charge which could be made for beer and
lodging still remained as before.[19]

At first, in order to encourage " honest and well-minded people " to keep an ordinary, a law was passed making it very easy
for them to collect their debts. If one of the ordinary keeper's
guests failed to pay what he owed, the man who maintained
" the victualling house " was not compelled to resort to the
courts in order to collect what was due him. His claims, however, had to be in writing and properly witnessed. A later
statute required the innkeeper to swear before one of the judges
of the provincial court as to the accuracy of his account before
he could have an execution for the amount due. The amount,
however, which the ordinary keeper could recover was limited.
His claim would not be allowed if he furnished a guest with
more than two meals a day, or more than a pottle (half a
gallon) of beer. The charge for the meals, beer and also a
night's lodging must be in accordance with the rates fixed by
law.[20]

Near the end of the proprietary period many ordinary keepers
seem to have been guilty of asking " excessive and outrageous
prices " for their wines and liquors. Tavern keepers would
refuse to let their guests have a statement showing the amount
of liquor which had been furnished for fear an overcharge
might be discovered. Sometimes when his guest was drunk the
innkeeper would make him sign a statement admitting the correctness of an overcharge. In order to break up such practices,
a law was enacted making it necessary for an ordinary keeper to
swear to the accuracy of his accounts before one of several
designated proprietary officials.[21]

In view of all the statutes which were passed regulating
ordinaries, one would expect to find many instances in which
ordinary keepers were accused of violating these laws. Comparatively few cases, however, could be found in the early records. One man was fined for selling wine and liquor in his
house without a license. An amusing case arose in Kent county,
where Edward Sweatnam ran an ordinary. It seems that Sweat-

nam served one of his guests a rattlesnake telling him that it was an eel. The guest later sued the ordinary keeper for having done this and recovered twenty thousand pounds of tobacco damages. The guest claimed that eating the snake not only made him sick but also caused him to lose his hair.[22]

At a session of the court leet and court baron of Thomas Gerrard, Lord of St. Clement's Manor, a jury presented Humphrey Willey for keeping " a tipling house and selling his drink without license at unlawful rates." For doing this Willey was fined " according to the act of assembly in that case made and provided." Two other men were presented and fined by the same court on a similar charge.[23]

There is one other case where a man, although not an ordinary keeper, was accused of asking too much for his liquor. This was when a complaint of " extortion " was laid against John Taylor for selling his drams at forty pounds of tobacco per bottle. Several planters, it seems, had sick wives for whom they wanted the drams. When one of them had to pay the amount Taylor asked, he remarked that "no ordinary keeper in the country sold so dear." Taylor's only comment was: " Well, well." Although the judges of the provincial court were of the opinion that Taylor had asked too much for his drams, they decided to dismiss the charge against him " upon his good obearance (behavior) hereafter." [24]

Most of the ordinaries were at St. Mary's. Indeed, as early as 1635, plans were made " for convenient houses to be set up at St. Mary's where all strangers may, at their first coming, be entertained with lodging and other fitting accommodations for themselves and their goods, till they can better provide for themselves." The village authorities were authorized to issue licenses to ordinary keepers and to require each of them to put up a bond of two thousand pounds of tobacco as security that they would maintain order in their hostelries. Ordinaries were required to have accommodations for twelve persons and "stable room" for twenty horses. Rates at which wines and liquors could be sold in St. Mary's were decided upon by the village officials. These rates varied from year to year. The following were the prices that could be asked during the year 1684:—

WINES AND LIQUORS	POUNDS OF TOBACCO PER GALLON
Brandy	120
Rum	100
Fayal and St. George's wine	80
Madeira wine	100
Claret and Porto Port	100
Passada wine	120
Sherry and Rhenish wine	120
White wine	100
Cider, perry or quince	30
Canary wine	150

Brandy punch, made of one quart of brandy, sold for fifty pounds of tobacco a bowl, while rum punch, made of the same amount of rum, sold for forty pounds of tobacco per bowl. Ordinary keepers in St. Mary's, like those in the counties, must observe these fixed prices.[25]

One of those who kept an ordinary in St. Mary's was an illiterate woman named Hannah Lee, widow of Hugh Lee. When, during the spring of 1662, the members of the assembly were discussing the necessity of providing " a house and place " for holding the meetings of the assembly and the sessions of the provincial court, the assemblymen finally decided that the house of Mrs. Lee would be suitable for this purpose. In consideration of the twelve hundred pounds of tobacco which she received, Mrs. Lee not only agreed to repair the roof of her house, but also to maintain an ordinary, or inn, in her house for the accommodation of the proprietary officials. In her house some of the burgesses, who came from distant parts, slept.[26]

Not long after this Mrs. Lee married William Price, her servant. Both were now summoned before the provincial court for their failure to repair the roof of Mrs. Lee's house, (used as state house) as had been agreed. Hannah was unable to appear, " being not able to travel so far." When, however, her husband did make his appearance, he was informed by the court that he must " forthwith cover the state house at St. Mary's," or remain in the sheriff's custody until he gave bond that he would make the necessary repairs.[27]

This was not the only litigation in which the Prices were involved. She and her husband were sued by William Hollingsworth, of Salem, New England, on some debts which Hannah, before her marriage to Price, had owed the New Englander. Hollingsworth said that although he " divers times gently required " the Prices to pay him a specified amount of tobacco for " the several goods and merchandises " which he had sold to Hannah, they had refused to do so. Mr. and Mrs. Price were also involved in litigation regarding the settlement of an orphan's estate.[28]

As a result of all these lawsuits, Hannah finally landed in prison. In a humble petition to the governor and council she stated—

That through a long and hard durance of imprisonment that necessary clothing, both of woolen and linen apparel your petitioner had, is now worn out so that your petitioner is in great distress of relief herein.

Mrs. Price asked that either the cases in which she was involved be brought " to a speedy trial," or if " a longer restraint " was necessary, then, in that case, some provision should be made for her, as to the proprietary officials " shall seem meet." After considering this petition, the judges of the provincial court instructed the sheriff of the county, in which William Price lived, to levy on his property in order to provide his wife with sufficient clothing.[29]

Others besides Mrs. Lee undertook to keep ordinaries at St. Mary's. James Jolly, an appropriate name for a tavern keeper, was one of them. At his inn he provided " drink and diet " not only for members of the lower and upper house of the assembly, but there, too, stayed some Susquehannock Indians when on a mission to St. Mary's. On one occasion, it appears, that Jolly's account book was stolen. The upper house at once ordered that no one should leave the state house until a search had been made for this book. The sheriff was also instructed to search the persons of any one whom Jolly suspected as having stolen the book.[30]

Although Jolly at one time agreed to keep an ordinary in

a building which was used as a state house and also to keep
the structure in repair, he did not live up to his promise. He
was still open for business, however, in his own house at St.
Mary's which was on a plantation known as "Kitt Martin's
Point." As a side line Jolly traded with the Indians in order
to keep his guests supplied with corn. The ordinary keeper did
such an extensive business that he appointed an agent to buy
liquors for him. Like Mrs. Lee he was too ignorant to sign his
own name. Jolly and his wife later moved to the Eastern Shore
where they settled on the south bank of the Pocomoke river.[31]

Another man who ran an ordinary at St. Mary's was William
Smith. When both Mrs. Lee and Jolly had failed in their
undertaking to keep an ordinary in the building used for the
meetings of the assembly and the provincial court, Smith entered
into the same agreement. He also promised to build a new
state house, but later asked to be relieved of this undertaking.
Smith appears to have had difficulty in collecting the debts due
him by those who patronized his ordinary. The rates which
he charged for wines and liquors were also subject to investigation.[32]

Complaints were raised against Robert Gellie, another ordinary keeper in St. Mary's, on the ground that the conditions
in his tavern were "prejudicial to the public." It was said that
during the time the provincial court was in session, jurors, attorneys, and others went to Gellie's inn, and were there "detained,
and disordered." It was also alleged that clerks and other
proprietary officials spent too much of their time at Gellie's,
the result of which was that attention to public affairs suffered
"by reason of the ordinary." In view of these complaints,
the governor and his council asked the village authorities at St.
Mary's to take steps to have Gellie's ordinary suppressed.[33]

Despite the arrangements made for having ordinaries at St.
Mary's, there were times when sufficient accommodations were
lacking, especially when the provincial court held its sessions
and other occasions of "public convention." People would ride
into the village on horseback only to find that there were no
lodgings available. To remedy this situation, several ordinary
keepers, including Henry Exon and Francis Catterson, were
told that they must at once furnish more accommodations.[34]

Catterson informed the governor and council that it would be almost impossible for him to comply with this order within a short time. As he was, he said, a young man and " newly settled in the world," he could not purchase as large an amount of household goods and provisions as those who had been in that business for some time. Even to maintain lodgings for ten people he had incurred many debts. While he wished to entertain " to the utmost of his power," he did not think it advisable to buy additional provisions and liquors, unless he had some assurance that his license as an ordinary keeper would be renewed. The provincial officials told young Catterson that if he put up sufficient security his license would be renewed.[35]

Anthony Underwood, a delegate, sought to recover for the expense which he had incurred by lodging a proprietary official at his house in St. Mary's. Although he was allowed to recover on his account, members of the council said that they would not allow this to occur again, as they thought it was unfair to ordinary keepers who were required to take out licenses. When, however, Charles Calvert entertained the members of his council at his house at St. John's, the lower house thought that his Lordship should be allowed thirty thousand pounds of tobacco to reimburse him for his expense. Evidently, the prominence of the man who did the entertaining made a difference, or possibly the lower house was more generous with public funds.[36]

When we come to consider the ordinary keepers in the different counties, we find that Thomas Belcher, in Anne Arundel county, received a license to keep an inn, or ordinary, in that county " for the entertainment and accommodation of strangers and others, as occasion shall require." Belcher was authorized to sell beer, wine, " strong waters, or any other fitting and wholesome drink, victuals, or provisions." Belcher's name does not sound like a recommendation of the wholesomeness of the provisions he dispensed.[37]

In Calvert county, Joseph Edloe in a petition to the governor and council, informed them that the situation of his house on the Patuxent river was " more fit and convenient for a public than a private house," and therefore asked permission to keep an ordinary. Edloe was told that he must first be recommended for such a license by the commissioners of his county.[38]

Philip Lynes kept an ordinary in Charles county near the court house. Lynes submitted to the members of the assembly an expense account for taking care of a man by the name of Paul Scurfield who had been sick at his inn. The account follows:

To his diet from April 2nd, 1688 until 28th of this instant November, three meals per day at 12 lbs. tob. per meal comes to..............................	8640	lbs.	tob.
To lodging for said Scurfield at 4 lbs. per night......	960	"	"
To diet for a nurse to attend him for said time comes to the same as above..........................	8640	"	"
To lodging for the nurse at 4 lbs. per night..........	960	"	"
To a pint of cider, or other liquor, each morning the nurse had before (she) went to dress Dr. Scurfield's sores comes to at 30 lbs. per gallon..............	900	"	"
To the nurse I pay her per the court's order of St. Mary's county 300 lbs. tob. per month, being eight months..	2400	"	"
	23700		

To two flock beds, blankets and other necessaries quite rotted and utterly ruined and fit for nothing by his lying in the same, I leave to the honourable houses of assembly to allow me for them what they think convenient.

To four sheets and a table cloth for rowlers and trusses had to dress his wounds.

May it please the honourable houses to well consider the great trouble and inconveniency that I have had with the abovesaid Doctor Scurfield and that I have only charged for his diet and lodging, and the like for his nurse, according as the law allows, and allow the same to your honour's petitioner.

Philip Lynes.[39]

From the number of instances of profanity in early Maryland one is inclined to agree with two visitors to the colony who said that the planters were " very godless and profane." In an attempt to break up this practice, the proprietary officials fined any one ten pounds of tobacco who was guilty of profane cursing or swearing. Should a person become known as " a common swearer, blasphemer, or curser, by any imprecations whatsoever against God or man," he might be punished as the members of any court thought fit.[40]

These laws were invoked in a number of cases. Thomas Hills for swearing " divers oaths " was ordered to appear before the members of the lower house and there acknowledge " his faults for swearing, expressing his hearty sorrow for the same." In Kent county, William Price and Jane Salter were accused of taking the name of God in vain " thereby transgressing both the laws of God and man." It was said that Price swore three oaths and Jane, the wife of John Salter, two oaths. Price was compelled by the Kent county commissioners to pay ten pounds of tobacco for each of his oaths, while John Salter had to pay the same amount for each of his wife's oaths. Clearly it was wise to keep one's wife from swearing.[41]

Michael Baisey was accused of swearing at James Veitch, a sheriff. When the latter asked Baisey to let him have his grindstone, Michael was alleged to have replied: " God's blood, I will be the death of that man that shall fetch it away." Baisey added that Veitch had better bring a strong guard with him if he wished to take the grindstone. For his swearing at the sheriff, Baisey had to pay ten pounds of tobacco and find " sureties " for his future good behaviour.[42]

There are two cases in which men were accused of being " common swearers." One of them was Captain Thomas Bradnox, a justice of the peace in Kent county. It was said that, on one occasion, Bradnox uttered at least one hundred oaths and that he called one man " loggerhead, puppy and fool." When the captain admitted he had sworn " in his passion," the members of the Kent county court debated as to the amount their fellow commissioner should be fined. One justice thought that a fine of two hundred pounds of tobacco should be imposed, while another advocated three hundred pounds of tobacco in view of the fact that this was the second time Captain Bradnox had been convicted of swearing. The latter was the amount which was finally imposed.[43]

The other man accused of being " a common swearer " was Gregory Murell. The occasions upon which Gregory swore were numerous. He had one argument with Thomas Ringgold, a justice of the peace in Kent county, about a steer. Murell swore " by God's blood " that if Ringgold killed the animal, he would claim half of it. Ringgold, on his part, called Murell a thief, said that he had stolen a boat, and that he was not fit to live in the colony. Murell had another dispute with a man by the name of William Elliott, during which Murell swore " a great oath," and told Elliott that he would throw a hammer in his face. In a controversy with Robert Knapp, Murell swore " very outrageous and many desperate oaths," and, on still another occasion, Murell boxed another man on the ears at the same time uttering " many bitter oaths." From all these instances, the Kent county commissioners were convinced that Murell was " a common swearer," but they only fined him two shillings, six pence, as this was his first conviction.[44]

Sunday, or the Sabbath, was not a day for recreation in early Maryland. Under the Roman Catholic proprietors, rigid views were held about the observance of this day. By the terms of the Toleration Act of 1649 any one was fined who profaned " the Sabbath, or Lord's day called Sunday, by frequent swearing, drunkenness, or by any uncivil or disorderly recreation, or by working on that day when absolute necessity doth not require it." If the offender did not have the means of paying the fine, he could then be compelled " in open court " to acknowledge " the scandal and offence he hath given against God and the good and civil government of the province." A man guilty of profaning the Sabbath a third time might be publicly whipped.[45]

Several years after this, an act was passed which forbade any one to hunt or shoot a gun on Sunday. About twenty-five years later, another law was passed " for keeping holy the Lord's Day." Members of the assembly thought that the new statute was necessary, since the day was being profaned " by working, drunkenness, swearing, gaming, unlawful pastimes and other debaucheries to the high dishonour of Almighty God . . ." In order to remedy this condition, an act was passed which stated:

No person, or persons, within this province shall work or do any bodily labour, or occupation, upon any Lords Day, commonly called Sunday, nor shall command or willfully suffer or permit of his or their

children, hired servants, servants, or slaves, to work or labour as aforesaid, the absolute works of necessity and mercy always excepted.

And for that through the wicked and profane licentiousness of several
persons inhabiting or travelling into this province it hath been customary for such sort of persons to spend the said day in the bodily
exercise or occupation of fishing, be it enacted that no person or persons whatsoever, inhabitant or foreigner, shall upon any Lords Day
presume to take fish in any bay, port, or creek of this province with
any nets, tramells, seines, hooks or lines, or other instruments of fishing,
nor shall command, willfully suffer, or permit, any of his or their
children, hired servants, servants, or slaves, so to fish as aforesaid.

And be it likewise enacted that no person, or persons, shall profane
or abuse the said Lords Day by drunkenness, swearing, gaming at cards,
dice, billiards, shuffle boards, bowls, ninepins, horserace, fowling, or
hunting, or any other unlawful sports or recreations.

And if any person, or persons, within this province shall offend in
all or any the premises, he shall forfeit and pay for every offence one
hundred pounds of tobacco.[46]

There are several cases of where men were accused of violating the laws passed for the observance of Sunday. In Charles
county, a jury of inquest presented several persons " for Sab-
both-breaking." Most of the violations of the Sunday laws,
however, occurred in Kent county. In this county, Henry Clay
was presented " for striking tobacco on the Sabboth Day," and
Captain John Russell, himself a Kent county justice, was presented for fighting with another man " on the Lord's Day." [47]

Edward Rogers was charged by the Kent county commissioners with shooting on Sunday. Rogers, it seems, had killed a
turkey on this day. In his own defence, Rogers said that he did
not know that there was any law prohibiting shooting on Sunday. He added that this was the first time he had hunted on
Sunday, that he was sorry he had done so and promised " never
to do the like." The Kent justices decided to accept these
excuses. They told Rogers that if he paid the cost of the suit,
they would not impose any penalty on him. Other men in Kent
county, including Matthew Read, a commissioner of the local
court, were accused of " misbehaviour, profaning the Sabbath,
and shooting off, or discharging their guns, unseasonably."
When Read and his friends promised " amendment," the Kent

court, considering that this was their first offence of this kind, decided not to punish them.[48]

Ordinary keepers had to be very careful about observing Sunday. They could not permit any disorder in their taverns on that day " by gaming or exorbitant drinking during the time of Divine Service." Members of the assembly deplored the fact that despite the existing regulations there was still much " drinking, tipling and gaming " at the ordinaries. This, they considered, was not only profaning the Sabbath, but it was also debauching youths and increasing vice. Accordingly, a law was passed which prohibited the sale of liquor in the taverns on Sundays and also forbade the playing of any games, such as dice or cards. An innkeeper who violated this law might not only lose his license, but also have to pay a fine of two thousand pounds of tobacco.[49]

The officials at St. Mary's also adopted regulations for the observance of Sunday which must be observed by all ordinaries in the town. They were led to take this step because of the " great debaucheries and disorders " occasioned by too much drinking, swearing and gaming on that day. The local authorities decided to prohibit the sale of intoxicants in the village taverns on Sunday to any one, " travellers, strangers, and sick people only excepted." Nor were card playing or dice games to be allowed. Twice on Sunday, once in the morning and once in the afternoon, the constable of St. Mary's was to make an inspection of all the ordinaries in the village to see that these regulations were being observed.[50]

Many Marylanders like to think of the colony during the proprietary period as a place where all faiths were tolerated. While the Toleration Act of 1649 was a step in this direction, there was not complete toleration under the provisions of this statute. If, for example, any person blasphemed or cursed God, or denied that Jesus Christ was the Son of God, or denied the Holy Trinity, " the Father, Son and Holy Ghost," or the Godhead of any of the said three persons of the Trinity, or the Unity of the Godhead, or even used any " reproachful words " about the Holy Trinity, all such persons could be put to death and their possessions confiscated. Nor could any one make any reproachful remarks about the Virgin Mary, or the apostles, or the evangelists, without either being fined, or whipped and imprisoned.[51]

Under the so-called Toleration Act of 1649 there was obviously no toleration for Jews. This is shown by the trial of Jacob Lumbrozo who held this faith. At a meeting of the provincial court held at St. Mary's during the winter of 1658, this man, who was a chirurgeon by profession, was charged with uttering words of blasphemy against "Our Blessed Saviour, Jesus Christ."

At his trial, the prosecution produced two witnesses, one of whom was John Fossett. The latter said that he and Lumbrozo were visiting Richard Preston when a discussion arose about Christianity. According to Fossett, when he remarked that Christ was " more than man, as did appear by his Resurrection," Lumbrozo replied " that his disciples stole him away." But how about the miracles which Christ performed, asked Fossett. " Such works might be done by necromancy, or sorcery," answered Lumbrozo. "Then do you take Christ to be a necromancer? " challenged Fossett. At this the Jew laughed but said nothing.

Richard Preston was then called to the witness stand, who testified that in his house he overheard the following conversation between Lumbrozo and Josias Cole, a Quaker.

Cole: Do the Jews look for a Messiah?
Lumbrozo: Yes.
Cole: Who was He that was crucified at Jerusalem?
Lumbrozo: He was a man.
Cole: How did He do all his miracles?
Lumbrozo: He did them by art magic.
Cole: How did His disciples do the same miracles after He was crucified?
Lumbrozo: He taught them his art.

When called upon to testify in his own behalf, Lumbrozo said that when answering the questions propounded to him by Fossett and Cole about the Christian religion he was merely expressing his opinion as a Jew and that he had not intended to say anything in a scoffing way, or " in derogation of Him, whom the Christians acknowledge for their Messiah." After Lumbrozo had finished testifying, the judges of the provincial court ordered the Jew to remain in prison until the next meeting of the court, at which time the accusation against him would be further considered. The case, however, never came up again for another hearing. When, upon the death of Oliver Cromwell, his son Richard was proclaimed Lord Protector of England, Josias Fendall, then Governor of Maryland, in honor of this event, pardoned and acquitted every one in the province who " stood indicted, convicted, or condemned to die " for some criminal offence. This pardon, issued in the name of the Lord Proprietor of Maryland, saved Lumbrozo from further trial and possible conviction for blasphemy.[52]

It may be wondered why witchcraft should be considered in connection with drunkenness and profanity. As a drunken man is likely to be profane, the connection between profanity and drunkenness is obvious. And so, it may be added, is the connection between profanity and witchcraft. Witches, sorcerers and sorceresses, were supposed to be persons who denied God and made a league with the devil.

People who lived during the seventeenth century were apt to by very superstitious. Passengers of a vessel named the *Johanna* imagined that they had heard something which looked like a dog, cry like a child. The belief in witches and witchcraft was quite prevalent. Even so learned a man as Father Andrew White, head of the Jesuit mission in Maryland, expressed a belief in witches.[53]

By the provisions of an early Maryland statute sorcery was made a capital offence. Any one convicted of this, the act stated, might be put to death, or have his or her hand burnt. The offender could also be exiled or imprisoned for life. Commissioners of all the counties in Maryland were authorized to investigate all cases of " witchcrafts, enchantments, sorceries, and magic arts." In such cases, however, the county justices were not permitted " to take life or member," but must send the accused, together with the indictment, to a session of the provincial court " there to be tried." During the fall of 1665, a grand jury was instructed by the provincial court to investigate cases of witchcraft. Apparently the jurors only considered one case, that of Elizabeth Bennett. When the grand jury refused to indict her as a witch, she was later cleared by proclamation.[54]

Not all who were accused of witchcraft were as fortunate as Elizabeth Bennett. At a meeting of the provincial court on the 29th day of September, 1685, Rebecca Fowler was indicted by a grand jury—

For that she, the said Rebecca Fowler, the last day of August in the year of our Lord, 1685, and at divers other days and times, as well before and after, having not the fear of God before her eyes, but being led by the instigation of the devil certain evil and diabolical arts, called witchcrafts, enchantments, charms and sorceries, then wickedly, devilishly and feloniously, at Mount Calvert hundred and several other places in Calvert county, of her malice forethought feloniously did use, practice and exercise, in, upon, and against one Francis Sandsbury, late of Calvert county aforesaid, laborer, and several other persons of the said county, whereby the said Francis Sandsbury and several others, as aforesaid, the last day of August, in the year aforesaid and several other days and times as well before as after, at Mount Calvert hundred and several other places in the said county, in his and their bodies were very much the worse, consumed, pined and lamed against the peace, etc., and against the form of the statute in this case made and provided.

To this indictment Rebecca pleaded not guilty. She was tried before a jury who rendered the following verdict:

We find that Rebecca Fowler is guilty of the matters of fact charged in the indictment against her and if the court find the matters contained in the indictment make her guilty of witchcraft, charms and sorceries, etc., then they find her guilty. And if the court find those matters contained in the indictment do not make her guilty of witchcraft, charms, sorceries, etc., then they find her not guilty.

In view of this finding of the jury, judgment was " respited " until the court had time to further consider the case. After the court reconvened a few days later, Rebecca was again brought to the bar and the judges having " advised themselves of and upon the premises, it is considered by the court that the said Rebecca Fowler be hanged by the neck until she be dead, which was performed the ninth day of October aforesaid." [55]

A man by the name of John Cowman almost met the same

fate as Rebecca when he was convicted " for witchcraft, con-
juration, sorcery, or enchantment, used upon the body of Elizabeth Goodale." Cowman begged the delegates of the lower
house to ask the governor for a reprieve and stay of execution.
The assemblymen did as they were requested and beseeched
the governor " that the rigour and severity of the law to which
the condemned malefactor hath miserably exposed himself may
be remitted and relaxed." The governor, after considering this
petition, said that he was willing that Cowman be reprieved
and his execution stayed, provided the sheriff take him to the
gallows, and, " the rope being about his neck, it be there made
known to him how much he is beholding to the lower house of
assembly for mediating and interceding in his behalf." After
this was done the governor wanted Cowman to remain in St.
Mary's, where he was to be employed as he and the members of
his council thought fit.[56]

While Rebecca Fowler was the only person to be executed in
Maryland for witchcraft, there were other cases in which this
was an issue. A woman by the name of Joan Mitchell, the wife
of Thomas Mitchell, appears to have figured in several witchcraft cases. We find her husband complaining of " the abuseful
reproaches " offered to his wife by Mrs. Hatch, the wife of
John Hatch. From the testimony of witnesses, it appears that
when Joan Mitchell asked Mrs. Hatch how she fared, the
latter replied that she thought that Joan had bewitched her
face. Mrs. Mitchell at once asked Mrs. Hatch if she had really
meant what she had just said. The latter replied that as she had
lately suffered much from a soreness in her mouth she did
" verily believe she was bewitched." Mrs. Mitchell said that if
that was the case, then she would hold Mrs. Hatch responsible
for her slanderous remarks. To this threat Mrs. Hatch replied
that if the suit was brought, she would endeavour to prove what
she had said was true, and if she could not, then she would be
willing in open court to acknowledge that she had wronged
Joan. As Thomas Mitchell died not long after this, no decision
was reached in the case.[57]

It was Mrs. Mitchell's turn to institute a suit of her own. In
an action of defamation against Francis Doughty, a minister,

12

Joan Mitchell claimed that she had been abused and her good name taken from her. Joan asked that an investigation should be made of her character, and, if she was proved innocent, then she might have " the law " against Doughty who had slandered her. The minister, it appears, had come to the colony from Virginia with his son Enoch. While in Virginia Francis Doughty had persecuted another woman on the grounds of " witchery."

In order to prove that the Reverend Doughty was responsible for the injury to her reputation, Mrs. Mitchell referred to some scandalous reports which he had circulated about her. According to Joan, his reverence had said that when she had spoken to a woman in church, the latter at once " fell a aching as if she had been mad." As a result of this and other stories which the minister had spread abroad about her, Joan maintained that her reputation had suffered in the neighbourhood. Indeed, on one occasion, Mrs. Mitchell said that while she was on her way to church on Sunday a man had thrown stones at her. Joan was certain, however, that as she trusted in God, he would plead her cause and would not suffer " the poor and innocent to perish by the hands of their enemies."

When this case first came up for trial, Francis Doughty, because of his sickness, did not appear in court to defend the suit. When, however, at a latter session, neither Mrs. Mitchell nor the minister appeared in court either in person or by their attorneys, the justices decided that " the action should die." For what reason Mrs. Mitchell failed to prosecute this case is not clear. Doughty, who disgraced the cloth he wore, should have been held responsible for his remarks.[58]

But it was not against minister Doughty alone that Joan Mitchell brought actions of defamation. She also instituted a suit against his son, Enoch. In order to prove her case against this twenty-two-year-old youth, Joan introduced several witnesses. Two of them, both women, testified how they had heard Enoch Doughty ask Mrs. Mitchell if she wasn't the woman who had taken a swim in a river. The third witness, a man, said that he knew nothing of " the above discourse." With this evidence before them, the county commissioners decided there was " no cause of action," and dismissed Mrs. Mitchell's suit. In the seventeenth century few, if any, women were able to swim and to accuse a woman or girl of being able to do this was equivalent to saying that she had the characteristics of a witch.[59]

A Mrs. Long was the next person to be sued by Mrs. Mitchell for slandering her and injuring her reputation and " good name." To support this charge, Mrs. Mitchell had a young man testify who said that he had heard Mrs. Long say that some chickens which she had received from " Goodie " Mitchell died in such a strange way that she thought " some old witch or other had bewitched them." The young man's wife gave testimony which confirmed that of her husbands. Mrs. Long, however, denied ever having spoken " any ill " of Mrs. Mitchell at any time, nor did she want to accuse her of anything now. It was probably because of this statement of Mrs. Long that the commissioners of Charles county again decided that Mrs. Mitchell had not shown a cause of action.[60]

Against James Walker, one of the Charles county justices, Joan Mitchell brought another and her last suit of defamation, alleging that Walker " hath spoken words tending to the taking away of her good name." Joan asked the Charles county court to question Walker and that if he had nothing " to tax her with," then she wanted to be vindicated. When Walker denied having spoken any ill of Joan and she, on her part, was unable " to prove anything against him," the county commissioners dismissed Mrs. Mitchell's case.[61]

We have seen how Thomas Mitchell, the husband of Joan Mitchell, resented his wife being called a witch. A similar suit was brought by Richard Manship against Peter Godson, a chirurgeon, for saying that he would prove that Manship's wife was a witch. When the case came up for trial, Godson said that he had grounds for making such a remark. When, on one occasion, he went to see Mrs. Manship she laid some straw on the floor and said in a jesting way: " They say I am a witch. If I am a witch, they say I have not power to skip over these two straws." Then, according to the chirurgeon, Mrs. Manship bade him " in the name of Jesus " to skip over the straws.

About a day after this conversation, Godson said that he became lame. The final result of this case was that Godson had to express his sorrow for having slandered Mrs. Manship by calling her a witch. Godson also had to pay all court costs.[62]

Although, as we have seen, there was only one case in which a woman was hanged for witchcraft in early Maryland, there were two other hangings, both women, on boats bound for Maryland from England. One of these occurred on the ship *Charity* of London. When the stormy seas which this vessel encountered did not abate, the passengers on the *Charity* decided that they were caused by " the malevolence of witches." Seizing an old woman, known as Mary Lee, they bound her and accused her of sorcery. She was hanged the following day and her corpse was thrown into the sea. When the ship reached Maryland, John Bosworth, her master, was summoned before the provincial authorities. Bosworth escaped being held responsible for the death of Mary Lee by being able to show not only that he was not present when the old woman was hanged, but also that he had tried to prevent the execution.

Among the reasons that led the seamen on the *Charity* to believe that Mary Lee was a witch was her " deportment and discourse," and also because they had found " some signal or mark of a witch " on her, which, on the following day, " was shrunk into her body for the most part." It was said that the old woman confessed to being a witch, but if this was true the confession was probably extorted under fear of bodily injury.[63]

The other case of hanging a woman as a witch occurred on a vessel also bound for Maryland of which John Greene was the master, and Edward Prescott, a merchant, the owner. It was Colonel John Washington, of Westmoreland county in Virginia, the great grandfather of the famous George, who entered a complaint against Prescott for hanging Elizabeth Richardson as a witch as he was " outward bound from England hither." Josias Fendall, who was governor of Maryland at this time, at once ordered Prescott arrested. At the same time Fendall wrote Colonel Washington to inform him of Prescott's arrest and of his coming trial before the provincial court on October 4th, at which time he expected Washington to be present to make good his charge. The colonel replied that it would be impossible for him to be present on the day set, " because then, God willing, I intend to get my young son baptized; all the company and gossips being already invited, besides in this short time witnesses cannot be got to come over." If, however, Prescott's trial could be postponed to the next meeting of the provincial court, Washington promised that he would make every effort to be present.

Prescott was brought to trial on the day originally set without Washington being present. The merchant did not deny that there was a woman by the name of Elizabeth Richardson hanged on his ship. His defence was that although he was the owner of the ship, John Greene, the master, was then in command. When he had protested against hanging the woman, Prescott said that Greene and his crew "were ready to mutiny." For this reason it was unfair, he maintained, to hold him responsible for the woman's death. As no one brought forward any evidence to refute these statements, Prescott was acquitted.[64]

CHAPTER VIII

ADULTERY, FORNICATION AND BASTARDY

One proprietary governor of Maryland, in an address to the members of the assembly, urged them to suppress "that most horrid and damnable sin of adultery, which, in these days, is grown to that height that with the prophet we may justly say the land is full of adulterers." Continuing, the executive said:

> This sin (adultery) in the old law was, as I wish it were by law now, made punishable by death for it was decreed that he that committeth adultery with his neighbour's wife, the adulterer and the adulteress shall surely be put to death, and with reason, for it not only brings our estates to ruin, as it is said that by means of a whorish women a man is brought to a piece of bread, but it also brings both body and soul to eternal death. For, it is said, that adulterers shall not inherit the Kingdom of God, so abominable and shameful it is for men, especially married men, to keep whores, as I hear some do not only abroad but even at home under their wives' noses, where strumpets rule and the wives obey to the scandal of all honest and good men.
> Wherefore I pray it may be taken into such consideration as for the future such villains may be excluded from all human, or at least Christian societies.

The members of the lower house were of the opinion that the prevalence of adultery and other crimes in Maryland might be due to the fact that the laws already in existence were not properly enforced. Although both delegates and members of the council expressed a willingness to enact other laws for the suppression of this and other offenses, no such laws were passed. By the provisions of the law already in effect, any person who had committed adultery or fornication was to be censured or punished, as any of the provincial justices before whom such cases came, thought fit, provided the punishment did not involve the taking of "life or member." [1]

There were instances of adultery and of fornication among those who held office in proprietary Maryland. Captain William Mitchell, a member of the governor's council, was accused of adultery. It was because Cecil Calvert, proprietor of Maryland, had confidence in Mitchell's "honour, worth, and good abilities" that he had appointed him a member of the council. Yet within a year after he had come to Maryland the captain was in disgrace.[2]

Sometime before he arrived in the colony Mitchell stayed in London on the Strand, "near the Savoy." While there, Mitchell, who was a married man, visited William Smith and succeeded in persuading him to allow his daughter, Susan, to go with him to Maryland. This girl, it appears, was the wife of Humphrey Warren. The status in which Susan accompanied Mitchell to the New World is not clear and was the subject of dispute. Mitchell claimed that in consideration of the one hundred pounds sterling which he had loaned Mrs. Warren, she had promised to act as his servant until this loan was repaid. Mrs. Warren, on the other hand, contended that she not only had paid the cost of her own passage, but that Mitchell had by "harsh and cruel usage" compelled her to give him a bond for one hundred pounds sterling, although she was not indebted to him. Susan told how Mitchell had said that unless she signed the bond he would make her fetch water for him and even clean "his foot boy's shoes." Apparently Captain Mitchell had even boasted to others about the deception which he had worked on Mrs. Warren.

After Captain Mitchell and Mrs. Warren arrived in Maryland, they lived together at a place called "The White House." Martha Webb, a servant in this house, who was later called upon to testify, spoke of finding Mrs. Warren and Captain Mitchell in bed together. Not long after this Susan was found to be with child. At first she denied having had intercourse with Mitchell and said that if she was pregnant "it was inspired by the Holy Ghost, and not by man." Later, however, Mrs. Warren admitted that Captain Mitchell was the father of her child.

The captain, having decided that something must be done to get rid of the baby, had a concoction made consisting of a very strong physic mixed with eggs which he forced Susan to drink. Soon after this was administered, Mrs. Warren became very ill. She was compelled to spend much of her time upon a closet stool "purging very strongly." The physic, which must have contained some kind of poison, caused the woman " to break forth into boils and blains, her whole body being scurfy, and the hair of her head almost fallen off." The poison in the physic killed the child within her: shortly afterwards it was still born.

When all of these facts, gathered from the testimony of witnesses, were considered by a grand jury Captain Mitchell was indicted for having committed adultery with Susan and also with having " murtherously endeavoured to destroy or murder the child by him begotten in the womb of the said Sarah Warren." The captain did not ask for a jury trial, but said that he wanted the provincial court to pass judgment without further delay. The court did as requested. Their judgment was that " for his several offences of adultery, fornication and murtherous intention," Mitchell should be fined five thousand pounds of tobacco. He also had to give a bond for his future good behaviour.

Mrs. Warren did not fare as well as her seducer. While admitting her " offence," she, too, waived her right to a trial by jury and asked that the judges of the court would be " favourable unto her." The provincial court, unmoved by this plea, decreed that as a punishment for her " lewd course of life," she should be " forthwith whipped with thirty-nine lashes upon her bare back." Mrs. Warren received this punishment " with some mitigation," upon the intercession of members of the council and others to the governor in her behalf. Possibly she was not given the full number of stripes. Mrs. Warren had some consolation for her lashing when the provincial court judges informed her that she would not have to act as a servant for Captain Mitchell any longer. To their honours it seemed that her relationship with the captain had been that of a companion and bedfellow rather than that of a servant.[3]

When Cecil Calvert received an account of the way Captain Mitchell had acted, he regretted having placed any confidence in a man who lived such " a scandalous life " and had also professed himself of " no religion." From the records it appears that Mitchell, while living with Mrs. Warren, had tried to persuade her that there was no God and that, on another occasion, when talking with some men, he told them that he was astonished that the world had been deluded by " a man and a pigeon," by which he referred to Jesus Christ and the Holy Ghost.

Lord Baltimore compelled Captain Mitchell to resign as a member of the governor's council in Maryland and forbade him thereafter to hold any public office in Maryland. Determined to prevent, if possible, the recurrence of " such high offences to Almighty God and such dishonour to us and our government," Cecil Calvert informed William Stone, then governor of Maryland, that if in the future any proprietary official should be convicted of living " scandalously and viciously with any lewd women," or of making atheistical remarks, then, in all such cases, the official should be suspended from office and the record of the trial sent to him. Calvert also asked Governor Stone to be careful to investigate all cases of " scandalous and evil comportments and misdemeanors " on the part of any provincial official.[4]

During the remainder of the proprietary period, there were no other cases of where a man holding the high office of councillor was accused of immorality. Thomas Hynson, Jr., a commissioner of the Talbot county court, was faced with a similar charge, however, when he was presented to the court, of which he was a member, for committing fornication with Anne Gaine. Young Hynson was a son of Thomas Hynson, Sr. who was a justice of the peace in Kent county.

When brought before the Talbot county court, young Hynson expressed his sorrow for what he had done, but added that he had lately made Anne " his lawful wife." In order, however, to express their disapproval of what Hynson had done, the Talbot justices suspended their fellow commissioner from the bench " for one year and a day." As for Anne, the justices directed Hynson to bring her before them at the next session. When Anne made her appearance, she acknowledged " her fault with extreme sorrow of the same," and submitted herself to the judgment of the court. In view of her " submission," the Talbot court declared that she would not be punished.

But the Hynson case did not end with this decision. It seems that thereafter, on several occasions, James Ringgold, also a commissioner of Talbot county, referred to Hynson's offense.

Unable to stand this taunting any longer, Hynson appealed to his fellow justices of the Talbot court. In his petition he said that he had already been punished for having had relations with Anne, an offense due to his " frailty and weakness." Now, however, continued Hynson, to make matters worse, Justice Ringgold added to his punishment by constantly referring to his offense, using " opprobrious and scandalous speeches." Hynson asked the Talbot commissioners that they should either force him to resign as a justice, or grant him some satisfaction against Ringgold. When questioned about the complaint which Hynson had lodged against him, Ringgold admitted having made some slurring remarks, for which, he said, he was " heartily sorrow." The Talbot commissioners finally succeeded in persuading Hynson to accept this apology.[5]

While there are no other cases in which a justice of the peace was accused of adultery or of fornication, the wife of Thomas Bradnox, a commissioner of Kent county, was involved in a rather unsavoury affair with another man. It seems that to Bradnox's house came one evening John Salter and John Gibson. Liquor was produced and soon the justice and his two guests were making merry. While the drinking was in progress, Salter got up and went to look for Bradnox's wife, Mary. When he found her, Salter told Mrs. Bradnox that he would make her go to bed with him. Mary replied that Salter need not use force since " by fair means " he could " do much." It did not take Salter long to act on this suggestion. His breeches were soon " down in his hand," while Mrs. Bradnox lay on the bed " with her coats up as high as her breast." After having intercourse with Mary, Salter left the room.

When, soon after this, Justice Bradnox found out what had occurred, he was naturally very indignant with his wife. Grabbing her by " the birth," he dragged her about " in an inhumane manner," finally pulling her from the bed to the floor. When Mrs. Bradnox cried for help, Salter ran back into the room, seized Mr. Bradnox by the throat, almost strangling him. While the scuffle was going on, the justice accused Salter of having had relations with his wife. Salter at once retorted that it was no more than what Bradnox had done with his wife and that he could prove she was the justice's " whore." John Gibson and others in the house finally succeeded in parting Bradnox and Salter and in persuading the latter to go home. As he was leaving, the justice followed Salter, threatening to shoot him, knock him on the head and break his legs. Nothing daunted, Salter exclaimed: " Old Tom, do thy worst, I fear thee not," and added that he would accuse Bradnox of rape.

Things had just about quieted down in the Bradnox household when, at " about two hours before sunrising," two prominent Kent Islanders arrived at the justice's home. They were Captains Joseph Wickes and Robert Vaughan. These two men had just crossed the bay bringing to Captain Bradnox " a rundlet of drams." After they had all had a few drinks, Justice Bradnox went to his wife's bedside and asked her " to drink with him, or to pledge him a dram, and to forget and pass by all malice or cause of discord that was betwixt them." Mrs. Bradnox was not so anxious to do this, just having been roughly handled by her husband. Finally, however, she put the bottle to her mouth, but whether she drank or not, Captain Wickes said that he was not certain.

Bradnox now insisted that every one in his house have a drink with him. It was not long before the justice was " so much disquised with drink that he was not able to go nor stand." Attempting to rise from his chair, he fell to the floor " by which fall his nose did bleed much." Mrs. Bradnox, with the aid of Captain Wickes, picked the drunken justice up, laid him on a bed and held him down while they dried the blood " until it left bleeding."

Although the testimony of many witnesses was taken in this case, no decision was ever reached by the Kent county court. Possibly the commissioners were of the opinion that it was a case of the pot calling the kettle black.[6]

Other colonists besides those holding office were involved in cases of immorality. When Walter Pake found Paul Simpson, a mariner, having intercourse with his wife, Pake drawing his sword, stabbed Simpson in the arm and side. The affair of honour did not end with this. Pake made so many threats as to what he intended to do to Simpson that the latter appealed to the provincial court to restrain Pake from making another attack on him. The court compelled Pake to take out a bond by the terms of which he agreed to keep the peace. The affair must have been finally settled out of court upon some basis agreeable to both parties, as we find Simpson later agreeing to release Pake from his bond of " peace and good behaviour." [7]

In most cases of immorality severe punishment was meted out, especially to the woman. The following case, however, was an exception. How Robert Harwood succeeded in having intercourse with Elizabeth Gary is described by the young woman:

> About three years since, Robert Harwood began his first pretended love towards me, and ever since, through his suggestions and delusions hath he followed me withall, till at a certain time about a year since he followed me to the garden, where my mother sent me to gather a sallet (salad). He never left his attempt till he forced me to yield to lie with him, and after he had obtained his filthy desire and lust upon me, said: 'Now, I should nor could have any other man but him.' Several times I told him that I would not have him, were it not for that filthy act he committed with me. And Harwood made me answer again that he did not know what he did, for he had no other way to keep me but by that in lying with me.

Although Elizabeth regretted that she had had intercourse with her suitor, she did not seem anxious to marry him. However, her stepfather, Peter Sharpe, decided that something must be done. Accordingly, a most unusual agreement was drawn up between Sharpe and Harwood.

By the terms of this agreement, Elizabeth was to go and live with a neighbour for a period of six weeks. While there her expenses were to be paid by young Harwood. During the six weeks, Harwood was to be free to visit Elizabeth at any time " and to use all fair and lawful endeavours " in an attempt to persuade her to marry him. At these meetings one or more neighbours must always be present. If during the six weeks Robert succeeded in making Elizabeth agree to marry him, then the marriage was to take place at once. Harwood must promise, however, that after they were married he would never " upbraid, or deride, or use any bitterness with Elizabeth in relation to any former passages between them." If he broke this promise, then, by the terms of the agreement, he was no longer to have any control over Elizabeth's possessions or property. But suppose Robert was not successful in persuading Elizabeth to marry him? In that case, it was understood and agreed that Robert would never see Elizabeth again, or try to induce her to marry him. Unfortunately, there is no record of whether young Harwood was successful or failed in his courtship. It seems likely, however, that he married Elizabeth. The " filthy act " Robert had committed with her would certainly tend to make her less desirable as a wife in the eyes of other men.[8]

As has already been stated, most women found guilty of illicit intercourse with a man were severely punished. Accordingly, when the commissioners of Charles county found Agnes Taylor guilty of " having played the whore," they ordered her tied to the whipping post and, " in the public view of the people," given twenty lashes on her bare back. It seems that Agnes had previously been accused " of the like crime." [9]

In the same county, Mary Hews and Katherine Budd were accused of being " loose livers." Curiously enough, a young man by the name of Thomas Shelton appears to have seduced both women. Witnesses were called to testify. One spoke of seeing Mary Hews in bed with Thomas and that both were naked, another testified that Shelton was " in his drawers," while still another man said that he was unable to see who was the woman in bed with Shelton as her face was " ktvered." In view of this evidence, the county justices ordered that Mary Hews publicly ask forgiveness for the scandal she had caused and that she should not see young Shelton again.[10]

Katherine Budd, the other " loose liver," did not fare as well as Mary Hews. As has been already stated, Katherine had relations with Shelton who had been responsible for Mary's undoing. Witnesses testified as to having seen Katherine and Shelton in bed together " in one another's arms." One witness spoke of seeing Shelton coming out of Katherine's room clad only in his shirt. Katherine, it seems, had come to Maryland from Accomac, Virginia, where it was said she had spent what her husband had earned " in evil company." As punishment for Katherine, the Charles county commissioners ordered that she have given her " forthwith " twenty lashes on her bare back.

As the evidence produced against Mary Hews and Katherine Budd was substantially the same, it is hard to justify the difference in punishment. It is interesting to note that twenty-eight-year-old Shelton, who had seduced both women, received no punishment for his " loose living." [11]

In some cases of adultery, or of fornication, the man was only fined, while the woman's back was lashed with a whip. In one instance, for example, Edward Coppage was brought before the Kent county court charged with having adulterous relations with Elizabeth Risby, a married woman. When convicted, Coppage was sentenced to pay a fine of five hundred pounds of tobacco. He was also warned that in the future he must not see Elizabeth again. As for Elizabeth, she was given fifteen lashes for her indiscretion and at the same time required to give a bond for her future good behaviour.[12]

In the following case the difference in the punishment received by the man and woman seems very unfair. This was when John Nevill, a seaman, and Susan Atcheson were brought before the provincial court " for suspicion of adultery and fornication." Many witnesses were called. Some spoke of having seen John and Susan together in a thicket. One woman was so much astonished at what she saw that she went to get some of her neighbours to come and watch the proceedings. Another woman said that Nevill had begged her not to tell Susan's husband what she had seen. When reprimanded for her actions, Susan told one neighbour that she went with Nevill because her own husband abused her so much that she could not love him.

A man named Thomas Plott gave the most incriminating account. Plott told the members of the provincial court how:

I did see Nevill and Susan upon a bed together, and Susan called out and desired Nevill to be quiet, but Nevill would not be quiet, but pulled up Susan's clothes so that I did see her nakedness . . .

Nevill, who was naturally somewhat embarrassed by Plott's presence, told him to get out as he wanted " to swive Susan." Plott did as he was requested, but he had not gone very far from the house when he heard Susan cry out: " For God's sake, help." Plott said that he then returned to the house, went into the room where Nevill and Susan were and told them " to come off the bed for shame." Although the couple did as they were told, Nevill soon afterwards threw Susan on the bed again.

While on the witness stand, Plott recounted how, at another time, an Indian came to tell him that upon hearing " a bustling in the loft," he went up a ladder and found Susan and Nevill " sack a sack." Plott said that he then followed the savage up the ladder and he, too, found Nevill " upon Susan . . . with his clothes all off, but his shirt." On still another occasion, Plott related how one night, during the past winter, being asleep on a bed in the chimney corner, he awoke to find Nevill " atop of " Susan on the floor in front of the fireplace. When he saw that he had been discovered, Nevill threatened to whip Plott " till the blood should follow," should he dare disclose what he had seen.

Another witness by the name of Mary Gillford testified that when she saw Nevill and Susan together she had her hand " in his breeches " and that Nevill's hand was " in Susan's placket."

After hearing all of this testimony, the judges of the provincial court were of the opinion that as Nevill and Susan had lived " in a notorious and scandalous course of life tending to adultery and fornication," both of them should receive twenty stripes with a whip upon their bare backs. Soon after this judgment was rendered, some of John Nevill's friends petitioned the court that instead of whipping Nevill a fine should be imposed upon him " in hopes of his future carriage and comportment." The judges finally decided that if Nevill's friends would agree to pay a fine of five hundred pounds of tobacco for his lascivious conduct, then, in that case, the court would remit the punishment of whipping. As for poor Susan, since she appears to have had no friends of means, her back was bared for the application of the lash.[13]

In the following case, however, both offenders received the same punishment. William Mullins, it appears, admitted to the justices of the Talbot county court that he had " lain with Sarah Spurdance." Their honours at once ordered that William should have twenty lashes on his bare back. As for Sarah, the sheriff was instructed to bring her before the next meeting of the court for trial. In due time the girl was presented for having committed fornication. Witnesses were summoned by the prosecution. They testified how Mullins and Sarah " did lie naked together in bed divers nights," and how Sarah had told them that she was with child by Mullins. As punishment for her conduct, the Talbot commissioners decreed that as soon as Sarah had delivered her infant she should receive twenty lashes on her back.[14]

It is hard to understand the decision of the court in the following case when they are discussing the punishment to be imposed on a man and a woman. It appears that Edward Hudson and Dorothy Holt, a married woman, were charged " with divers lewd, incontinent, and scandalous actions and practices." When the case came up for trial, several witnesses testified as to having found Hudson and Dorothy in bed together. Apparently, Dorothy did not entertain a very high opinion of her husband, Robert Holt. On one occasion, she was heard to say that she hoped " he might rot limb from limb." At times Dorothy had even threatened to kill her husband.

After hearing all the evidence, the court found Hudson and Dorothy guilty of adultery. Hudson was sentenced to receive thirty lashes, while Dorothy was to be given fifty stripes. In order to prevent Edward from having further intercourse with Dorothy, he was told to leave St. Mary's. If he returned without the permission of the governor, he might receive another whipping. As for Dorothy, she was told that she must live at least five miles away from her husband's house, and if, in the future, she soiled his reputation by loose living she would be severely whipped. After having delivered itself of this gory ultimatum, the court suddenly reversed itself. This must have been a relief to Hudson and Dorothy. At any rate, the judges now informed them if they made a humble submission, acknowledged their offense, asked for a pardon and promised " amendment," then the governor might be willing to remit the punishment which the court had ordered. As for Robert Holt and his erring wife, Dorothy, they were told that there was nothing to prevent them living together as man and wife should they come to some agreement.[15]

During the summer of 1657, it was discovered that a man named Alexander King and a woman by the name of Mary Butler had " lived and bedded together as man and wife " for about three months. John Butler, Mary's husband lived in York, Virginia, whom Mary had left " without his leave or knowledge." Both King and Mary Butler were arrested, but King " broke prison " and made good his escape. Mary was not so lucky. The provincial court decided that twenty lashes should be given Mary on her bare back, " ten immediately at the court door and ten at the river side of the Potomac." After these whippings had been administered, the sheriff was directed to take Mary across the Potomac to the Virginia side of the river where she was to be delivered into the hands of the provincial authorities there.[16]

It is interesting to compare the decision in the case just discussed with one in Virginia where a Maryland woman had relations with a Virginian. In the Virginia case, it appears that a married man named Carline, who was an inhabitant of that colony, had persuaded a maidservant, who lived on Kent Island, to run away with him to Virginia. Upon their arrival there, the couple stayed with Captain Fleet as his guests. Just before bedtime, Carline addressing the girl, said: " Honey, thou art a cold. Wilt thou go to bed? " To this the young woman replied: " No sweetheart." But Carline being a passionate youth took the girl in his arms and threw her on a bed. During the night, Fleet heard the young couple " make a noise." Fleet, indignant, called on some of his servants to turn the illicit love makers out of doors.

After this, Carline and the young woman were taken before a court held at Rappahannock, Virginia. For her one night of love, the girl servant received thirty lashes with a whip. As for young Carline, he was fined " for keeping the servant away so long and for disowning his own wife." He was also banished from Virginia.[17]

One case came before the Maryland courts, where a woman, named Mary Taylor, a native of the colony, was suspected of having a child by George Catchmey, a Virginian. The illicit intercourse appears to have occurred while Mrs. Taylor, the wife of Robert Taylor, was visiting in Virginia. When Taylor heard of the birth of the infant, he was said to have remarked that he would " turn his wife and the bastard out of doors." On her knees Mary begged forgiveness, but Taylor, adamant,

13

only exclaimed: " O! thou wicked, base woman, how can I forgive thee, I cannot forgive thee."

Mary Taylor appears to have resented the insinuations made by some of her neighbours. When a Mrs. Johnson told her that she knew that George Catchmey was the father of her new-born baby, Mary replied that the Virginian was no more the father of her infant than he was of any child of hers. Resenting this remark, Mrs. Johnson called Mrs. Taylor " an impudent . . . brazen faced whore." Whereupon Mary Taylor retorted: " It hath pleased God to make us both alike." This was too much for Mrs. Johnson who struck Mary Taylor. When the fight was over, Mrs. Johnson, as a parting shot, said that any rate she knew Mrs. Taylor's " night's work " with Catchmey was going to cost her " a whipped back."

George Catchmey was finally persuaded to come to the Taylor home. Inside locked doors he was accused of being the father of Mary's baby. At first he denied this, but when Robert Taylor took down his gun threatening to shoot him, Catchmey changed his mind on the question of parentage. In the conversation which ensued, Catchmey asked Taylor how many pounds of tobacco he would have to offer him in order to make him forget the matter and accept the child as his own.

After hearing all the testimony in this case, and there was much of it, the provincial court came to the conclusion that as the offense with which Mary Taylor was charged occurred in Virginia, the Maryland court had no " cognizance " of the matter. The judges dismissed the case against Mary, at the same time informing her that she was now at liberty to prosecute her accusers, " if she see cause." [18]

There were a number of instances where men had intercourse with women servants. John Bowles appears to have been quite promiscuous in his relations with servant girls. It was reported that he " lay " with two maidservants in the employment of Thomas Hynson, Jr. and that he also had relations with a woman servant of another man. When the charge was proved in the latter case, the girl was given twenty lashes on her bare back. Bowles escaped with the payment of a fine of five hundred pounds of tobacco.[19]

William Robisson, in a petition to the Charles county court, alleged that William Wennam had dishonored his house " by committing fornication " with Anne Mardin, his maidservant. At the trial of this case, Robisson called witnesses in an attempt to prove his contention. The first witness, a youth by the name of Richard Smith, said that Wennam had confessed to him that he had " lain with Anne Mardin once and that he knew not what to do to procure a pair of shoes and stockings to be married in." Smith added that Wennam had also said he was afraid that Francis Fitzherbert, the minister, would excommunicate him and that none of his friends " would abide him." Joan Nevill, the next witness, testified how Wennam had told her that if Robisson forced him to marry Anne Mardin he would, on each day of their married life, bind her to a tree and give her a whipping.

After Joan and several other witnesses had finished testifying, Wennam maintained that from the evidence produced there was no proof of " a carnal copulation." No witness, Wennam added, had seen him and Anne " rem in rem," that is, in the act of copulating. The county commissioners agreeing with Wennam, dismissed Robisson's case against him.[20]

In the cases which have been just discussed about maidservants, no mention was made of there being a child born as the result of the illicit intercourse. There were, however, a number of instances where a bastard was born. During the early years of the colony, indeed until 1658, there was no law in effect which sought either to prevent the practice of having intercourse with women servants, or which made any provision for the care of the bastard child. Each master or mistress of a maidservant must seek to recover what he or she could from the man who had seduced the girl. Thus we find Robert Taylor, owner of a woman servant, bringing action against John Hambleton claiming that he was the father of his servant's child which she was soon to deliver. Taylor claimed that Hambleton had been caught acting " suspiciously and uncivilly with her and that he obscures himself from the sheriff." In a petition to the provincial court, Taylor asked that an attachment might issue against such property of Hambleton's, as shall be found in his possession, until he put up security to answer the suit

" and make good your petitioner's sufferings." The court so ordered.[21]

In still another case, instituted before 1658, Roger Grosse sued Roger Scott " for getting his maidservant, Frances Smith, with child." Grosse said that Frances had sworn that Scott was the father of her child. In view of these allegations, the provincial court ordered Scott committed to prison until he gave security for the maintenance of the child. The judges also decreed that Scott must make good " the damages " which Grosse had sustained by his " getting his maid with child." [22]

In 1658 the proprietary officials began to be alarmed by the number of unmarried woman servants who had " been gotten with child," during the time of their servitude, " to the great dishonour of God, and the apparent damage to the masters, or owners, of such servants." The colonial officials decided to take steps to protect the masters, or owners, of such servants against the loss of their services occasioned by child bearing and nursing and raising the infant. Accordingly, the assembly passed a law which provided that if the servant woman, who was the mother of a bastard, could not show who was the father of her infant, then she must make good to her master, or mistress, by additional servitude, or " other ways," as the court should determine, the time lost in serving her owner by having the baby.

If a maidservant was able to show who was the father of her child, then the father, if a servant, must make good half " the damages," or the time lost in service by the girl servant bearing a child. If the man who had seduced the servant was a freeman, then he must make good the whole damage caused the owner of the servant by serving himself as a servant of the same master, or the freeman could be made to recompense the owner of the maidservant in other ways, as the court before whom the case was brought thought fit.

The maidservant was also allowed to produce evidence to show that the man with whom she had intercourse had promised to marry her. If this was done, to the satisfaction of the court, then the man, if unmarried and a freeman, must marry the girl, or " recompense her abuse," as the court should decide, " the quality and condition of the persons considered." [23]

After the passage of the Act of 1658, many cases of bastardy came before the provincial courts. First, as to these cases in which the maidservant could or would not give any proof of the man who had seduced her. Penelope Hall, a servant who lived on Snow Hill Manor, was indicted in 1661 for having a bastard, or as the indictment stated:

That the said Penelope having had carnal knowledge with some person or other of evil behaviour was sometime in December last brought to bed of a bastard child to the evil example of others and the form of the law, or statute, in that case provided and against the peace of the Lord Proprietor, his rule and government.

When Penelope, at her trial before the provincial court, admitted her guilt, the judges decided that she must serve her master additional time.[24]

Some years after the decision in this case, a similar case came before the Talbot county court commissioners. This was when Sissly Rogers was ordered to serve her master, Trustrum Thomas, an additional year for having had a bastard child.[25]

In the two cases just discussed, the decision of the court in requiring the maidservant to serve additional time for the time lost by childbearing was in accordance with the provisions of the Act of 1658. As the servant did not name the man by whom she had been seduced, she alone had to bear all the punishment. But what of the cases where the servant girl named her seducer? In order to prove this charge, if denied by the man, it was necessary to produce the testimony of witnesses, or the statement of the woman while in " the pains and throes of travail, etc." First, to take the case of where the seducer was like the girl a member of the servant class.[26]

Frances Shambrooke was presented by a constable of Kent county " for having a bastard contrary to the law of God and man." Although the court decided that she should have twenty lashes given her, at the request of her master and some women friends, the corporal punishment was remitted and Frances was only required to pay the costs of the suit. During the time she was " in travail," Frances said that James Phillips was the father of her child, which was affirmed by the testimony of two women witnesses appearing before the Kent county court commissioners. Phillips was an indentured servant. The court ordered that he should give sufficient security to pay half of the cost of the child's maintenance. The two men who employed Phillips were required " to enter bond for him that the said child be not chargeable to the county." [27]

Frances Smith, a servant of Abraham Bishop, when examined by the Talbot county commissioners, declared that Thomas Mawman, another servant, was the father of her bastard child. Mawman did not deny this charge. Therefore, in accordance with provisions of the Act of 1658, the court justices ordered that Frances must pay one half of the damages caused her master by having the child and Mawman the other half. Mawman also undertook to look after the child, who was a girl, until she was twenty-one. When Mawman told the county commissioners that he could not be put to " extraordinary charge " in raising the child, the justices said that as long as the girl was " carefully looked after," he could make whatever provision for her that he wished. Mawman finally succeeded in persuading a man by the name of Edward Fuller to take the custody of his illigitimate offspring. The Talbot commissioners, approving of this plan, ordered Fuller to provide for Mawman's child until she was twenty-one. The girl, no doubt, paid for her keep by becoming indentured to Fuller.[28]

In another case in Talbot county, a jury of inquest presented George Thirle and Mary Barnett for having a bastard child. For this offence, Mary and George were each to be given twenty lashes on their bare backs. After they had received this punishment, the boy and the girl made a proposition to the county commissioners. They said that if their master, Colonel Henry Coursey, would give them permission to marry, they would " make satisfaction " by way of additional servitude to their master for every child they might have while in his service in order to make up for the loss of time, trouble, and expense incurred in raising children born during the period of their servitude. While there is no record of how the Talbot justices acted on this suggestion, it is hoped that they approved of it.[29]

So much for the cases in which the begetter of a bastard child was a servant. Next, to consider those cases where the seducer was a freeman, and although single did not promise to marry the girl. In Somerset county, John Cooper had a bastard child by Susanna Brayfeelds. For the expense to which he had been put for the care of the child, Randall Revell, the master of Susanna, sued Cooper for one thousand pounds of tobacco. Apparently, Cooper was unable to pay this amount as soon afterwards he agreed to serve Revell for a period of two years. This service, it was understood, was to be in lieu of the service which Revell could have required of Susanna. Cooper also agreed to be responsible for any expense thereafter incurred in raising the bastard child he had by Susanna. It was also agreed that in consideration of Cooper's serving Revell, Susanna was to become the servant of Cooper for the unexpired term of service still due Revell.

Susanna appears to have been rather promiscuous in her sex relations. Soon after this, while Cooper's servant, she became pregnant again but by another man, whose name was John Griffith. It was Cooper's turn to complain against Griffith for having " got " his servant with a bastard child. The Somerset commissioners ordered Griffith to pay Cooper thirteen hundred pounds of tobacco for the loss of time, etc., caused Cooper by his servant having the child. As for Susanna she was required to pay five hundred pounds of tobacco as a penalty for her offense. If she failed to do so, she was to be publicly whipped. Later, it appears, that Cooper married Susanna. The facts in this case leave one rather dizzy.[30]

In Talbot county, there were several instances of where a girl servant had a bastard child by a freeman. In these cases, as in the one just discussed, there is no evidence that the freeman promised to marry the girl. A grand jury in Talbot county presented Andrew Twotley and Elizabeth Wharton for having a bastard infant. For this offence Elizabeth received thirty lashes on her bare back. Twotley was compelled to pay her master five hundred pounds of tobacco for the loss of Elizabeth's services while she was " having a bastard child." [31]

At another session of the Talbot county court, Jane Harrison, a maidservant, was presented for having a bastard. When Jane declared under oath in open court that one Roger Somers was the father of her illegitimate offspring, the county justices required Somers to give bond with sufficient security for the maintenance of the child. Apparently no witnesses were called to confirm Jane's statement on the question of parentage. In a similar case where a maidservant of William Gary accused Richard Chapling of having begotten her bastard child, the Talbot commissioners ordered Chapling to give bond that he would " keep the county from the charge of a child." Possibly neither Somers or Chapling denied the accusations made by the two women.[32]

At a meeting of the Kent county court, John Ereckson complained that his maidservant, Mary Stedhed, was " with child " by Matthew Read, one of the county commissioners. Mary alleged that the child was begotten on Candlemas Day during the night. The sheriff of the county was instructed to summon a jury of women " to search the said servant " in order to find out, if possible, whether Mary was pregnant. After the women, who were nine in number, had made " the search," they reported that it was impossible for them to determine whether Mary Stedhed " be with child or not." To be on the safe side, however, the Kent county commissioners ordered Read to give a bond for five thousand pounds of tobacco " to save the county harmless from the said woman's child in case it proved to be his." As in the cases previously discussed, there was no evidence produced to show that Justice Read had promised to marry the girl.[33]

Suppose a man after having sexual intercourse with a servant which resulted in the birth of a bastard, then fled from the province. This happened when Denis White, who was a horse trader from New England, had relations, while in the colony, with Anne York, a maidservant of John Anderton. As Anderton was left with the responsibility of raising the bastard child, he asked for and received authority from the county commissioners, where he lived, to levy an attachment on White's estate to the extent of three thousand pounds of tobacco.[34]

No case could be found in the early records of where an unmarried freeman promised to marry the servant girl with whom he had intercourse. There is one instance, however, of where the woman claimed that a married man had made this promise to her. This was when Elizabeth Lockett was presented to the Kent county court for having a bastard child. Elizabeth had come to the colony when only sixteen and had been indentured to Matthew Read, a county commissioner. From the testimony of witnesses, it appears that Elizabeth " began her labour on Tuesday night and so remained till Wednesday night and about cock crowing she was delivered." When her baby was born, the girl claimed that Thomas Bright was the father and that there was " a piece of money broken betwixt them and that Bright promised her marriage before the child was got."

Bright, probably because he was a married man, only had to pay the expenses that " did ensue by their unlawful doing and the cost of the suit." Elizabeth, however, besides receiving twenty lashes on her back " well laid on," also had to pay for the trouble and expense which another woman had been put to in taking care of her bastard child.[35]

There was one other case in which a servant girl had relations with a married man which resulted in the birth of a bastard. The girl's name was Jane Palldin and the man John Norton. When the case came up for trial, one witness said that he had seen Jane and Norton sitting before a fire and that Norton had his hand " in the slit of Jane's coat."

Norton appears to have been very angry with his wife for telling neighbours about his affair with Jane. On one occasion, drawing a knife, he exclaimed: " God's wounds, you whore, if it had not been for you, this would never have come out." When a neighbour intervened to prevent him stabbing his wife and asked him if he was " minded to go to the devil," Norton replied that he did not care. As for Mrs. Norton, she referred to the servant girl as that " damned jade."

Jane Palldin described how Norton used to come to see her, while his wife was away and tried to induce her " to be dishonest with him." Jane added that Norton never " gave over his importune suit and fair speeches, until he had obtained her consent to lie with him, and yield unto him the carnal use of her body." The servant girl insisted that Norton was the father of her child as she never " did know any one's body since she came into the country but his." On one occasion, Jane struck Norton with a knife telling him that she had been " undone " by him.

With these facts before them, the judges of the provincial court compelled Norton to put up security both for bringing up the bastard child and also for his good behaviour in the future. William Dorrington, Jane's master, was told that he must " do his endeavour " to see that Jane and Norton were kept apart. As for Jane, the court ordered her stripped to receive thirty lashes on her back.[36]

An unusual case involving a bastard came up for trial in the Baltimore county court. It appears that Thomas Marsh had sold to George Utie a woman servant whose name was Hannah Bowen. After the sale, Hannah was " brought to bed of a bastard child " of which she was later delivered. It was Utie's contention that Marsh was the father of the child. For this reason, and also because he had been put to considerable expense in caring for the woman and her infant, Utie refused to pay Marsh for the servant. Marsh, on his part, claimed that this was untrue. He said that a man by the name of Edward Winwood was really the father of the child, and that Utie had invented the story about his relations with Hannah Bowen not only to defraud him of the tobacco which he owed him for the servant, but also to ruin " his credit and fortune." As there is no record of a judgment, the case was probably settled out of court.[37]

In all the cases so far considered it was women of the servant class who were accused of giving birth to bastards. While it is true that most of the illigitimate children seem to have been born of women in servitude, there were cases of where other women had bastards. To take, for example, the case of Major Edward Inglish. Some colonists wanted him punished " for a bastard child laid to him by Gundry's wench, besides that he had by Wroth's wife." Inglish was the presiding justice of Cecil county.[38]

In Charles and also in Talbot counties, women were lashed at the whipping post for having bastards. In Charles county, Ann Cooper " in the public view of the people " was given twelve lashes, while in Talbot county, Sisilly Johnson received twenty lashes for the same offense. In Talbot county, another woman by the name of Martha Chessill received thirty lashes. When Martha took an oath that her infant was begotten by John Rice, the Talbot justices ordered Rice " to father the child and pay five hundred pounds of tobacco." Although the records do not so state, it is possible in the cases just discussed, as in the following ones, the women may have been servants. Indeed, the punishment meted out in all these cases, as compared with the cases previously discussed, makes one feel that this might be true.[39]

Anne Barbery, during a session of the Calvert county court, was accused of having a bastard. When the child later died, the commissioners of that county, "upon suspicion of felony," sent the case up to the provincial court for a hearing. It appears that after the child was born, Anne hid the infant in a tobacco house. Later she planned to take the baby to Joseph Edlow, whom, Anne claimed, was the father of the child. As no one was present when the infant was born, its navel cord had not been given the proper attention. Those who saw the baby after its death, noticed that "the navel string was untied," that there was a blue spot on the infant's stomach and that its lips were black. However no blood was seen which would indicate that the baby might have been murdered by its mother. It seemed more probable that the child had died from lack of proper care during its concealment. With these facts before them, the provincial court ordered Anne to receive thirty lashes on her bare back.[40]

At a meeting of the county court of Charles county, Lucy Stratton appeared holding a baby in her arms. When asked who was the father of the child, Lucy said that it was Arthur Turner, her former master. When Turner denied this charge, the county commissioners decided to punish the girl, first, for having been "brought to bed of a bastard," and, secondly, for having "unnaturally dried up her milk through which action the infant's life might have been in danger." For these offenses the sheriff was ordered to give Lucy thirty lashes.

Lucy Stratton was not, however, satisfied with this disposition of the case. Knowing that Arthur Turner was the father of her child, she again brought the case before the county court in order to make him pay for its maintenance. When the case was re-opened for a hearing, Lucy Stratton produced witnesses who testified that they had heard Turner admit that he was the father of Lucy's child. The same witnesses said that they had heard Turner offer to marry the girl and thus, in a measure, right the wrong which he had done her. Lucy apparently had rejected this offer saying that Turner was " a very lustful man." Turner, objecting to this remark, had asked Lucy to recall the time she had without his solicitation come to his bed, put her hand under the bedclothes and taken him " by the private parts." Other witnesses were then called whose testimony made it seem doubtful whether Turner was really the father of Lucy's child. Nevertheless, the jury before whom the case was tried, decided that he was and that he must provide for the bastard's maintenance. Turner was told that he must either pay so many pounds of tobacco a year to Lucy for the child's support, or he could, if he wished, bring the child up himself.

Arthur Turner, dissatisfied with this decision of the county court, appealed to the provincial court. The members of this court reversed the decision of the lower court. As Lucy Stratton had refused the offer of marriage made to her by Turner, she must provide for the child herself without any allowance from him.[41]

Because Peter Calloway had a bastard child by Elizabeth Johnson, the justices of Somerset county compelled Peter to pay a fine of one thousand pounds of tobacco and Elizabeth five hundred pounds. Calloway was also required to give one hundred pounds of tobacco to Elizabeth " for her abuse " and to put up security for the child's maintenance " that the county may come to no charge." When Peter and Elizabeth failed to pay the fines demanded of them, they were both summoned before the Somerset court. At the hearing, it was reported that Elizabeth " wandreth to and again amongst the Indians and layeth in the marshes." The county commissioners ordered the woman apprehended and turned over to a magistrate who should give her correction " for her idleness and also provide her a service that she may work for her living." [42]

Anthony Purse, a constable of Talbot county, and Anne Mungummory were brought before the Talbot commissioners " upon suspicion " that Purse " had got Anne with child." Anne claimed that Purse had gotten the child on New Year's Day. In view of this information, the Talbot court ordered Purse to put up a bond, obliging him to maintain the bastard, should it prove to be his child. Anne was told to come before the court after she had delivered her baby.

When later, Anne and Purse, her alleged seducer, appeared before the Talbot justices, evidence was produced which showed that the bastard had been born seven and a half months after New Year's Day, the day on which Purse had intercourse with Anne. At the time of its birth the child was perfectly formed. For this reason the Talbot commissioners were of the opinion that Constable Purse could not be the father of the infant and the charge against him was dismissed. As for Anne, their honours ordered that she should receive thirty lashes on her bare back.[43]

During a session of the Charles county court, Anne Williams asked for maintenance for a child, which, she said, Richard Smith " hath got by her." At the trial, Smith in his own defense said:

That whereas this impudent woman hath most scandalously cast aspersions upon me, and I, having taken into my consideration the injury, I do think it most meet for me to let her run on in her own perdition as she hath begun.

If so be that you, the county commissioners, will be pleased to permit her to take her deposition concerning the allegations she hath alleged against me, I am contented thereupon to take the child and to maintain it, trusting in the severe judgment of God against perjured persons.

After this statement, the county justices ordered that Smith should maintain the child. This seems rather hard on Smith to be compelled to support a child on the statement of the woman alone. Anne Williams received, of course, the usual lashing, thirty stripes " well laid on." [44]

Before closing this chapter, mention should be made of the steps which it was necessary to take before entering into the state of matrimony and also, not less important, the chances of escape from this bond should the conditions prove unfavourable for its continuance.

By the terms of a law passed in 1640, persons planning to marry must do one of two things. Either the banns must be published three days before the ceremony in some chapel, or other place, for posting public documents so that any one who had some reason for objecting to the marriage could do so, or else a sworn statement must be filed in a county court that neither the man or the woman was " an apprentice, or ward, or precontracted, or within the forbidden degrees of consanguinity, or under the government of parents or tutors," or, in other words, that there was no impediment to the marriage. The man and woman must obtain from the judge, or the register of the county court, a certificate stating that such a sworn statement had been made. Unless the banns were published or the sworn statement filed, no one could " solemnize the marriage." [45]

Richard Thompson taking advantage of this Act of 1640 in the year following its passage, took an oath in court " that neither himself, nor Ursual Bish, to his knowledge, is apprentice, or ward, etc., etc.," using the same words as were in the act. At the same time he made these statements, Thompson put up two thousand pounds of tobacco as " caution " money.[46]

During the spring of 1658, an act regarding marriages was passed, which, with minor changes, remained in force until the end of the proprietary period. By the provisions of this law, all persons intending to marry, must apply for a marriage certificate either at the county court, or church or chapel near where they lived. The application must be made either when the court was in session or a congregation was present in the chapel or church. This was in order that people might have notice of this application. This was known as posting the banns, as the application was called. If, after the banns had been posted for three weeks, no one raised any objection to the proposed marriage, then the county commissioners, or the minister or pastor of a church, had power to issue a certificate authorizing the marriage. If persons wishing to wed, failed to post the banns, or to obtain a marriage certificate, or license, then, in that case, they were liable to a fine of one thousand pounds of tobacco. Any magistrate, or minister, joining persons in wedlock before the terms of the law had been complied with was liable to pay a fine of five thousand pounds of tobacco, half of which went to the person who acted as " informer " on the magistrate or minister.

Having obtained the necessary marriage certificate, or license, then the young couple, or old, as the case might be, must apply to some minister, pastor, or civil magistrate, to solemnize the marriage. In order that "the consent of the parties" might appear, a certain "form of words" was used which are given here in full for comparison with the ritual used today, almost three hundred years later. The man must first take the woman by the right hand and say:

I, John Smith (or whatever the name), do take thee, Alice Adams (or whatever the name), to my wedded wife, to have and to hold from this day forward, for better, for worse, for rich, or for poor, in sickness and in health, till death us do part, and thereto I plight thee my troth.

After he had finished saying this, the man released the woman's hand and she, in turn, taking the man's right hand, said:

I, Alice Adams, take thee, John Smith, to my wedded husband, to have and to hold from this day forward, for better, for worse, for rich, or for poor, in sickness and in health, till death us do part, and thereto I plight my troth.

After the woman had finished, then the minister, or magistrate, as the case might be, must say:

I being hereunto by law authorized do pronounce you lawful man and wife.[47]

Besides going through this ceremony, all marriages must be witnessed. At first only two witnesses were required, but near the end of the proprietary period five were necessary. The magistrate, or minister, who performed the marriage ceremony was entitled to receive a fee of one hundred pounds of tobacco for his services, or sixteen shillings sterling. The secretary of the province asked an additional one hundred and twenty pounds of tobacco, or twenty shillings for the marriage license. In other words, it cost nearly two pounds sterling to get married in early Maryland.[48]

In the early records, there is only one instance of where after the names of a man and woman had been posted showing their intention to marry, another man made an objection to the coming marriage. This was in the case of Elenor Empson, who, in a petition to the commissioners of Charles county, said that Richard Watson had written forbidding "the banns of matrimony" without any reason for doing so. It appears that Watson had written a note to Francis Doughty, a minister, asking him to forbid her marriage on the ground "that she was his wife before God." Watson, who was blind, denied having sent or signed such a letter. If his name was on it, this was only because some one had taken advantage of his blindness. Watson added that he never had "any interest" in Elenor. In view of this statement, Elenor's suit against Watson was dismissed.[49]

There are, however, several instances of where it was alleged that a marriage had taken place without the posting, or publication of the banns, or obtaining a marriage license, as the law required. This was when Richard Smith declared that Captain Thomas Manning, who must have been a magistrate, had joined in marriage William Chaplin and Mary Richardson without complying with the usual formalities. When Manning admitted having done this, he was compelled to pay five thousand pounds of tobacco, one half of which went to Smith who had informed the proprietary officials of Manning's offense. Chaplin and Mary Richardson were also compelled to pay a fine of one thousand pounds of tobacco.[50]

Jacob Lumbrozo, a Jew doctor, also reported to the proprietary authorities that John Legatt, a minister, had married two couples without complying with the provisions of the law. At his trial Legatt pleaded: "Not Guilty." When Lumbrozo was unable to offer any evidence to support his charge, his suit against Legatt was dismissed.[51]

Colonel Henry Coursey, a prominent colonist living on the Eastern Shore, had Archibald Burnet, a Scotchman, committed to prison for marrying a girl who was only eleven years old. It appears that Burnet was not qualified by civil or ecclesiastical law to perform the ceremony, and also that he had not obtained the consent of the girl's guardian.[52]

In a similar case, William Parker brought suit against Henry Mitchell "in an action of felony" for stealing away Grace Molden, his daughter-in-law, who was under age. In taking away Grace, it was charged that Mitchell violated a penal statute which protected orphans while under the care of guardians. Mitchell denying that he had stolen Grace, said that she willingly consented to be his wife. At the trial, witnesses testi-

fied that Mitchell did not appear to have used force in taking
the girl away with him. When Grace herself said that she went
away with Mitchell of her own volition, because of " love and
affection," intending to become his wife, Parker's case was
dismissed.[53]

In Charles county, during the fall of 1665, Giles Tompkinson
was accused of begetting a bastard. Tompkinson did not deny
that he was the father of the infant, but maintained that the
woman was and is " before the getting of her with child his
lawful wife." Tompkinson added that his marriage to the
woman was " as good as possibly it could be made by Protestants, he being one, because before that time and ever since
there hath not been a Protestant minister in the province." For
this reason, argued Tompkinson, in order to constitute a valid
marriage it was only necessary for them to consent to be married
and have " publication thereof before a lawful Churchman."
As it was obvious that they had wanted to be married, Tompkinson said that " for the world's satisfaction they here publish
themselves man and wife till death them do part." [54]

During the proprietary period, Maryland seems to have enjoyed the reputation of being a " Gretna Green." People would
come from Virginia and the other neighbouring colonies to be
married in Maryland. This " private and clandestine marrying
of strangers and others " was considered " a reflection " upon
the local government. Apparently, the Maryland magistrates, or
ministers, while in most instances complying with the law in
marrying inhabitants of the province, they did not, however,
adhere to the terms of the law when marrying couples from
other colonies. In order to prevent such marriages in the future,
the proprietary officials issued a proclamation forbidding magistrates, or ministers, to marry persons coming from a neighbouring colony unless they had first obtained the same license required of residents of the province.[55]

Although, while Maryland was under the Catholic proprietors, there were no cases of absolute divorce, there were
several instances of what amounted to a separation. When,
for example, the relations between Cornelius Cannaday, a brickmaker by trade, and his wife, Susan, became strained, the question as to what should be done was referred to Richard Preston

14

and Michael Brooke, who were to act as arbitrators. When these two men came to the conclusion that the Cannadays were determined " to live asunder," the arbitrators drew up an agreement by the terms of which Susan was to be paid one thousand pounds of tobacco by her husband and also to retain her own household goods. After this was done, Susan promised that she would not in the future demand or expect any further " allowance, maintainance, or subsistence " from her husband.[56]

At a session of the provincial court held during the fall of 1680, Elizabeth Tennisson, of St. Mary's county, entered a complaint against the treatment which she had received at the hands of her husband, John Tennison. In her petition to the court Elizabeth said:

> That she could not live peaceably and quietly with him, but in danger and hazard of her life, and therefore craved the allowance of a competent maintenance to be awarded her apart from her husband.

Replying to his wife's complaint, John Tennisson denied the allegations made by her. As he was ready and willing to live with his wife and afford her maintenance, the statements which she made, he said, were " false, malicious, and altogether unreasonable." Tennisson admitted, however, that he could never entertain " that love and respect for his wife, or afford her that countenance in his house, as is properly due from a man to his wife."

In view of these facts, the members of the council acting as a provincial court delivered the following judgment:

> That the said John Tennisson forthwith deliver unto his wife, or her order, one good bed and furniture called by the name of her own bed, all her wearing apparel, and allow her yearly for the time to come three hundred pounds of meat, three barrels of corn, and one thousand pounds of tobacco, for and towards her maintenance during her natural life, to be paid and allotted her in such part and place of this province as she shall desire and appoint, and that the said John Tennisson give good security for his true performance of what is hereby required of him . . .[57]

As is shown by the sordid facts of the following case, adultery was a ground for a separation between man and wife. Many witnesses were called in the case in which Robert and Elizabeth Robins were involved. William Marshall was one of these witnesses who said that when he told Robert Robins that he thought it was a pity he and his wife did not get along together, Robins replied: " What would you have me do, for she is a common whore, and I have good witness that William Herde rid her from stump to stump." Apparently Herde was his wife's own brother. One day meeting with him, Robins exclaimed: " Herde, you rogue, you swived your sister from tree to tree and I will have you to the court." Nothing daunted, the incestuous Herde replied, " I will see her hanged before I come to court about her." Another witness told how, when he was on board of a vessel anchored near Newtown, Elizabeth Robins came on board and slept with a man by the name of Hunniford.

As a result of Elizabeth's promiscuous sex relations, she was found to be pregnant. A jury of women instructed " to search the body " of Elizabeth Robins reported that they had found her " in a very sad condition." She had confessed to taking savin twice. Savin, or savine, is a juniper berry with a bitter acrid taste sometimes used as an abortifacient. Elizabeth claimed that she took the drug for worms not knowing that she was with child. When later the child was born, Robins asked his wife whether it was " begotten of his body." Elizabeth replied that she " would not swear it."

Knowing as he did all this about his wife, Robert Robins accused his wife of adultery and sought a separation. The Charles county commissioners, before whom the case was tried, did not, however, feel that the " divers depositions " which Robins produced substantiated the charge of adultery. Accordingly, Robins was ordered to take Elizabeth as his wife again and provide for her and her children. The justices, however, told Robins that if, later, he was able to prove that the infant to which Elizabeth had recently given birth was the child of another man, then, in that case, he would not be obliged to take care of either his wife or the child. Apparently Robins was successful in later being able to prove that the infant was not begotten by him. This is shown by the records of a subsequent meeting of the Charles county court where we find the following entries:

I, Robert Robins, do hereby disclaim my wife, Elizabeth Robins, forever to acknowledge her as my wife . . .

Elizabeth also disclaimed Robert as her husband and added:

I do hereby oblige myself and every one from me never to molest or trouble him any further for maintenance or any other necessaries.[58]

Daniel Clarke, who was a delegate in the assembly, was charged by his wife with " divers inhumane usages and beastly crimes." The records do not, however, disclose what was the nature of his crimes, or whether his wife was later separated from him. In another case of the same kind, the woman was successful in securing a separation. This was the case of Elizabeth Leshley who accused her husband of sodomy. Robert Leshley, her husband, submitted to the governor a petition which gives an account of the affair:

That your petitioner's wife being a woman of implacable, turbulent spirit has at several times unjustly and wrongfully made divers complaints and accusations against your petitioner before several of his Lordship's justices of the peace of Calvert county, but finding they took not the effect she hoped for, and that her design of procuring a considerable part of your petitioner's estate to be allowed her for a maintenance with which she might live at her pleasure where she listed and quite desert the family of your petitioner . . . she hath since very falsely, injuriously and maliciously accused your petitioner of buggery to the great scandal and irreparable prejudice of your petitioner's credit and reputation which he does and always shall esteem of greater price and value than his life.

Wherefore your petitioner humbly prayeth, that your Honour will be graciously pleased to grant him a warrant, that she may be brought before your Honour and her complaints and accusations against your petitioner heard and examined before your Honour whereby your petitioner may purge and cleanse himself of that grievous scandal cast upon him and that for the future he may live more peaceably and comfortably either with his said wife or by allowing her such a maintenance as your Honour shall order and command.

And as in duty bound he shall ever pray, etc.

Upon the receipt of this petition, the governor ordered the constable of Calvert county to bring Elizabeth, the wife of Robert Leshley, to the next meeting of the council. When, at this session, it appeared that both Robert and Elizabeth had had several differences between them which could not be composed and that neither of them wished to cohabit together again, the proprietary officials ordered Robert Leshley to pay his wife two thousand pounds of tobacco a year for her maintenance.[59]

Having more than one wife was a serious offense in early Maryland. The offender whether principal, or accessory before the fact, could be put to death, or burnt in the hand, etc., etc. There are two cases of bigamy recorded in the proprietary records. In one of them Anne Thompson, the wife of Robert Thompson, was convicted of this offense and " adjudged to be burnt in the hand for having two husbands." Anne appealed from this judgment of the provincial court to the council and prayed—

Their honours' commiseration of her condition, humbly acknowledging her error therein, and faithfully promised amendment of her life and to be more wary and diligent for the future.

When Mrs. Thompson also promised that she would separate from the man that she had last married and " cleave " to her first husband, and with him " soberly and honestly to live," the members of the council decided to pardon her.[60]

The other case of bigamy was when Robert Holt, a cooper of St. Mary's county, was accused of having married a woman by the name of Christian Bonnefeild, while his own wife, Dorothy, was still living. William Wilkinson, who was both a court clerk and a parson, was accused of having been accessory to this felony in that he married the couple after he had " divorced " Robert Holt from his first wife.

While Pastor Wilkinson said that he had married Robert Holt and Christian Bonnefeild, he denied that he had done anything " by way of divorce " between Holt and his former wife Dorothy. Wilkinson admitted, however, that he drew up and signed as a witness a paper which released Robert Holt from " all claim of marriage " by Dorothy. It is hard to see the distinction. This agreement he had drawn up, it seems, at their mutual request after Dorothy had confessed that she had had two bastards by another man and refused " to be reconciled " to Robert Holt.

In view of the evidence which had been produced, a grand jury indicted Holt and Christian Bonnefeild for bigamy and the pastor for being accessory to this felony. A petty, or trial jury, was then summoned. Holt and Christian did not want Catholics to be members of this jury which decided " so weighty a business as life and death," but wanted a jury composed of Protestants to consider the case. This request was granted. Before the case was actually tried, however, the proclamation of an amnesty freed both parties. Not satisfied with this lucky turn of the scales of justice in their favour, Holt continued to live with Christian Bonnefeild. For this reason, Holt was again presented for continuing to cohabit and living " incontinently " with Christian. Before this second case came up for trial, Holt died.[61]

Chapter IX

DEFAMATION

To slander another is not a criminal offense. However, having considered theft, hog stealing, adultery, etc., it is not inappropriate to discuss the punishment which was inflicted on those accusing others of these offenses.

From the number of cases of defamation in early Maryland, it is quite obvious that the colonists were very zealous in guarding their reputations. One took the chance of being sued if he called another a knave, rogue, or rascal. When, for example, Thomas South called Matthew Read " a knave," the latter brought an action of slander against South. At the trial, when South was unable to prove what he had said about Read was true, he was compelled " to confess and acknowledge his offence in open court." [1]

In Charles county, Thomas Baker, a justice of the county court, sued William Robisson for referring to him as " a rogue and a rascal." As Baker said such a remark was prejudicial to his reputation, he asked that Robisson must prove the truth of the words which he had spoken, otherwise he should be punished. After witnesses had testified that they had heard Robisson call Baker a rogue and rascal, the commissioners of the Charles county court ordered Robisson to acknowledge in court that he had injured the reputation of Baker. Robisson was also compelled to pay the cost of the suit. [2]

An amusing exchange of slanderous words took place in the same county. It appears in an argument about a bottle Margaret Codwell called John Wood " a rogue, rascal and knave." As Wood started to leave the house, he said to Margaret, " God be with you," to which the girl replied, " The devil go with you." Wood's answer to this was to clap his hand " on his britch and bid her kiss it." No action of defamation grew out of this incident, possibly because each had insulted the other. [3]

In most of the cases of defamation in proprietary Maryland

a far more serious charge was made than calling a man a rogue, rascal or knave. There are several instances of where a man was accused of having perjured himself. This, of course, was a grave charge to make. It usually meant that a man was accused of making false statements while testifying under oath as a witness. The punishment of those guilty of perjury was severe. As we saw when discussing this question in a previous chapter, the offender's ears might be nailed to the pillory and then cut off.[4]

Francis Brooks sued Edward Commins in an action of defamation for having slandered him " with divers scandalous speeches," especially in calling him " perjured " and in saying that he had taken a false oath. When this case came up for trial, Commins admitted having uttered the words in question, expressed sorrow for having done so, and asked Brooks to withdraw his action against him. Brooks agreed to do so. In another case, Thomas Mathews sued William Finney for slandering him by saying that he had perjured himself when testifying as a witness, and that he could cause him " to loose his ears," which was the punishment for perjury. The jury, before whom this case was tried, awarded Mathews three thousand pounds of tobacco damages and also required Finney to ask forgiveness of Mathews on his knees in open court.[5]

Hubert Patee was accused by George Bussey and James Veitch of having " notoriously slandered them in their reputations," by saying of them that when testifying before the provincial court they had " forsworn themselves to the pit of hell." When Veitch and Bussey " proved their complaint," the court ordered Patee to ask the forgiveness of the two men whom he had defamed and also to pay a fine of five hundred pounds of tobacco. This fine was later remitted upon Patee's " good abearing."[6]

William Price complained against Elizabeth Martin, the wife of Robert Martin, for defaming him by calling him "a perjured rogue." It appears that in some previous litigation Price had given testimony which cast some doubt upon Mrs. Martin's morals. It was for this reason that Mrs. Martin had in anger said that Price had perjured himself. When, at the trial, Mrs.

Martin did not deny having made the slanderous remark about
Price, she was told that she must either pay Price three hundred
pounds of tobacco " for the reparation of his credit," or else
acknowledge her offense and ask Price's forgiveness in open
court. No matter which of these penalties she chose, Mrs.
Martin had to pay an additional three hundred pounds of
tobacco " as a fine."

Not long after this judgment was entered, Robert Martin,
Mrs. Martin's husband, begged the Kent county commissioners
that, in accordance with their " wonted clemency," they would
not exact the three hundred pounds of tobacco demanded as a
fine. Martin gave as the reasons for his request, first, that his
wife's defamatory remark about Price was due to her " weakness and passions," and, in the second place, that he himself
was in such straightened circumstances that he was unable to
pay the fine. The Kent justices lending a willing ear to Martin's
" humble request," decided that he should not be compelled to
pay " any part, or parcel " of the three hundred pounds of
tobacco.[7]

There are several instances of where a man defamed another
by accusing him of cheating or some such form of dishonesty.
Thomas Hynson, a justice of the Kent county court, brought an
action of slander against Matthew Read, a fellow commissioner,
on the ground that Read " did vilify him in saying he either
got his living by cheating and cozening, or sharking and cozening." After Justice Hynson had introduced several witnesses,
whose testimony supported his contention, Read was compelled
to acknowledge his offense in open court and promise " amendment." He also had to pay five hundred pounds of tobacco
" towards the reparation of Mr. Hynson's credit." [8]

In a similar case, Hugh Stanley sued Thomas Paggett in an
action of slander for calling him both a knave and a cheating
knave and saying that he could prove it. Stanley said that as
he had been for many years a justice of the peace, his reputation as such was injured by these false and malicious words.
He asked thirty thousand pounds of tobacco damages. At the
prevailing rate of two pence per pound of tobacco, this was
two hundred and fifty pounds sterling, a tidy sum in those
early days.

Paggett's first step was to file a demurrer to Stanley's declaration. By this he meant that he neither denied or admitted the truth of the facts stated in the declaration, but maintained that even if true the language he used about Stanley was not actionable. The judges of the provincial court overruled Paggett's demurrer. This decision meant that if Stanley could prove that Paggett had called him a cheating knave as a result of which he had suffered injury to his reputation, then a cause of action did exist.

Paggett now pleaded "a justification and travers," or, in other words, that he was justified in calling Stanley a cheating knave. Witnesses were summoned to prove Paggett's contention. A man and his wife declared that Stanley had refused to give Paggett "a certain bill for tobacco" which he had promised to do when requested. Two men testified that Stanley had killed a calf, without the permission of the owner, and afterwards had denied doing this. Evidence was produced which showed that Stanley "did knavishly cause" Paggett to pay some tobacco to one man, and that, on another occasion, Stanley had demanded tobacco of another man which he "knavishly denied" having in his own possession.

After the witnesses had all been heard, the case went to the jury. Before they retired they received the following charge:

Whether Thomas Paggett did falsely and therefore maliciously call Mr. Stanley, knave and cheating knave, and say he would prove him a cheating knave—

Whether in case he did falsely and therefore maliciously call him knave, cheating knave, and say he would prove him a cheating knave, he be damnified thirty thousand pounds of tobacco or anything at all.

In a short time the jury returned and rendered the following verdict:

We of the jury do find no actual damage sustained by the plaintiff (Stanley) by the defendant (Paggett) calling him a cheating knave—

Neither what hath been proved of the defendant's charge against the plaintiff to be so valid as to repute him a cheating knave.

The judges of the provincial court at once ordered the jury's verdict "entered for the judgment." Stanley not only lost his

case, but he also had to pay all the costs of the litigation, including fees for witnesses, attorneys, jurors, and even the cost of
Paggett's attendance in court for six days.[9]

To accuse a man or woman of theft was, of course, a very
serious charge to make. There are several instances where this
accusation was made in proprietary Maryland. Edward Commins complained against George Manners for having defamed
him " publicly " by accusing him of taking from him " feloniously " an iron pestle and also for having upbraided him with
" many injurious words." Commins produced witnesses who
substantiated his accusation. The jury, before whom this case
was tried, found Manners guilty, awarded Commins three hundred pounds of tobacco damages and made Manners apologize
to Commins.[10]

John Deare brought an action of slander against Nicholas
Brown for having said that he was a thief and had stolen some
household goods. Brown, it seems, had also made some defamatory statements about Deare's wife. For his slanderous
remarks about Deare, Brown was compelled to ask Deare's
forgiveness in open court, pay him five hundred pounds of
tobacco " for the reparation of his credit," and an additional
five hundred pounds of tobacco " for public use." Brown must
remain " in the sheriff's hands " until he paid these sums. For
the defamatory statements he had made about Deare's wife,
Brown had to acknowledge his offense, ask her pardon and pay
an extra three hundred pounds of tobacco.[11]

Elizabeth Jolly, wife of an innkeeper, slandered two women
neighbours when she accused them of stealing some linen.
When Mrs. Jolly was unable to prove this charge, the court
ordered her committed until she should in open court acknowledge that she had wronged the two women in question and also
put up security for her future good behaviour.[12]

James Lee brought an action of defamation against John
Meekes, a chirurgeon, for having called him a thief. The testimony of the witnesses whom Lee called did not make out a
very strong case for him. Although two of the witnesses said
that they had heard Meekes call Lee a thief, the other two declared that they had never heard Meekes make any such statement.

After Lee's witnesses had finished testifying, Meekes brought forward his witnesses. From their testimony we are able to piece together an account of what occurred. It appears that Meekes, John Lewger, the colonial secretary, and Lee all met together in Meekes' house where the latter kept a store as well as practiced chirurgery. The three men had a very merry evening drinking together. When Lewger decided to retire, Meekes offered him his bed while he slept as well as he could on a chest. Lee, however, stayed up all night consoling himself with draughts of rum. A boy servant employed by Meekes said that when he arose early the next morning in order to beat corn, he noticed Lee at a rum cask filling up a bottle by drawing it off from the cask. The youth said that Lee had asked him to have a drink with him but that he had declined. Soon after this, Meekes and Lewger awoke and made preparations to attend a session of the Charles county court. Meekes deciding that he would like a morning draught, looked for the bottle which he used " to draw drink in." Apparently this was the same bottle which Lee had used for this purpose. Although a careful search was made about the store and all the servants questioned, the bottle could not be found. Lee did admit, however, that he might have drawn a bottle of drink " in my drunken humour."

The jury that considered the facts in this case brought in the following verdict:

As it was a drunken business, the charge of the suit shall be equally divided and that Mr. Meekes shall acknowledge in open court that he hath injured James Lee by calling him ' thief.'

The commissioners of Charles county lost no time in confirming the verdict of the jury. The decision in this case seems reasonable. Meekes was hardly justified in calling Lee a thief for having drawn a bottle or two of liquor from his cask while he was intoxicated.[13]

The decision of the Charles county commissioners in another defamation suit is not so easy to understand or to defend. This was when John Courts sued Hannibal and Elizabeth Spicer in an action of slander. The witnesses, whom Courts produced,

testified that they had heard the Spicers say that Courts was a
slanderous man and had slandered " the whole country." The
same witnesses said that the Spicers had told them how Courts,
on one occasion, had said of a man named William Empson
that he was a thief " from his cradle."

In view of this testimony, the justices ordered the sheriff to
give Hannibal and his wife six lashes apiece. When Elizabeth
said that she was pregnant, the commissioners ordered that she
should not be whipped " until such time that she be delivered."
It is hard to understand the decision in this case. How could
any one slander " the whole country," and, even if he could
such remarks would hardly be actionable.[14]

During a fall session of the court of Calvert county, Thomas
Hooton brought an action of defamation against Demetrius
Cartwright. In a petition to the commissioners of this county
court, Hooton alleged:

That Demetrius Cartwright hath falsely slandered and defamed your
petitioner by saying your petitioner stole goods from Michael Catterson
and others by which report your petitioner hath been very much defamed
and blemished in his credit. Now, so it is that your petitioner being
found guiltless, humbly craveth of this worshipful court redress against
the said Cartwright for the said defamation, according as this worshipful court shall adjudge, with cost of suit, and your petitioner as in
duty bound shall pray, etc.

Several witnesses were called by Hooton. Two of them testified that they had heard Cartwright accuse Hooton of theft.
This case was finally appealed to the provincial court. For
some unknown reason Hooton did not come to St. Mary's to
prosecute the case. Accordingly, Cartwright asked for " a dis-
mission from the cause of action with cost of suit." This request
the provincial court judges granted. It would appear that
Hooton had a good case which he might have won.[15]

Besides accusing their fellow colonists of theft, some slandered them by referring to them as hog stealers. Richard
Wollman, a commissioner of Talbot county, brought two actions
against a man by the name of Thomas Wilkinson. In one case,
Wilkinson was charged with hunting and injuring the commissioner's swine. The other suit was an action of defamation

in which the justice said that Wilkinson had accused him of
hog stealing. Being a magistrate this remark, said Wollman,
was a reflection on his good name and " disenabled " him to
execute his office. He therefore asked that either Wilkinson
prove the charge he had made, or else that he himself might
have satisfaction according to law, " whereby his good name
may be vindicated." When this defamation case came up for
trial, Wilkinson, kneeling before the commissioners of the
Talbot county court, admitted that he had " abused and wrongfully defamed " Wollman. The justice was so pleased with
this admission that he agreed to withdraw both suits. Wilkinson was ordered to pay the cost of the litigation.[16]

Seth Foster, another commissioner of Talbot county, brought
an action of defamation against Robert Knapp for spreading
abroad the report that he was " a hog stealing fellow from his
cradle." Just as Wollman had said, Foster declared that this
slanderous statement was harmful to his reputation as a justice
of the peace. When Knapp, at the trial of this case, admitted
that he had " wrongfully scandalized " Foster, the Talbot commissioners ordered him to pay the cost of the suit. Later,
Robert Knapp sued Seth Foster in an action of assault and battery, alleging that Foster and his wife had " most violently set
upon him and beat him so much that he was forced to keep
to his bed." Apparently Foster was not satisfied with a mere
apology.[17]

Edward Burton instituted an action of defamation against
Samuel King and Eleanor, his wife, because she had called him
" a hog stealer," and had claimed that she could prove it was
true. When this case came up for trial, Eleanor in open court
asked forgiveness of Burton, saying that she had made the
slanderous remark " out of passion." Burton accepted the
apology.[18]

In the Somerset county records, there were several actions of
slander entered, but none of these cases came to trial except
one. This was when Randall Revell, a commissioner of that
county, sued Phillip Barre for having vilified and defamed him
" in a most scandalous manner " by saying that he had killed
one of his hogs. The case was tried before a jury who decided

that Barre was guilty of slander and should upon his knees acknowledge himself " culpable of the false accusation." Barre also had to pay the cost of the suit.[19]

In the provincial court, Walter Beane sued John Ballance in an action of defamation because he had accused him of altering the marks " of a parcel of pigs." Although Ballance denied having made such a statement, Beane was able to prove the truth of his charge. The court thereupon ordered Ballance to ask Beane's forgiveness in open court.[20]

Looking through the records of the Charles county court, we find that Arthur Turner instituted a suit against James Bowling (or Bouline) for having defamed him by saying that he had slaughtered a calf which did not belong to him. In order to prove that Bowling had made this slanderous statement about him, Turner called eight witnesses to testify in his behalf. It would be neither interesting nor profitable to give in detail the testimony of all these people, especially as the county commissioners, after hearing all the evidence, decided that Turner had no cause of action. None the less, the case does show how carefully the Maryland planter tried to protect his reputation when he was accused of killing or injuring another's livestock.[21]

A case in which the charge of being a hog stealer was made came before the Kent county justices. It involved Captain Joseph Wickes and Thomas Ringgold, both members of this court. Between these two prominent colonists relations had been strained for some time. Ringgold had only recently had Wickes arrested " in an action of account for a barrel of corn." As Ringgold had lost this suit he was, no doubt, looking for a chance to square his accounts with Captain Wickes. Accordingly, when shortly after this Captain Wickes entered the court room, Ringgold remarked that he did not think it " fitting that any whore master should sit at the table there." Wickes at once retorted that " it was better to be a whore master than a thief as he (Ringgold) was."

The result of this exchange of pleasantries was that Ringgold brought an action of slander against Captain Wickes for having called him a thief. At the trial, Wickes tried by " several depositions " to prove that Ringgold really was a hog stealer. After

" a long debate in the cause," Captain Wickes' attorney asked that the case might be appealed to the provincial court. The Kent county commissioners granted this request, provided Wickes was willing to pay " double damages" if he lost the case on appeal. When neither the captain, or his attorney, appeared at the next session of the provincial court, the judges of that court ordered the sheriff of Kent county to take Captain Wickes " into safe keeping, until he give bond with security for his appearance at the next provincial court . . . to answer to the suit of the said Ringgold." No decision was ever reached in this case. Possibly it was settled out of court.[22]

But to return to the slanderous remark Ringgold had made about Captain Wickes being " a whore master." Certainly this was worse than calling a man a hog stealer. We may be sure that Wickes would have brought an action of defamation against Ringgold had it been possible to do so. That there was some basis for Ringgold's remark is shown by Constable John Ellis' presentment found in the Kent county records. It stated:

> I present unto this court Mary Hartwell, the now wife of Mr. Joseph Wickes, for bringing up a bastard upon the Island (Kent), and charging Mr. Wickes with the begetting of the said child.
>
> By me: John Ellis

After this presentment was filed in court, the Kent justices suspended Captain Wickes from sitting on the bench with them until this charge was either proved or disproved. Soon after this the Kent commissioners, after considering all the evidence against Wickes, came to the conclusion that he was not guilty " in the matter of fact." Therefore, they reversed the order suspending him from the county court and placed him " in statuo quo prius to his place and judicature." What influenced the Kent justices in reaching this decision? While it was true Captain Wickes had a child by Mary Hartwell, he had later married her.[23]

John Wickes, possibly a brother of Joseph Wickes, was also involved in a defamation suit. This was when John Wedge brought an action against him because of " a scandalous libel against the reputation of his wife and family." Wedge asked that John Wickes, whom he termed "a notorious fellow," should be severely punished. When the case came up for trial before the Talbot county commissioners, it appeared that while with a group of friends Wickes had sung a defamatory song about the Wedges, although he himself had not written it. The song was read in court and found to be "scandalous" to Wedge and his wife. For this reason Wickes was ordered to pay one thousand pounds of tobacco "for breach of his Lordship's peace," and also compelled to ask the forgiveness of the Wedges "for the abuse done them." [24]

Captain Joseph Wickes was not by any means the only man accused of having immoral relations with women. John Meredith sued William Daynes in an action of defamation for spreading about a report that Meredith "had got one of his negroes with child and that he had a black bastard in Virginia." These statements, said Meredith, tend "much to his disgrace and defamation." In view of what Meredith had said, the court ordered Daynes to appear at the next session. He would then be given an opportunity to deny the statements attributed to him. If, however, he had slandered Meredith he must, in that case, ask his forgiveness, or pay a fine.[25]

There are several instances of where a woman slandered a man by accusing him of having sexual intercourse with her. Thomas Bradnox, a commissioner of Kent county, brought an action of defamation against a maidservant of Thomas South whose name was Margaret Mannering. This woman, it seems, had formerly been in the employ of Bradnox. When with her new master, it was Captain Bradnox's contention, that Margaret had made slanderous remarks about the treatment which she had received as his servant, saying among other things that the captain "had the use of her body." Margaret was found guilty of slander, ordered stripped to the waist and given ten lashes with a whip.[26]

When Mary Willkin's slandered John Vicaris, another Kent county justice, by saying that he was the father of her bastard child, the court ordered that Mary should have thirty lashes on her bare back "well laid on." In Talbot county, William Bag-

15

ley sued Elizabeth Smith for having defamed him by saying that he had "carnal use of her body at his pleasure." When Elizabeth later admitted that she had "wrongfully abused" Bagley, the Talbot justices ordered her to receive thirty lashes on her naked back "and also to pay the cost of suit." The Charles county court commissioners compelled a woman named Mary Grub to ask the forgiveness of John Cage upon her bended knees in open court for having "maliciously wronged him in laying a child to him." [27]

In one case a bachelor, named Robert Bryan, brought an action of defamation against Teressa Arnald for spreading about a report that he had had intercourse with her on several occasions. As a result of this slanderous remark, Bryan said, not only was his "credit and good name impaired," but also his chances of "attaining a wife." When, at the trial of this case, Bryan proved that his charge was true, the county commissioners ordered the sheriff to give Teressa fifteen lashes on her bare back.[28]

William Berry accused John Little of slandering his father and mother by saying that Berry's mother was already married to another man when his father took her for his wife. Indeed, Little was alleged to have said that Berry's father always kept a boat on hand to escape with, in case "his wife's tother husband did come in." For these defamatory remarks, Little had to pay a fine of five hundred pounds of tobacco and put up security for his good behaviour. Little was also compelled to stand by the whipping post "stripped naked from his waist upwards for the space of one hour with a whip over his head." When Little's wife and "divers neighbours" of his called attention to his age and "the unseasonableness" of the weather, the court shortened the time that Little must stand by the whipping post, requesting that it be made "as short as may be." [29]

In all of the cases just considered, except the last, immorality on the part of the man was alleged. Far greater in number were the cases where a woman's morals were questioned. In some instances, it was a single woman who objected to the defamatory remarks made about her. In an early case, Anne

Avery, a maidservant, sued Richard Cox in an action of slander
for saying that " Daniel, the Governor's Irishman, had lain with
her two moonshiny nights under a walnut tree." When at his
trial Cox was found guilty of slander, the court ordered that
he should acknowledge that he had wronged Anne and " publicly " ask her forgiveness in court. Cox was also required to
pay one thousand pounds of tobacco " in reparation of her
fame," and was to be imprisoned in irons until he " performed
this judgment." [30]

Sarah Taylor, another spinster, brought a suit of defamation
against Thomas Courtney for boasting to other people " that
he had lain with her." According to Sarah, Courtney had made
love to her and she had treated his attentions seriously until
she found out that he was talking behind her back and reflect-
ing on her morality. For " the impudent and false abuses "
which he had circulated about her, Sarah asked that her honour
as " a poor abused maid " might be vindicated. A jury awarded
Sarah five hundred pounds of tobacco damages and also compelled Courtney to ask Sarah's forgiveness upon his knees in
open court for having made slanderous remarks about her.[31]

In several instances, the charge was made that a married
woman was guilty of adultery by having sexual intercourse
with another man. John Little, who, as we have seen, slandered William Berry's father and mother, also " notoriously
scandalized " Mrs. Elizabeth Potts by saying that he saw her
having intercourse with an Indian youth. By way of punishing
Little, he was compelled to pay a fine of five hundred pounds of
tobacco and to stand " for the space of one hour at the door of
the court with a paper in his hat written in capital letters
signifying that he hath scandalized Mrs. Potts." [32]

Thomas Emerson in an action of defamation alleged that
Jane, the wife of Leonard Daniel, had slandered his wife by
saying that she had had sexual intercourse with two other men.
At the trial of this case, Emerson having proved that his case
was well founded, Jane was ordered to receive thirty lashes.
After hearing this sentence imposed, Jane came forward and
kneeling before the Talbot county commissioners begged forgiveness of the Emersons. Jane said that she was sorry she had

" wrongfully abused " Emerson's wife. In view of this apology, the county justices decided that they would not have Jane whipped.[33]

As the decision in the following case shows, it was sometimes dangerous to accuse a married woman of adultery, unless the person making this serious charge had very good grounds for doing so. Bridget Johnson, wife of David Johnson, entered a complaint against John Clymer and Elizabeth Madberry that must have shocked the inhabitants of Talbot county. According to Mrs. Johnson, she was sent one day by her husband to the house of John Madberry to fetch Elizabeth, Madberry's wife. Upon her arrival at the Madberry home she was surprised to find John Clymer " pulling down his breeches " at the bedside of Mrs. Madberry who was asleep. The following conversation then ensued:

> Mrs. Johnson: Thou impudent fellow, what dost thou with the woman that is fast asleep?
> John Clymer: Thou damned whore, what business hast thou here?

Whereupon, according to Mrs. Johnson, Clymer threw a pestle at her which hit her " at the bottom of her belly."

Mrs. Johnson now left the Madberry house but returned not long afterwards. On her second visit, Mrs. Johnson found Clymer in bed with Mrs. Madberry. Clymer's breeches were down as before, while Mrs. Madberry, though asleep, had " her coats up and her arms about him with her knees around his middle." Mrs. Johnson said that she pulled Mrs. Madberry by the coat to awaken her, and then asked her what " she did in that posture." Mrs. Madberry, on her part, asked Mrs. Johnson what she meant by tearing her clothes, as she thought she had been lying in bed with her husband. As for Clymer, he proceeded to strike Mrs. Johnson about the head and face with his fist.

Witnesses summoned in this case gave further details as to what had occurred. According to some of them, Mrs. Madberry said that Clymer had come into her house " unknown to her," and while she was asleep. According to others, Mrs. Johnson had said that both Clymer and Mrs. Madberry were asleep when she found them in bed together and that she had pulled

Clymer's " prick out of her cunt with her two fingers," and then tried to rip away a piece of Mrs. Madberry's dress, but only succeeded in tearing her clothes. Mrs. Johnson also tried to cut off a piece of Clymer's breeches but without better results.

Other witnesses were called to give an account of what they had heard Mrs. Madberry say about the incident. Apparently Mrs. Madberry denied having intercourse with Clymer and claimed that the whole story was invented by Mrs. Johnson out of " her jealous head." This jealousy was due to the fact that Mrs. Johnson had at one time been very much in love with Clymer. It seems that Clymer had even promised to marry her upon the death of her husband. When she was asked by some friends why she did not tell her husband about her love for another man, Mrs. Johnson had replied that she dared not do so and begged no one to disclose her secret. Now, however, Mrs. Johnson felt differently towards Clymer and one witness had heard her say that she hated him " as bad as a toad."

After the Talbot justices had heard all the evidence, they decided that Mrs. Johnson's complaint against Clymer and Mrs. Madberry was grounded on malice. She had loved Clymer once, now she evidently hated him. In view of this finding, the commissioners ordered the sheriff of the county to fasten Mrs. Johnson to the whipping post and give her twenty lashes " well laid on her bare back." As for John Clymer, he was compelled to give a bond for his future " good behaviour." [34]

About the most serious charge that could be made against a man's wife was to accuse her of being " a whore." There are several instances of this in early Maryland. When in Talbot county Richard Austin, a servant, was charged with having called his mistress a whore " with other malicious words," the justices of this county ordered that Austin should receive thirty lashes on his bare back.[35]

Edward Harwood brought an action of defamation against Elizabeth Greene for having grossly abused his wife, Olive, in calling her a whore and saying that she could prove that it was true. Olive, it appears, was the daughter of John Williams. For her defamatory remarks Elizabeth was compelled to ask Olive's forgiveness in open court and also had to pay the costs of the suit.

For some unknown reason, Elizabeth Greene seems to have had a grudge against the Williams family. She said that John Williams was " the king of thieves " and his wife the queen. In an action of defamation Williams said that as he and his wife had been " cleared by law from those scandals," he asked that Elizabeth should make some satisfaction for injuring their reputation and " credit." In this case, however, the judges of the provincial court did not consider the words actionable as there were " no scandal " to Williams and his wife. Accordingly, the case was dismissed.[36]

Henry Clay sued Thomas Ward, a chirurgeon, in an action of slander for calling his wife " a burnt arse whore " who had " the pox." This case was appealed from the Kent county court to the provincial court. After hearing the evidence in this case, the higher court decided that Ward must return to Kent county, and, at the next meeting of the court there, acknowledge " in open court " that he had done the wife of Henry Clay a great wrong in using such slanderous and abusive words about her. He must beg Mrs. Clay's pardon, who was also to be present in court, and promise never " to wrong her so again." If Ward refused to do this " in a submissive, due, respective manner " within a specified time, then he must pay a fine of one thousand pounds of tobacco, and, if his " estate " was not sufficient to pay this amount, he would be publicly whipped " with one and thirty stripes." [37]

One important case involving a married woman's honour came before the council, or upper house of the assembly, on appeal from the provincial court. This was the case in which Henry Spinke sued Dr. Luke Barber for slandering his wife, Ellinor. It appears that Barber made the defamatory remarks when he was being sued by one Benjamin Hunton at a session of the St. Mary's county court. During this trial, Ellinor Spinke was summoned as a witness for Hunton. Doctor Barber said that the oath Mrs. Spinke took as a witness was " false and malicious." Barber also called Mrs. Spinke a whore, adding as proof of this charge that he had, on one occasion, found her in bed with a man named Tom Hughes. At that time, said the doctor, Ellinor had " her coats up " and Hughes had his

" breeches down," and both were engaged in " uncivil actions not fit to be named." Barber then explained how Ellinor had previously been one of the servants whom he had brought from England, all of whom were of low character, " rogues and whores from Newgate, some from Bridewell and some from the whipping post," and that Ellinor was " the impudentest whore of them all."

In suing Doctor Barber for having made these slanderous remarks, Spinke stated that his wife had been in the colony for almost five years and that during that time had enjoyed a good reputation, " without the least blemish of immodesty that carping envy could suspect either in her behaviour, speech, or carriage." Spinke added that his wife's former life in England was known to have been " honest, modest and civil." Now, however, because of Barber's remarks his wife's reputation and honour had been injured and all the virtues of her former life " blasted and she blazed for a whore and strumpet." For this reason, Spinke asked " reparation " against Dr. Barber not only for slandering his wife, but also for impugning her testimony as a witness " in contempt of government."

Henry Spinke won his case before a jury in the provincial court. He was awarded thirty thousand pounds of tobacco damages. Dr. Barber at once appealed the case to the upper house of the assembly in order that he might have an opportunity " to show the errors." Among the grounds for his appeal were that the writ and the declaration in the proceedings instituted by Spinke " doth differ and this is error." In the hearing of this case on appeal before the upper house, Spinke maintained that the difference between the writ and the declaration was not error but only " a matter of form," while Dr. Barber contended that the difference was " a matter of substance." The members of the council, or upper house, agreed with the doctor that the error was substantial and that the judgment of the provincial court was for this reason " erroneous." Both parties were ordered to divide the cost of the litigation.[38]

Three important cases of defamation in which the term " whore " was bandied about came before the Charles county court commissioners. In one of these cases John and his wife

Margery Gould brought an action of defamation against Giles and Elizabeth Glover. As neither John nor Margery could read or write, they made the Jew doctor, Jacob Lumbrozo, whom they referred to as their " trusty and well-beloved friend," their attorney to represent them in the trial of the case. Acting in this capacity Lumbrozo offered the following petition to the justices of the Charles county court:

To the Worshipful Commissioners of Charles County the humble petition of John Gould humbly sheweth:

That whereas Giles Glover hath much defamed your petitioner's wife in calling her whore and in saying that he could prove her a whore, which is the greatest infamy that a malicious tongue can cast on a woman seeing that—

> She lives forever in eternal shame
> That lives to see the death of her good name.

Therefore your petitioner humbly craveth a jury of able men to consider the premises and to bring in their verdict for the reparation of your poor petitioner's wife's wounded fame as they shall think fit, etc.

The above petition, it will be noticed, is only concerned with the abuse which Giles Glover had heaped upon Margaret Gould. In a separate petition, the Jew doctor set forth the slanderous statement which Elizabeth, the wife of Giles Glover, had made. After the usual formal opening words the petition went on—

That whereas Elizabeth Glover, wife to Giles Glover, hath much abused your petitioner's wife in saying go, you whore, play the whore in the cornfield again.

Therefore your petitioner humbly craveth a jury for the reparation of the said wrong, humbly considering—

> That though the speech be near so false an ill
> That one believes it not, another will,
> And so their malice very seldom fails,
> But one way, or another, still prevails.

Dr. Lumbrozo now called his first witness, Joseph Dorrosell. On the witness stand, Dorrosell related how he had heard Elizabeth Glover say to John Gould's wife: " Go into the cornfield

and play the whore again." After Dorrosell had finished testifying, the hearing of the case was postponed until the next
meeting of the county court. For some unknown reason a decision was never reached in this case.[39]

Perhaps the reason that we hear nothing further of this case
is that not long after this the Jew doctor, Jacob Lumbrozo,
brought a suit of defamation against John and Margery Gould,
his erstwhile clients. Lumbrozo, in order to prove his case
against the Goulds, produced as a witness, Joseph Dorrosell,
the same man who had testified in the previous case. Dorrosell
related how Margery Gould had told him that the doctor
" would have lain with her " and had promised her that if she
would do this, he would give her and her husband half his
plantation and half his stock of hogs. According to Dorrosell,
John Gould had told him that Lumbrozo had thrown his wife
upon a bed and tried in vain to force her to have intercourse
with him. When his wife asked the Jew if he was not ashamed
of himself, the latter replied that he was not, as he could " show
her scripture for it."

The Goulds were defended in this case by Daniel Johnson.
He criticized Lumbrozo for bringing such a suit. Johnson said
that the Goulds themselves, who were servants of the Jew, had
intended to ask the county commissioners to be released of their
service because of the Jew's attempt to force his wife to let him
have carnal use of her body, or to be his " whore," as Johnson
expressed it. When Johnson dared the doctor to prosecute his
action of defamation against his clients, Lumbrozo, his bluff
called, " withdrew himself." The county justices immediately
dismissed his case.[40]

This decision of the county court seems just. As Johnson
had pointed out, Lumbrozo had tried to anticipate the suit
which his servants, the Goulds, had planned to bring in order
to end their service with him on the ground that he sought to
have intercourse with Margery Gould. By withdrawing from
his action of defamation, the Jew admitted his own guilt.

It is interesting to note that about a year before this dispute
between Doctor Lumbrozo and the Goulds, John Hammond
had brought a suit against the doctor for spreading abroad a

slanderous report. Hammond said that Lumbrozo was telling people that Hammond had offered to let him lie with his wife in satisfaction of a debt which Hammond owed the Jew. According to Hammond, Lumbrozo had made this remark so often that it had become a matter of " general discourse " and had so harmed him " that he is the by-word and scoff of many and hath so irreparably injured him both in this province and Virginia that it hath wholly taken away his hope of livelihood." A jury awarded Hammond five thousand pounds of tobacco damages.[41]

As we shall see in the following chapter, Jacob Lumbrozo was accused of performing an abortion. In a previous chapter, it will be recalled, Lumbrozo was indicted but not convicted of blasphemy. The doctor was at another time accused of receiving stolen goods. This Jew, who wore a black " ruff caster," or cloak, and " a shagged caster hat," must have been quite an unusual character. A native of Lisbon, Portugal, he was probably the first of his faith to come to Maryland. He was naturalized in 1663. He engaged in many other activities besides the practice of chirurgery. He hunted wolves and traded with the Indians. At his house on Nanjemoy creek, in Charles county, he kept an ordinary or inn. Lumbrozo also acted as an attorney on several occasions. In one instance, we find him acting in this capacity for Giles Brent, a prominent colonist. The Jew chirurgeon also served on several juries. Not the least of his accomplishments was, as we have just seen, his ability to quote poetry and the scripture.[42]

The last two cases of defamation to be considered are those in which John Nevill and his wife, Joan, of Charles county, were involved. In one case they were the defendants, in the other the plaintiffs. First, to consider the case in which the Nevills were being sued. Richard Dodd claimed that Joan Nevill did " falsely and maliciously utter, publish, declare and express several scandalous words of and against Mary Dodd (his wife) much to the scandal, prejudice and defamation of the said Mary." Among the slanderous words which Joan was alleged to have uttered, were that Mary Dodd was " the whore of Captain William Batten " and she could prove it. As a

result of such statements, Richard and his wife, Mary, said that
they were both " infinitely damnified in their reputations and
impaired in their credits," for which they asked reparation of
Joan and her husband, John Nevill.

In order to prove the truth of this allegation, the Dodds
called several witnesses, among them Mary Roe. From her
testimony and that of several others we are able to piece together the story of what happened. It appears that one day
Mary Dodd came to see Mrs. Nevill. The following conversation then ensued:

Mrs. Nevill: Thou jade, get thou out of my ground, for what busi-
            ness hast thou here.
Mrs. Dodd: I am in the path, I will go when I please.
Mrs. Nevill: If thou wilt not get out of my ground, I will set
            thee out.

When Mrs. Nevill walked towards Mrs. Dodd, the latter turned
on Mrs. Nevill and exclaimed: " Stand off from me, or I will
strike thee." With that Mrs. Dodd struck Mrs. Nevill in the
face. " Thou jade," said Joan Nevill, " dust thou strike me in
my own ground," and with that Joan grabbed Mary Dodd.

Now it appears that as Mrs. Dodd was holding a baby in
her arms, she told Mrs. Nevill to let go of her for fear that
the child might fall to the ground. Mrs. Nevill tried to get a
man who was standing nearby to take the infant from Mrs.
Dodd's arms, but she refused to give the baby up. But this did
not deter Mrs. Nevill from giving Mrs. Dodd " a good blow on
the chops," at the same time exclaiming, " By God, you shall
have one for the other blow, thou jade, I will have my revenge
of thee yet." To this remark Mrs. Dodd retorted: " Do not
threaten me, for threatened folks live long." To continue with
with the dialogue:

Mrs. Nevill: Bald eagle, get thee home and eat some of Gammer
            Belaine's fat pork.
Mrs. Dodd: If I did eat fat pork, I do not eat rammish boar.
Mrs. Nevill: Who did?
Mrs. Dodd: I do not, thou troge.
Mrs. Nevill: Thou whore, who is that thou callest ' troge '?

Mrs. Dodd: I am no Scotchman's whore.

Mrs. Nevill: Neither Scotch, Irish, or English come amiss to you.

It was now Mrs. Dodd's turn to tell Mrs. Nevill to go and eat some of Mrs. Belaine's fat pork, to which Mrs. Nevill replied by spitting on Mrs. Dodd. The latter, discouraged perhaps by this last mode of attack, decided to leave the scene. As a parting gesture Mrs. Nevill holloed at Mrs. Dodd, at the same time " holding forth her fingers, to wit, her forefinger and her little finger."

The fingers held in this manner made what was known as " the sign of the horns." This gesture had two meanings according to the way in which the fingers were held. If the fingers were pointed towards a person suspected of practicing witchcraft, it was supposed to ward off the baneful influence of the black magic, or the evil eye. If, on the other hand, the fingers were held pointing to one's own chin, this was a gesture showing great contempt and an insinuation of a woman's unfaithfulness to her husband. As Mrs. Nevill had called Mrs. Dodd " a whore," the former must have held her fingers pointing to her chin. Indeed, Mrs. Nevill was heard by several witnesses to have said that she could prove that Mrs. Dodd was Captain William Batten's whore and that, on one occasion, she " lay with him in the sight of six or eight men, with her coats up to her mouth." Mrs. Nevill was also heard to say that a child of Mrs. Dodd had black eyes like Captain Batten's.

After all the witnesses had finished testifying, Richard Dodd and his wife asked the Charles county commissioners that they might appeal this case to the provincial court, stating as the reason for this request " that his wife's reputation being so far ruinated that the reparation thereof is beyond the cognizance of this court." The county justices granted his request. As there is no record of the case subsequently coming before the provincial court, the suit may have been settled out of court.[43]

As has been already stated, the Nevills themselves were the plaintiffs in another case which came before the Charles county court. In a petition to the justices of this court, John Nevill said that his wife had been much defamed by Thomas Baker's " slandering tongue." Baker was a commissioner of Charles

county. Nevill said that if the remarks attributed to his wife
by Baker were true and she had meant them, she would, in that
case, be " a creature not modest enough to keep the brutes of
the forest company." If, on the other hand, the remarks attributed to her by Justice Baker were made because of " over-
much joy of her safe delivery of a child," then Baker should
not have repeated them, since by doing so " a most malicious
and injurious infamy " has been cast upon her. Nevill asked
the county commissioners to allow him satisfaction for the
injury done to his wife's reputation.

George Thompson was the first witness called by Nevill.
Baker objected to Thompson's testifying, saying that in order
to injure him Thompson might even perjure himself on the
witness stand. Thompson's reply to this accusation was that it
was well known that Baker had been " a common defamer "
of almost all of his neighbours and that he was willing to prove
this by the testimony of his neighbours.

Over Baker's objection, Thompson was finally sworn as a
witness. Thompson testified that Baker told him that Mrs.
Nevill was to have been a witness in a case between him and
another man and that Baker had said that if she had ever taken
the witness stand he would have questioned her right to testify
" for she was a common whore." The reason for his making
this statement, Baker said, was that just after Mrs. Nevill had
given birth to a baby girl, she invited a man by the name of
William Empson, " that accidentally happened to be there, to
come to bed with her and get her a boy to her girl." Empson's
disgusted reply to this was, according to Baker, to call his dog
Trojan " to perform the office," as being more fitting for a
dog than a man.

Baker also objected to the next witness summoned by Nevill
who was William Robisson. As in the case of Thompson, Robisson was finally allowed to testify. His testimony was similar
to Thompson's. He said that he had heard Baker speak of Mrs.
Nevill as " an impudent quean (slut) . . . and notorious whore,"
and that Baker had then repeated the conversation, which, he
claimed, had taken place between Empson and Mrs. Nevill
after the birth of her infant. Robisson said that, on another

occasion, Baker told him how he had "jeared" John Black-wood about Nevill's wife. Baker told Blackwood that he was certain that he could never forbear her company, for he knew that he "fucks her oftener than John Nevill himself."

James Fox, another witness called by Baker, gave substan-tially the same testimony as had Robisson and Thompson. After Nevill had finished presenting his case, Baker then called wit-nesses to testify in his behalf. All three of Baker's witnesses said that they had heard the story about Mrs. Nevill, of how she had asked William Empson to have intercourse with her and how he had contemptuously suggested his dog as a substi-tute. In other words, Baker attempted to show by the testimony of his own witnesses that what he had said about Mrs. Nevill was common gossip.

After the county commisioners had heard all the evidence, they delivered an opinion in which they declared that Baker had maliciously defamed Mrs. Nevill because it was "against nature" that she could have made the remark attributed to her, that is, that she wanted intercourse with a strange man imme-diately after the birth of her baby. For this reason, the court said, Baker should not have "reiterated" such a story "as an infamy unto her." By way of punishing Baker, the justices said that he must ask forgiveness for his "verbal injury" in open court and upon his bended knees and that he must also pay the cost of the suit. John Nevill and his wife also wanted Baker to ask God's forgiveness for his remarks.[44]

It will be recalled that in the case just discussed, George Thompson had declared that Baker was "a common defamer" of many colonists and that he could prove this by the testimony of Baker's neighbours. Because of this statement, Baker sued Thompson in an action of defamation. The latter asked that witnesses might be sworn who would testify as to the truth of his accusation. The county commissioners granted his request.

Edmond Lindsey was the first witness to take the stand. He said that he had heard Baker call Job Chandler "a spindle-shanked dog" and had added that he would like to stab Chand-ler. Lindsey also testified that ever since he had lived "in these parts" Baker had been considered a common hog stealer by almost all his neighbours.

Richard Roe then came to the witness stand. Roe said that when, on one occasion, Baker had called him " a thievish rogue," he had, in reply, accused the justice of being a hog stealer. Indeed, added Roe, during the ten years he had known Baker he had the reputation of being a hog stealer. Roe then testified as to some of the defamatory remarks which Baker had made about some of his women neighbours, and how Baker had said that he had seen Pope's wife's " cunt which is like a shot bag," and that Mistress Hatch had " a cunt big enough to make souse for all the dogs in the country."

William Robisson, who had testified in the previous case, was the next witness. Robisson confirmed what Lindsey and Roe had said about the reputation which Baker had of being a hog stealer. In addition to hearing the same slanderous remarks about Pope's wife and Mistress Hatch, Robisson said that he had heard Baker say that he thought " a Mr. Batcheler must get a swan's neck to put upon the end of his prick, or else he could not get his wife with child."

After all these witnesses had testified, it is not surprising that Baker decided to " let fall " his defamation suit against Thompson. The later had certainly good grounds for calling Baker " a common defamer." The Charles county court compelled Baker to pay all the costs of the litigation.[45]

Because, as we have seen, Thomas Baker had called him " a spindle-shanked dog," Job Chandler brought an action of defamation against the Charles county magistrate. When two men testified that they had heard Baker use these words, the county court ordered that Baker should be bound " unto his good behaviour," and that if he should be found guilty of making similar remarks in the future, he would then " stand liable to the censure of the court." [46]

CHAPTER X

SICKNESS, CHIRURGERY AND BURIALS

Except for the one instance where a chirurgeon is accused of being an abortionist, there are no other criminal cases considered in this chapter. However, it is hoped, the interest of the subject matter will justify the inclusion of the chapter in this book.

Sickness appears to have been quite prevalent in early Maryland. Councillors would allege their illness as an excuse for not attending a meeting of the council. One councillor said that he was "sick and could not pass the river." A member of the lower house, or delegate, who could not attend a meeting of the assembly "either through sickness, or other urgent occasion," had the right "to constitute another his proxy in his room during such his absence." [1]

Judges of the provincial court would sometimes fail to appear at a session, "being indisposed in health." Witnesses and litigants would plead various ailments as their reason for not coming to a meeting of a court. One man said that he could not come "without great hazard and endangering of his health," another that he was "sick and not able to travel so far as to court." Other excuses alleged by some were "a great fit of sickness," lameness, or that their family was sick. One woman said that she was unable to administer an estate because she was so sick and weak that she could not rise "when set or scarce lift her hands to her mouth to feed herself." [2]

One colonist, an indentured servant, took his illness philosophically. When troubled "with a throng of unruly distempers," he consoled himself with the thought that "body diseases do but mellow a man for heaven, and so ferments him in this world, as he shall need no long concoction in the grave, but hasten to the Resurrection." [3]

From the instances given and many others can be found in the early records, it is evident that sickness of one kind or another

232

was prevalent in Maryland. What was the cause and nature of these ailments? During the winter of 1663, a meeting of the provincial court was postponed " by reason of the present distemper now reigning in the country." Unfortunately, there is no indication given as to the nature of this " distemper." In one of the annual letters of the Jesuit priests, mention is made of " the common sickness prevailing in the colony," although nothing is said as to the character of this sickness. At one time " ague " was prevalent on Kent Island.[4]

A number of colonists seem to have suffered from ague and fever at various times. One man had " a tertian quartan ague," another man who had been swimming in a river came out of the water " sick of an ague and vomited, and of the said sickness died." A Jesuit priest contracted a fever and was sick with chills. Ague is a fever of malarial character attended by paroxysms. Each paroxysm is marked by chill, fever and sweating.[5]

Father Andrew White, head of the Jesuit mission in Maryland, in a letter to Lord Baltimore gave an explanation for the sickness then prevalent in the colony. He said:

This year (1638) indeed hath proved sick and epidemical and hath taken away sixteen of our colony rather by disorder of eating flesh and drinking hot waters and wine by advice of our chirurgeon, rather than by any great malice of their fevers, for they who kept our diet and abstinence generally recovered.

Really, my Lord, I take the cause of the sickness to be the overgoodness of the land which maketh viands too substantial that if due regulation be not used the time of summer when the heat of stomachs is commonly weakest, either they lie undigested and so breed agues, or are thoroughly digested and so breed great quantities of blood and vital spirits, which, taking fire either from the heat of the season, our buildings being far unfit for such a climate, or from some violent exercise, begets fevers troublesome enough where we want physic, yet not dangerous at all if people will be ruled in their diet, which is hard for the vulgar, unless we had a hospital here to care for them, and keep them to rule perforce, which some worthy persons of this place do think upon.[6]

There were a number of cases of flux and scurvy in early Maryland. Flux is a flowing or fluid discharge from the bowels

16

or other part, especially an excessive discharge, as the bloody flux, or dysentery. Scurvy is a disease characterized by livid spots, especially about the thighs and legs, and by spongy gums, and bleeding from almost all the mucous membranes. It is accompanied by general debility caused by the lack of nutritious food, especially vegetable food. It is prevalent among those confined to a limited range of food.[7]

James Wilson, a servant of Captain Thomas Bradnox, died of " an intermitting fever with the dropsy or scurvy." Dropsy is an abnormal accumulation of watery fluid in the body often due to stagnation of the circulation. Thomas Watson, another servant of Captain Bradnox, was thought to have scurvy. William Hemsley, who was probably a chirurgeon by trade, went to see Watson if he could cure him. Hemsley decided that it was not scurvy but dropsy and that the servant had such a bad case that he was incurable. On the other hand, John Dabb said that when Watson came to his house he seemed to be much " swelled with the scurvy." Dabb said that he and another man stripped Watson naked " to view him." They found his " members . . . broken and the skin gone off them." Before dressing Watson, Dabb said that he held a candle while his friend anointed the members.[8]

Abscesses, ulcers, and cankers are mentioned in the early records. One servant suffered from "a most rotten, filthy, stinking, ulcerated leg," and his hair seemed to be rotted with ashes. In one instance, a negro's hand was sore and one of his fingers so " mortified " that it was thought necessary to cut it off " to save the hand and arm from gangrene." Jacob Young, Maryland's most experienced interpreter of aboriginal tongues, was, on one occasion, so sick and weak, " being all over his body broken out with boils and sores," that he was in no condition to travel " without apparent danger of his life." [9]

Apoplectic fits with " rattling in the throat and drabbling at the mouth " were known in early Maryland. One man appears to have died of a heart attack.[10]

Servants newly arrived in the colony were expected to be sick before they were acclimated. One servant, who had a sore leg, was so lame that he was unable to work during the first six weeks of his employment. Of another servant it was said that although he was strong enough to carry four bushels of salt, yet when he took cold he was " troubled with the stone " for an hour or so at a time.[11]

There are several instances of insanity in early Maryland. The person would be spoken of as " non compos mentis." A man named Francis Armstrong said that during " his extremity of sickness " he was so troubled with a violent fever that his condition bordered on " lunacy." During this time, he said, he was not responsible for what he did, as he was always " insensible in such distempers." [12]

Jeremiah Hasling before his death was so ill that he was entirely deprived of his reason and senses. It was said that several designing persons influenced him to make a will by the terms of which his daughter was deprived of her rightful share of her father's estate. A man named Henry Grace was such " an infirmed man " that he was considered incapable of making any agreement. Any bargain or contract which he entered into was of no effect, unless it was made in open court or before two justices of the peace.[13]

William Wennam was not permitted to testify in one case when he was found by the court to be " extraordinarly simple." It appears that when Wennam was asked by a magistrate how many months there were between March and April his answer was three.[14]

Venereal diseases were known in early Maryland. It was sometimes referred to as " the country (or cuntry) duties," or the " pox." Joseph Wickes, a commissioner of Kent county, brought a suit against Richard Owens on the ground that he had failed to deliver him a maidservant that was " sound and in perfect health." Wickes claimed that the girl he actually received, whose name was Anne Gould, had the French pox, or syphilis, sometimes called " the cuntry disease." The Kent county justice maintained that Owens should be compelled to live up to his agreement by delivering to him a maidservant who was in good health.[15]

Witnesses were called in this case and from their testimony we learn the following facts. Not long after Anne Gould came to live with Captain Wickes she complained that she was very sick. Apparently she was so sore all over her body that she could not " well turn herself," or lift her arms or even walk. Her condition was described by those who saw her as " lothsome, stinking and perishing . . . . that smells and scents continually came from her." Being in this condition made it dangerous for the health of the Wickes family, as it appears that Anne had charge of " the ordering and dressing of all their provisions and washing their linen," and was thus in a position to spread her disease.

Although the poor girl finally died, efforts were made to cure her. Thomas Ward, a chirurgeon, and Captain William Fuller were consulted. Ward " let her blood " and gave her physic. Captain Fuller gave her some medicine for internal use and also salves to dress the sores with which her body was covered. When Fuller was asked whether Anne Gould did not have " the cuntry disease," he replied: " You may call it the Cuntry disease, or French disease, or what you will, but it is no better nor worse than the pox." Chirurgeon Ward was of a similar opinion, saying after he had examined her body that " the wench had got the biggest pox that could be got for the money."

Anne Gould blamed Richard Owens for her condition, saying that he had " undone her . . . and given that disease unto her . . . and that Owens did make use of her body, after a very inhumane manner, and kept her down upon her face, that she could no ways help herself." It appears that Owens also beat and abused his maidservant because she had " a peremptory tongue."

After hearing all the evidence, the judges of the provincial court were of the opinion that as Owens had failed to deliver to Wickes a maidservant that was " sound and in perfect health," he must deliver another woman servant to the justice of Kent county.[16]

The findings of the coroner's inquests which follow throw light on the knowledge people of the seventeenth century had about human anatomy. A chirurgeon often acted as the foreman of a coroner's jury. George Binx, for example, a chirurgeon, was made foreman of a coroner's jury to investigate the shooting of a young Indian. As foreman of the jury he declared that the youth was killed by a bullet which entered " the epigastrium near the navel on the right side, obliquely descending and piercing the guts, glancing on the last vertebra of the back, was lodged in the side of the ano." [17]

When Richard Morton, who was visiting the province was shot by a Maryland colonist, Stephen Clifton, another chirurgeon, was requested to go to see Morton and report on his condition. Clifton's report follows:

I found him wounded in his left arm with small burshot, so that from the elbow to the upper part of the os humeris there were eight orifices. The greatest orifice was upon the musculus part, near unto the musculus biceps, where a quantity of the shot had entered, making a large orifice to the head of the os humeris, with several cavities missing the bone, and penetrating into the center of the body. Likewise two other shots were placed, the one just above the bastard ribs, penetrating into the lungs, the other between the third and fourth ribs, into the body likewise, by means whereof his pulse was weak. His body, as he explained, was extremely cold: he talked very idly and was vexed with shortness of breath and spitting of blood. Thereby is gathered that certain and speedy death is at hand, which followed on the 17th of the said instant. In witness whereof, I have hereunto set my hand,

Stephen Clifton, Chirurgeon.[18]

On another occasion, Stephen Clifton, together with John Brooke, another chirurgeon, were ordered by a jury of inquest " to open two suspicious places " that had been discovered upon the corpse of Thomas Simmons, a servant, whose master was suspected of having beaten him to death. It appears that Simmons had been sick for about a year before his death. Clifton and Brooke did as they had been requested and reported that after the cutis and cuticula (cuticle) had been laid bare they found no contusion upon " the musculous (muscular) part, or fleshy panicle." Then, probing further, the chirurgeons said that after they had cut open the body they had found it clear of inward bruises, either upon the diaphragm, or within the ribs, and, they added:

The lungs were of a livid bluish colour full of putrid ulcers, the liver not much putrid, although it seemed to be disaffected by reason of its pale wan colour: the purse of the heart was putrid and rotten, by which we gather that this person by course of nature could not have lived long, putrefaction being got so near unto that noble part, the heart, even at the door.

> In witness whereof we have hereunto set our hands,
> this 13th day of July, 1664.
>
> Stephen Clifton,
> John Brooke.[19]

Thomas Goddard was a chirurgeon who lived in Talbot county. When a servant by the name of Samuel Youngman was found dead, Goddard was made foreman of a coroner's inquest to investigate the cause of Youngman's death. Many thought that Youngman had been murdered by his master, Francis Carpender. The finding of the coroner's jury follows:

> We of the jury having viewed the corpse of Samuel Youngman, found a depression in the cranium in one place, and another wound where all the muscle's flesh was corrupted, and withal finding corrupt blood between the dura (i. e. fibrous membrane) and pia water (pia mater?) and the brain, and several bruises in the head and body. Therefore our verdict is that the want of looking after the abovesaid wounds was the cause of his death.

> Thomas Goddard, Chirurgeon, Foreman.[20]

We come now to a consideration of the practice of chirurgery in early Maryland. It is interesting to note that by the terms of an act passed in 1640 the county courts were given power " to moderate the bills, wages, and rates of artificers, labourers and chirurgeons." This grouping together of these trades or professions is evidence of the low plane occupied by the practice of chirurgery. Indeed, the trade of barber and chirurgeon was often combined and we have as in the case of a colonist by the name of John Robinson, a " barber-chirurgeon." Robinson was also a carpenter.[21]

If any one was sick in seventeenth-century Maryland, the chances were that he would either be given a physic, or cathartic, to purge him, or he would have his blood let. Sometimes both treatments would be given the same man. George Binx, who is spoken of as " a gentleman, licentiate in physick," was constantly suing patients for the physic which he had administered. He brought an action against one man for " labour and physic extraordinary " given to his servants. Peter Sharpe, another chirurgeon, was accused of killing his patient " by taking too much blood from him." [22]

In one instance when an Indian was sick one of the Jesuit priests gave him a powder of " known efficacy mixed with holy water, and took care the day after, by the assistance of a boy, whom he had with him, to open one of his veins for bloodletting." After this treatment, it was claimed, that the savage recovered. It is interesting to note that the colonists themselves hoped to learn from the Indians the medicinal value of certain plants. It was said that—

The savages have a root which is an excellent preservative against poison, called by the English, snake root. Other herbs and roots they have, wherewith they cure all manner of wounds; also sassafras, gums, and balsam. An Indian seeing one of the English much troubled with the tooth-ache, fetched of the root of a tree, and gave the party some of it to hold in his mouth, and it eased the pain presently.[23]

The early colonists thought that Maryland " affords naturally, many excellent things for physic and surgery," and that there would be no lack of " those things that can be made useful to apothecaries . . . " Robert Wintour, member of the governor's council, said that there was in the province " sassafras, coprose, and clabore and a world of other commodities for medicine, druggists and dyers." [24]

Various drugs were used in the colony. John Wade, a chirurgeon, sought to recover at law for the medicines which he had administered to Thomas Medwell during his illness. The drugs included " a vomitive (emetic) potion, also the usual physic, and " phlebotomy (blood-letting) with diaphoretics and sudorific cordials and corroboratives for his stomach." Diaphoretic and sudorific cordials were medicines used to cause sweating. Medwell was also given five " dormitive (sleeping) cordials." Other treatments tried on Medwell including " breathing a vein," or blood-letting, and " astringent means," that is, drugs causing contractions and arresting discharges. Needless to say Medwell did not survive.[25]

Mention is made in early records of salves, Angelica root and mithridate. The latter was a medicine, compounded of several ingredients, which was supposed to act as a remedy, or preservative, against poison. It was so called from King Mithridates VI, its reputed inventor. Opium was in use as a drug during the early days of the colony. Pills which smelt very strong were given a woman to kill the child in her womb. They probably contained some kind of poison. Taking a physic called "savin" was the drug another woman quick with child used to get rid of her baby. Savin, a drug made from the oil of evergreen shrubs, was a poisonous stimulant used to produce abortions.[26]

Jacob Lumbrozo, the Jew chirurgeon, was " clapped up in prison " when the charge was made against him that he had committed an abortion on Elizabeth Wild, a maidservant in his employ. To support this accusation, which was made by James Lindsey, the latter offered in evidence the testimony of several witnesses. From their testimony we learn that the Jew doctor had intercourse with Elizabeth without her consent. He would throw the girl on a bed and when she tried to cry out, Lumbrozo would cover her mouth with his hand or a hankerchief. In this way he had " the use of her body." It was not long before Elizabeth was pregnant. In order to bring about an abortion, or miscarriage, the Jew chirurgeon gave his servant girl some strong physic to purge her. The treatment was successful as Elizabeth soon afterwards passed " a clod of blood as big as a fist." When Lumbrozo saw this in a chamber pot, he told the girl that her " body was clear."

When this case came before the commissioners of Charles county, the members of the grand jury were instructed to determine whether from the evidence there was ground for a presentment or indictment. After they had taken the matter under their consideration, the foreman of the jury made the following statement:

We find by her own public confession that she, Elizabeth, was with child by Jacob Lumbrozo and that he did give her physic to destroy it. For these reasons we do present them.

As the offense of which Elizabeth and the doctor were accused was a felony, the case was referred to the provincial court for trial. No statement of the outcome of this case can be found in the printed records of this court. It does appear from the county records, however, that Lumbrozo later married Elizabeth. Possibly his marriage to this girl, the principal witness against him, disqualified her from testifying. This may be the reason that we hear nothing further of the case.[27]

The fact that the Jew chirurgeon had performed an abortion apparently did not affect his professional standing, as later we find him attending Mary Gordian in order to help her recover " the perfect use of her limbs." Lumbrozo's treatment in Mary's case seems to have been a frequent administration of " physic." [28]

There is only one case in Maryland of where a chirurgeon was stopped from the further practice of his profession. This was when Edward Husbands was suspected of attempting to kill the governor and some of the members of the assembly by putting poison in a duck pie which they ate. Husbands was told that if thereafter he attempted to practice chirurgery in the colony, he would, " for every such offence," pay a fine of two hundred pounds sterling. Some time later Husbands fled from the province.[29]

The litigation in which a chirurgeon by the name of John Meekes was engaged throws light on the treatment which a sick patient received during the sevententh century. Meekes entered a suit against the estate of Humphrey Haggate for nearly two thousand pounds of tobacco " for physic, time and pains for his wife and himself," and at the same time submitted an itemized account of the drugs, etc., which he had administered to the Haggates during their illnesses.

It appears that Mrs. Anne Haggate was taken sick during the winter of 1662. Chirurgeon Meekes first had notice of her illness when he received the following letter from her husband, Humphrey:

Mr. Meekes I would desire you out of all love to take my mare from the messenger and come to my house for my wife is desperate ill with a pain in her thighs which doth remove into her knees and from there into the small of her back. She hath not taken any rest this week, but is like one distracted. She hath not had her courses since she weaned her child which maketh me think that may be some cause of her pain. Hoping you will come, yours to use in any other service whilst I am,

December 6th, 1662.   Humphrey Haggate.

Meekes at once answered this urgent request. During the time he treated Mrs. Haggate while she was sick and lame, these are " the means " which he used to effect a cure:

First day . . . . One dose of purging pills.
Second day . . . . . . One dose more of the like pills.
Third day . . . . One potion of mixture.
Fourth day . . . . . . Blood let in the foot.
Fifth day . . . . One dose of purging pills.
Sixth day . . . . One large plaster for the pain in her hip.
                           One parcel of ointment to embriate for her disease.

Meekes said that before he had administered these medicines, he had, at the request of the Haggates, sent them the following drugs:

Euecroticem cum duplix slipticon paracilue emplaister adherna and diapalma and ointment. Also three doses of troches of mir.

As to the meaning of some of these medicines. " Emplaister " means plaster, " diapalma " is a dessicant plaster formerly used in drying up sores, and " troches of mir " are, no doubt, tablets or lozenges of myrrh. " Adherna " may mean adhesive. But what is the meaning of " embriate," and of " euecroticem cum duplix slipticon paracilue?" Perhaps some twentieth-century physician can solve the question. Despite their terrifying sound, the application of these and the other remedies proved effective. Mrs. Haggate actually recovered. Possibly she got well in spite of her treatment.

Not long after his wife had recovered, Humphrey Haggate himself was taken sick. Again an urgent message was dispatched to Chirurgeon Meekes which read:

Mr. Meekes, I would desire you to come to me with all speed as

possibly you can, for I am very dangerously sick of a violent vomiting
and a looseness. Bring means along with you. I pray fail me not of
your coming speedily, or else it will be too late. I rest your loving
friend,

Humphrey Haggate.

P. S. I was taken with this looseness a
Friday last, and I have a stool every
quarter of an hour.

When a few days later Meekes arrived at Haggate's house, he
diagnosed the case as one of " a violent bloody flux (dysentery)." He proceeded to treat his patient as follows:

First day........ Two restringent potions. One suppository at
                  night to cause rest.
Second day...... One cordial potion.
                 Two restringent boluses.
                 One cordial bolus more.
                 One cordial potion at night.
                 One suppository at night as before to cause rest.
Third day....... One restringent clysters.
                 Two cordial boluses.
                 One suppository at night used as before to cause
                 rest.
                 One cordial given him in the night.
                 One parcel of ointment for his hips.
Fourth day...... One restringent potion in the morning.
                 Two potions of restringent means left with him.
                 One parcel of cardamoms and some left to be
                 used in his drink.

The " restringent " potions unquestionably had an astringent
effect on the bowels, " suppositories " were some fusible prepa-
ration for introduction into the rectum, while " clysters " were a
liquid injected into the rectum like an enema. " Boluses " were
medicated pills and " cardamoms " were an aromatic capsular
fruit used as an aid to the other stimulants for the stomach.
Humphrey Haggate did not survive this treatment. Anne Haggate, as administratrix of her husband's estate, was compelled
by the court to pay Meekes' full claim for drugs and services.[30]
That John Meekes, who hailed from London, was not above

reproach is shown by the next suit in which he was involved. Before coming to Maryland, a young man by the name of John Helme had agreed to serve Meekes " in the way of chirurgery," or, in other words, as an apprentice in that trade or profession, if it can be dignified with the latter name. Although Meekes promised to provide the youth with " meat, drink, apparel and lodging," he did not live up to this agreement. When he asked his master for some clothes, Helme said that Meekes not only refused to give him any, but told him that he might go " whither he would and be damned." Appearing in person one winter's day before the commissioners of the Charles county court, Helme called attention to his shirt, which was the only one he possessed, and to his clothes which were very " bare and thin for the time of the year." The apprentice begged the justices that—

Your worships will be pleased to take his sad condition into your serious consideration, being in a strange country and destitute of friends, that you would be pleased to order Mr. John Meekes to find and allow your petitioner clothing sufficient, or to set me free in court, whereby your petitioner may provide for himself, for he is quite naked, and the time of year too far spent for to get employment, and your petitioner shall in duty bound pray, etc.

Young Helme did not plead in vain. The county commissioners at once ordered Meekes to clothe the youth " from top to toe fit for an apprentice." [31]

Other chirurgeons were not as successful as John Meekes in recovering for their services. Adam Stavely had a dispute with Peter Sharpe, a chirurgeon, claiming that the latter had failed to cure the lameness in one of his legs which had been caused by the limb of a tree falling on it. While the cure was being effected, Sharpe went away for about ten days. During this time Stavely had another chirurgeon come to see him. Upon his examination of the wound, this man said that the sore was much " putrified " from neglect. Stavely was finally compelled to make " a tedious and chargeable journey to Manhattan for his cure."

There seems to have been a difference of opinion as to the best way to heal the wound. Sharpe prescribed " baths and oils," while another man thought that guaiacum was the best remedy, and that the remedies which Sharpe used would not only deprive the patient of his leg, but also of his life. Guaiacum was used for gout, rheumatism and skin diseases. In this case the judges of the provincial court decided that Sharpe had failed to earn the one thousand pounds of tobacco which he was to have received for healing the wound.[32]

Peter Godson was another chirurgeon who was not successful in all the litigation in which he was involved. He must have been very ignorant not being able to sign his own name. Thomas Iger asked to have returned to him the six hundred pounds of tobacco which he had paid Godson for a cure on the ground that the chirurgeon " hath left him worse than he found him." Iger won his case.[33]

Godson was, however, successful in another suit when he recovered from Bartholomew Herring nearly six hundred pounds of tobacco " for physic and surgery impended on his wife." It appears that when Godson arrived at Herring's house his wife was in a " violent fever." Apparently, she had been assaulted by one Philip Hyde, who had made a complete job of it. First, he had beaten her with a log of wood taken from the garden fence, then he had kicked her, and rubbed her face in some oyster shells that were in the garden walk. When he had finished with Mrs. Herring, her head and face, her sides, and even her " private places " were all bruised.[34]

Robert Ellyson, a barber-chirurgeon, complained against Nicholas Hervey, a planter, for his failure to pay him twelve hundred pounds of tobacco for the cure of his servant, Henry Spim. Ellyson said that he did " follow the cure for divers months and brought it to a good state and was ready to perfect it, till he was hindered and put off by Hervey." The latter, however, maintained that he did not hinder Ellyson, but that the chirurgeon neglected the treatment of his servant " to the endangering of the man's life." Indeed Hervey claimed that he was compelled to employ Henry Hooper, another chirurgeon, to complete the cure of his servant. The jury, before whom this case was tried, awarded Ellyson eight hundred pounds of tobacco.[35]

Encouraged, no doubt, by the success of his first suit, Ellyson, during the following year, brought another action against Hervey for physic given his wife when ill. Ellyson sued no less a person than Sir Edmund Plowden, Knight, for " chirurgery and physic " administered to two of his maidservants. As Plowden was in Virginia at that time, his right to these servants was attached until he should put up security to answer Ellyson's suit. The barber-chirurgeon also sued one of these servants in person " for account of physic." [36]

Ellyson was involved in litigation with Giles Brent. The latter claimed that one of his guns had been given to the chirurgeon by a Mr. Wyatt. Ellyson denied this. The evidence disclosed that Wyatt gave the gun to Ellyson when he knew the chirurgeon was drunk. For this reason the court decided that Wyatt " ought to impute to himself what happened from the fault of Ellyson in the negligent keeping of the gun during his indisposition." Although Ellyson was not held responsible for the gun, he was, however, fined one hundred pounds of tobacco for allowing himself to become non compos mentis through drink.[37]

There seem to have been a number of lay practitioners of chirurgery in early Maryland. Thomas Hebden was one of them. He sued Colonel Francis Trafford for having treated his servants with physic. The fact that he did not cure one patient of ague and fever did not prevent Hebden from bringing a suit for his services against the administrator of the deceased's estate. Hebden asked twenty-five pounds of tobacco for a purge and the same amount for " stopping of his blood." The lay practitioner brought another suit against Francis Ottoway, who was a chirurgeon, for not importing certain medicines " this shipping." But as the agreement failed to specify which "shipping" Ottoway was not held liable.[38]

Quite often a colonist of prominence would take care of a sick person. Thus we find Colonel William Stevens asking the provincial court for five hundred pounds of tobacco out of the estate of James Allen for the trouble and expense he had been put to in taking care of Allen during his last illness. Henry Morgan, a commissioner of the Kenty county court, made a similar demand when he asked his fellow commissioners to reimburse him " for the time of eight weeks he harboured in his house, cherished, and kept with meat, drink, and attendance in the time of the sickness of Valerus Leo." Morgan said that he also had to pay Leo's funeral expenses. The justice recovered six hundred pounds of tobacco.[39]

There are cases of where a servant would agree to serve his master for a certain time with the understanding that he would be cured of his disease. Thomas Watson, in a humble petition to the Talbot county justices, alleged that he had agreed to serve John Edmundson for a period of two years, provided his master would cure his sore leg. Before the two years were up, however, Edmundson " assigned " Watson to Richard Holland who thereupon undertook to carry out Edmundson's promise about curing the servant's leg. Watson said that although he had been in the service of Holland for over ten months, his leg was " never the better, nor never a cure." In view of these facts, Watson asked the county commissioners that Holland should either be required " to take some speedy course " for the cure of his leg, or else to set him free " for he worketh in great misery."

When this case came up for trial, Holland admitted that he had promised to cure Watson's leg, just as Edmundson, his former master, had done. Holland added that his own wife had even tried to do what she could to effect a cure. The Talbot magistrates then called Richard Tilghman, a prominent chirurgeon, to give his opinion of the condition of the servant's leg. When Tilghman said that he thought the leg was "very bad and required speedy help," the county justices informed Holland that he must either cure the leg at once or set Watson free. Holland replied that he would set the servant free. This was done.[40]

So much for the treatment afforded the sick. But suppose some one did not have the means to employ a chirurgeon, or had no one who would take care of him while sick. By the provisions of a law passed in 1650, all " maimed, lame or blind persons," in St. Mary's county, who were unable " to get their living by working or otherwise," were maintained out of the proceeds of a county levy. Frances Lucus, of this county, who was a widow, asked the governor and council that an allowance be made her out of the levy for the maintenance of her crippled child. Mrs. Lucus said:

> That your petitioner's husband died about three years since, and that since his decease your petitioner hath been compelled to satisfy his debts which amounted to more than the effects of his estate.
>
> That your petitioner hath two children, one whereof a girl about eleven years of age is a cripple and goes on her hands and knees, and your petitioner being a very poor woman and nothing to maintain herself and children withall doth humbly request some allowance out of the public for the maintenance of the said helpless and crippled child.
>
> And as in duty bound she shall pray, etc.

After considering this petition, the governor and council referred it to the commissioners of St. Mary's county who were requested " to consider the necessity and indigency of the petitioner so as to award her a charitable and competent maintenance out of the county levy." [41]

In a similar case, the justices of St. Mary's county were required to provide " a being, livelihood and maintenance " for Martha Crab, a lame woman, and charge it to their county levy.[42]

The same practice was observed in Charles county. This is shown by the case of Alexander Howell who asked for a maintenance being " disabled by divers diseases to provide for himself." The governor and council instructed the commissioners of Charles county to provide for Howell. In Somerset county, John Waerum, described as " a decrepid lad," was assigned by the magistrates of that county to Robert Hart in order " to save the county harmless from any further charge." In consideration of Hart's curing the youth, the latter was to serve him for three years " in such service as Hart shall employ him." [43]

In a petition to the commissioners of Charles county, Richard Watson said that although he was blind and unable to work or to look after what little he had, yet his name was still entered on the list of taxable persons of that county. Watson asked the justices to have his name erased from that list " to the end he may not with his child be forced to come upon the charge of the county." Watson's request was granted. In a similar case, in Talbot county, because James Smith was " lame and decrepid " he was exempted from all tax levies in that county.[44]

Chirurgeons, or men with some knowledge of chirurgery, accompanied military expeditions. Francis Stockett went with a company of men sent to assist the then friendly Susquehannock Indians. John Lemaire accompanied a troop of rangers sent against some marauding savages. Chirurgeons would take plasters and medicines with them on a military expedition. All chirurgeons on duty with companies or troops were paid out of the public levy " for all such chirurgery means they shall expend in the country's service." One chirurgeon received four hundred pounds of tobacco for each month he was on duty with his company. This was as much as a lieutenant received.[45]

Chirurgeons, or those familiar with chirurgery, took care of soldiers wounded in war with the Indians. Oliver Sprye received two thousand pounds of tobacco for his " charge of diet and curing of wounded men." Sprye also petitioned the governor and council " touching the charge of wounded men." John Wallcott was given seventeen hundred pounds of tobacco " for attending and dressing thirty-two prisoners." Bandages and trusses made of sheets and table cloths were used to dress wounds and sores.[46]

Henry Hooper sued Mistress Margaret Brent, then acting as the proprietor's attorney, " for his salary and chirurgery in the fort of St. Inigoes." While Margaret Brent admitted that Hooper's claim was just as regards " the surgery and physick to the soldiers during the time of the garrison," she refused to recognize Hooper's demand for his salary. The jury seems to have agreed with her. Hooper entered into an agreement with the governor of Maryland by the terms of which he was to serve his lordship for a period of twelve months " in the quality of a chirurgeon." During this time, the governor was to furnish Hooper with all the drugs he needed, with food and lodging, and to allow the chirurgeon to retain two-thirds of all he earned by his general practice in the colony during the twelve-month period.[47]

17

Richard Tilghman was another chirurgeon who attended wounded soldiers. On one occasion, he received one thousand pounds of tobacco from the commissioners of Talbot county for his attendance " in the war time." One man, who had been wounded during a campaign against the Indians, claimed that he had been treated unfairly by Tilghman. This man, whose name was Nicholas Brooke, said that although Tilghman had been paid by the proprietary authorities for curing his wound, the chirurgeon tried to make him pay for the same treatment. Brooke asked that Tilghman's suit to collect for his services should be dismissed.[48]

Dr. Tilghman himself brought a suit claiming to have healed one of George Hayes' arms which had been " lacerated and torn by an accident." As Hayes had no funds of his own, a man named William Smith promised to pay Tilghman for the cure. Although Smith, when summoned to the county court, maintained that the chirurgeon had not yet healed the arm, he was, nevertheless, willing to pay for " a full cure " of the arm. He did this, said Smith, because he did not wish Hayes, who was very poor, to become an object of charity and " burthensome " to the county. Smith added that he was also willing to pay for Hayes' maintenance. The county commissioners naturally took advantage of Smith's generous offers.[49]

Richard Tilghman lived on the Chester river, on the Eastern Shore of Maryland. He lies buried on his plantation known as the Hermitage. The tombstone bears the following inscription:

Vale
Ita Dixit
Richardus Tilghmanus B. M.
In artique chirurgi Magister qui
sub hoc tumulo sepultus est
Obiit, Janu. 7 mo. Anno 1675.[50]

In view of a letter, written in behalf of wounded soldiers, which they had received, the Talbot county justices ordered two injured men, Daniel Jones and William Smith, " to remain where they now are until their wounds be cured." Later, when it appeared that Smith might be permanently disabled, the Tal-

bot court granted him fifteen hundred pounds of tobacco. At a
still later date, the county commissioners allowed Howell Powell, a Quaker, twelve hundred pounds of tobacco for " the cure
and charges" of William Smith.[51]

Many women in early Maryland acted as nurses, or healers
of the sick. One nurse received three hundred pounds of
tobacco a month, which, at the prevailing price of tobacco at
two pence a pound, was over two pounds sterling. Mary
Doughty, a daughter of the Reverend Francis Doughty, treated
a number of sick persons. Her first marriage was to a man by
the name of Adrian Vanderdonck whose death soon afterwards
left her a widow, but not for long, as she soon took unto herself
a second spouse, Hugh Oneale. Mary appears quite often in the
Charles county court records. We find her, when still Mrs.
Vanderdonck, suing no less a person than Josias Fendall, one
time governor of Maryland, for cures she claimed to have effected on three of his servants. One man had a bad leg,
another a canker in the mouth and the third a sore mouth.
Although Mary maintained she had cured all three servants, she
said that Fendall had not paid her for her services.

In order to prove her case, Mary Vanderdonck produced in
court a note signed by Fendall in which he promised her " such
satisfaction," as she thought fit, for the cure of one of his servants. Witnesses, too, were summoned who testified that Mary
had succeeded in her " cures " and how in one instance she
" had brought the sore to the bigness of a groat (fourpence),
or a sixpence." Fendall appealed this case to the provincial
court, but as it never came up for trial there, it must have been
settled out of court.[52]

Mary Vanderdonck also brought a suit against Christopher
Russell, alleging that when he was very sick she had administered physic to him. Russell, defending the suit, said that he
had never sent for Mary and that no one " hath command of his
purse but himself." To prove her case, Mary had an elderly
man by the name of William Smoot to testify. Smoot, on the
witness stand, declared that he "doth verily think she (Mary)
saved his (Russell's) life under God and further sayeth not."
But this case, like the one against Fendall, seems never to have
been finally settled.[53]

Mary, however, was not deterred by the unsuccessful outcome of these suits. She was now the wife of Hugh Oneale. We find him, in his wife's behalf, bringing an action against William Heard, as the administrator of the estate of Samuel Parker. Oneale sought to recover one thousand pounds of tobacco for his wife's services in taking care of Joan, the wife of Samuel Parker, during her illness. From the testimony of witnesses we learn the following facts. Although Parker appears to have been willing to give Mary a heifer " for her pains," there had never been any promise made to pay her a stated amount of tobacco. A woman witness, Anne Gess, described the treatment which Mrs. Parker had received when under the care of Mary Oneale. Said Anne:

Mary brought Joan Parker something in a pot and something in a paper, and that she gave Joan a portion of that upon the point of a knife out of the paper, and when she had given it to her, it did her little good for the present. Towards night Joan cried out and said this woman hath given me something to mischief me for I will never take any more of it, for she thought it had poisoned her.

Anne Gess also recounted how Mary Oneale had given Joan Parker a clyster (enema), but that it "did not work with her and none that she gave her did her any good at all."

William Heard now came forward and said that as the administrator of Joan Parker's estate he would not pay the one thousand pounds of tobacco demanded by Mary Oneale, but that he would agree to pay four hundred pounds of tobacco for the physic which Mary had administered to Mrs. Parker during her illness. As for the claim of one thousand pounds of tobacco against the estate of Samuel Parker, Heard said that he could not be held liable for this as he had never been appointed administrator of his estate but only of his wife's estate. After the county commissioners had heard this statement, they decided to dismiss Mary's claim for one thousand pounds of tobacco.[54]

But Mary Oneale, as we have seen, was not the kind of person to be easily discouraged. She brought another suit against William Heard, this time an action of defamation. Mary sought to hold Heard personally responsible for the slanderous remarks which he had made about her, or, as she alleged in her petition to the Charles county court commissioners:

> Whereas William Heard hath utterly defamed your petitioner in saying that he would prove that the widow Parker did say upon her death bed that your petitioner had poisoned her and that he could bring your petitioner upon her twelve godfathers, wherefor your petitioner humbly craveth redress according to law in such cases provided, and your petitioner shall as in duty bound ever pray, etc.

Mrs. Oneale now introduced several witness " in confirmation of her petition," all of whom testified that they had been present when Heard had made remarks about Mrs. Oneale poisoning Joan Parker and that Heard had also said that he would bring Mary before her twelve godfathers, which was another way of saying before a jury of twelve men.

After the witnesses had finished testifying, William Heard informed the county justices that he had been at fault and that he had been " very indiscreet in speaking in so unreserved terms." He added:

> If her (Mrs. Oneale's) credit may be thereby any ways stained he doth humbly desire her and her husband to forgive him, he being contented to pay the cost and charge of the suit.

After Heard had made this admission, Hugh Oneale declared that he and his wife were satisfied with his apology and that they would release him forever " from all trouble and molestation " for his slanderous statements.[55]

Encouraged by the successful outcome of this suit, Hugh Oneale and his wife, Mary, decided to sue William Heard again for the medical services which Mary had rendered Joan Parker during her illness. This time the Oneales only asked for six hundred pounds of tobacco. It will be recalled that they had already recovered four hundred pounds of their original demand of one thousand pounds of tobacco. To this claim of Hugh and Mary Oneale, Heard pleaded " the act for the payment of debts by which there appears no cause of action, it being an account upon a dead man's estate." The county justices sustained this plea and nonsuited the Oneales.[56]

The Hugh Oneales were the defendants in one case. William Marshall, a county commissioner, sued them claiming that Mary had bought of him a maidservant, cattle, corn, etc., for which she had agreed to pay nearly fifteen hundred pounds of tobacco. At the trial of this case in order to offset this claim of Marshall, the Oneales offered a counter-claim by which, they maintained, the justice was indebted to them for nearly thirteen hundred pounds of tobacco for " physic " which Mrs. Oneale had administered to Marshall when sick. In order to support their claim, the Oneales offered in evidence the following statement:

To a purge	30	lbs. tob.
To a dose for sweating	40	" "
To two portions for the fever	100	" "
To a cordial	100	" "
To one thousand pounds of tobacco which Marshall agreed for his cure when he was sick	1000	" "
	1270	" "

In this case none of the litigants were able to prove the correctness of the full amount of their respective claims. After the accounts between them had been balanced, it was found that the Oneales owed Marshall about three hundred pounds of tobacco. The county court ordered them to pay this amount.[57]

Other women besides Mary Oneale took care of sick people. Mary, the wife of Thomas Bradnox, a commissioner of Kent county, made two claims, one against William Cox, and the other against his wife. Mary Bradnox alleged that when she attempted to cure the injured hand of William Cox, he had promised to give her the cow calf, which, incidentally, had caused the injury, as a reward for her services. Mary further claimed that Frances, the wife of William Cox, had promised to give her a yearling heifer for the trouble she took in curing her child's mouth and attending her during her sickness.[58]

Isabella Barnes sued John Winchester for her care and attendance on his wife during her last illness. Many witnesses were called, all of whom declared that Mrs. Winchester had received very good care and attention. They were all of the opinion that Mrs. Winchester lacked nothing during her sickness. People were in constant attendance upon her. Some attended her during the early part of the night, others watched over her " till almost cock's crowing." In the way of food and drink Mrs. Winchester apparently received all the colony had to offer. She was given sack, drams and beer and her food consisted of sugar, spice, preuons (prunes?), poultry stewed with butter and currants, and wild duck. Mrs. Winchester, however, was a difficult patient to handle. When near death, she got up, threw off all her clothes, and attempted to go to her husband, then sleeping on a bed before the fire. Although from the testimony of witnesses, Mrs. Barnes seems to have shown that she took very good care of Mrs. Winchester, there is no record of her having recovered for her services.[59]

Mary Gillford, a widow, in a petition to the judges of the provincial court asked them to recall how she had been asked by them the past April to take care of a sick boy in her house. And, continued Mrs. Gillford, the lad had " tarried there until the beginning of July following, and then went from me, blessed be God, in health." During the time the youth was with her, Mrs. Gillford said that she took good care of him so that he should want for nothing "that was requisite and fitting." The bed-ridden boy had " rotted two blankets and a bolster." Taking into consideration the time the youth had been at her house and her care in looking after him, Mrs. Gillford asked the provincial court that she might be paid for her trouble. The judges directed that Mrs. Gillford's petition should be referred to the commissioners of Calvert county, who should see that she had " satisfaction." [60]

There are several instances of where a man and his wife were lay practitioners. Each would offer their services in taking care of the sick. Oliver Sprye claimed that his wife had cured another man's wife of some " distemper." Sprye, it will be recalled, took care of wounded soldiers. The wife of John Cherman undertook to heal the sore leg of a servant, while he undertook to cure sick people. Thomas Hebden has been mentioned as a lay practitioner. His wife, Katherine, can also be so classified. To one of her patients she administered physic and to another she " did chirurgery " on his leg.[61]

Some women, of course, acted as midwives. After Mrs. Symon Overzee had died in childbirth, a suit was entered against her husband by Daniel Clocker, acting in behalf of his wife, Mary, for full satisfaction for his wife's " bringing Mrs. Overzee to bed, her lying in, and his wife's charges and pains in tending Mrs. Overzee's child." Clocker asked five hundred pounds of tobacco for his wife's services, as they had been rendered " in the busiest time of her dairy." He only recovered about half this amount.[62]

In Talbot county we find William Smith asking over two hundred pounds of tobacco of Thomas Vaughan for the twelve days during which his wife had attended Mrs. Vaughan " in child bed." [63]

It is appropriate to consider burials in connection with the practice of chirurgery, as the chances were that the chirurgeon's treatment of his patient would result in his death and burial. Unless a colonist happened to live at St. Mary's, he was usually buried on the plantation which had been his home. One minister deplored this fact and hoped that more churches and church-yards might be consecrated " to the end that Christians might be decently buried together, whereas now they bury in the several plantations where they lived." With plantations scattered far and wide, it was some time before churches were erected in the different counties. Indeed there would have been little justification for building them. With the difficulties of communication and travel such as they were, it would have been very hard for those living any distance away to reach the church.[64]

The proprietary officials disapproved of excessive funeral expenses and in the settlement of an estate such expenses would not be allowed. What was a reasonable expense depended on " the proportion of the estate and the quality of the person." If, however, the expenses were reasonable they would, in the settlement of an estate, be " first defrayed." Included in the burial expenses of a colonist were digging the grave, the coffin or canvas bag, and a winding sheet for the body. The friends or relatives of the deceased must be provided with black mourning ribbons. A funeral sermon was usually preached and a funeral dinner given at which beer or liquor were served. As a tribute to the dead, muskets were generally fired as the body was lowered into the grave.[65]

A case which came up for trial in the Charles county court throws light in an amusing way on the customs observed at a colonist's funeral. It appears that James Lee had been made one of the executors, or overseers, of the estate of Mrs. Joseph Lenton, a widow. Upon her death, Lee made arrangements for the funeral. In accordance with the usual custom, Lee bought two pounds of gunpowder to be used in firing off muskets at her interment. Lee also bought three barrels of beer which the mourners were to consume at the wake usual with burials during the seventeenth century. In addition to this, the executor distributed " black ribboning " among the friends and relatives of the departed. Altogether Lee spent over twelve hundred pounds of tobacco for powder, beer, ribbon, etc., used at Mrs. Lenton's funeral. When Francis Pope, who was the administrator of Joseph Lenton's estate, refused to allow Lee this amount, the latter brought suit in the county court to recover it. This suit was dismissed, as Lee failed to show that Joseph Lenton, Mrs. Lenton's deceased husband, had made him one of the executors, or overseers, of her estate.

In the second case which he brought to recover for his expenses at Mrs. Lenton's funeral, Lee took pains to show that he had been appointed one of the executors. At the trial, Lee also had witnesses testify that the beer had been consumed at the funeral and the black ribbon distributed among the mourners.

George Thompson acted as attorney in behalf of Francis Pope, the administrator of Joseph Lenton's estate. Thompson asked that the jury should be given the following instructions in the form of " interrogations."

First, to inquire whether at an ordinary planters wife's funeral it be not ridiculous to shoot, as they do at a young soldier's death, or other commanders in war.

Second, to inquire whether at the time of a funeral it be Christian-like for some few neighbours to be gathered together, and, instead of showing a mournfulness for the loss of their friend and neighbour, to turn to their carousing cups to the quantity of three barrels of beer to the value of nine hundred pounds of tobacco.

Third, to inquire whether it be not most unreasonable that James Lee for this same merrie meeting should charge three hundred pounds of tobacco for boat and hands to fetch this same drink.

Fourth, to inquire whether it be not absurd that the said Lee should charge the administrator of the deceased with thirty-six yards of black ribboning at twelve pounds of tobacco per yard . . . when the whole world may imagine that it was but a dolorous countenance to disguise his rejoicing heart.

The attorney for Lee now came forward with his list of instructions for the jury which were, in brief, a defence of the amount of tobacco spent on the burial of Mrs. Lenton on the ground " that those that be dead the living must bury, and the estate of the deceased must defray the charge. All the law allows it." Also that there was no carousing at the funeral and that the " black ribboning " could not have been bought more cheaply, etc., etc.

The jury now retired to consider the facts of the case in connection with their instructions. They soon returned and their foreman, Thomas Lomax, said that their verdict was for Lee and that he was entitled to recover the amount of tobacco which he had paid for the funeral expenses of Mrs. Lenton. Among the reasons which they gave for their finding was that Lee as one of the executors, or overseers, was acting within his powers when he saw to it that Mrs. Lenton had " a Christian burial," and so long as " a sufficient estate " was left, the amount expended at the funeral " only redounded to the credit and memory of the person deceased."

When Thompson, the attorney for the administrator, heard this verdict, he at once requested the county justices to suspend their judgment and grant him an appeal to the provincial court. This request was granted, but soon afterwards, it appears, this case was settled, or, as the records reads, " Pope compounded the business." [66]

BIBLIOGRAPHICAL NOTES

FOR

CHAPTERS ONE TO TEN

CHAPTER I—BIBLIOGRAPHICAL NOTES

NOTE

In the following citations all references are to the volumes of the Archives of Maryland unless otherwise indicated. For example, in the following citation XLIX, xxi, 18, means volume forty-nine of the Archives, page twenty-one of the introduction and page eighteen of the text. The notation, CMS, is a reference to Captains and Mariners of Early Maryland, by the same author.

* * *

[1] XLIX, xxi, 18; III, 475 (Tilghman); XLIX, 389, 399 (Utie). For additional information regarding Utie, see index, CMS.

[2] XVII, 135, 139, 140. For other examples of attempts to uphold the dignity of the proprietor or governor, see CMS, pp. 123-135. In re Charles Calvert, see index CMS. For a brief account of the political history of Maryland during the seventeenth century, see Historical Background in CMS.

[3] X, 35, 36. In re Price and Gardiner, see index, CMS. Robert Stack and William Southerly were indicted by a grand jury for disturbing a minister during divine service. To support this indictment, it was alleged that on one occasion when the minister was conducting services both Stack and Southerly " maliciously and purposely " disturbed the congregation and that, on another occasion, Stack alone attempted to disturb and " disquiet " the congregation. Stack was ordered to find sureties for his good behaviour in the future. (XLI, 522, 523; XLIX, 244).

[4] XVII, 288, 289, 321 (Saxon); IV, 419 (Brent). Later, because of sickness Saxon was relieved of his duties as doorkeeper. For further information regarding Stevens and Brent, see index, CMS.

[5] II, 253, 254 (Lewis); II, 86 (Bretton).

[6] VII, 104, 105 (Husbands); II, 55-57 (Erbery).

[7] XV, 16; III, 547. For information regarding Cecil Calvert, see index, CMS.

[8] I, 463; LIV, 139, 146. There was a good reason for Carline's attitude. Although he had served as a commissioner of the Kent county court when the province was under the Commonwealth commissioners, he had not, however, been retained as a member of this court after the Restoration. (LIV, 24, 25, 56, 141). It appears that members of the upper house, or councillors, resented it when members of the lower house, or delegates, appeared before them with their hats on. (XIII, 50, 81).

[9] III, 352, 362. Thomas Thurston was one Quaker who found himself in difficulty with the provincial authorities. Like others of his faith Thurston was opposed to taking an oath, even the oath of fidelity to the proprietary government. For this and other reasons Thurston was ordered to leave the colony. He was told that if he was found in the province seven days after he had received notice of banishment, or if he ever returned to the colony again, he would be forthwith arrested and whipped with thirty lashes by a constable. This constable would then turn him over to the next constable who would inflict the same punishment, and so whipped from constable to constable " till he be conveyed out of the province." (III, 348-353, 362, 364). For what happened to Thurston when he did return to the colony, see XLI, 104, 105, 268, 269, 286, 287, 319, 322, 331, 339. For more in re oath taking by Quakers, see II, 355, 356, 492; VII, 184.

261

262   CRIME AND PUNISHMENT IN EARLY MARYLAND

[10] X, 202 (Bushnell) ; X, 202, 221, 558; XLI, 35 (Cornelius) ; X, 221 (Empson). Roger Scott was also fined for swearing in the court room (X, 558), as was Henry Hooper (XLI, 35).

[11] XVII, 279, 280, 283, 287, 288, 325, 326, 382 (Higges) ; XVII, 279, 280, 283, 325, 326, 413, 414, 437, 438; V, 464 (Wells).

[12] LIV, 416, 417 (Rawlings) ; LIV, 459 (Davis). Jonathan Sibery was at one time a justice of Talbot county.

[13] LIV, xvii, 24, 25, 102, 194, 197. Hynson had been a commissioner of the Kent county court during the Commonwealth period, but had not been retained as a member of this court after the Restoration.

A man by the name of Valentine Hues was charged with " divers scandalous words spoken against the commissioners of Kent county in a very unseemly manner." When Hues acknowledged in open court that he had been at fault and asked to be forgiven, promising " never to do the like," the Kent justices because of this humble submission and " out of their clemency did remit his offence." (LIV, 286).

In Charles county complaint was made by Captain John Jenkins, one of the justices of that court, that some one had called him " Captain Grindingstone," obviously intending to refer to his harsh decisions. (LIII, 49, 51).

[14] III, 271, 276, 277; LIV, 1, 4, 9, 13-16. Vaughan's attitude can be explained. Recently the colony had come under the control of commissioners sent out from England by the Commonwealth government established by Oliver Cromwell. Many of those who held office under the Maryland proprietor were compelled to resign, including Captain Vaughan who had held the rank of commander of Kent Island. Having always been a loyal supporter of Lord Baltimore, Vaughan naturally resented the change of administration. This explains his contemptuous attitude towards the men who had been placed in charge of Kent Island affairs.

[15] III, 197, 198; IV, 439, 440, 459.

[16] LIV, 2.

[17] LIV, 122. In 1661 Captain Vaughan was called before the provincial court " to answer unto a contempt against his Lordship's government." What constituted this " contempt " we are not informed. At any rate, since Vaughan said that he was " sorry for the same," the judges of the court decided that he should be forgiven. (XLI, 526).

It is quite evident that Captain Vaughan was the type of man that would tolerate no opposition. This very quality made him extremely useful to the Calverts, as was shown by the important part he played in reducing Kent Island, when under William Claiborne, to proprietary control, and, still later, his loyalty to Lord Baltimore was proved during the Ingle Rebellion. (See index CMS, and also LIV, xiii-xv, xix, xxiii, xxvi).

[18] LIV, 350. In re Wickes see index, CMS.

[19] LIV, 352.

[20] LIV, 316. The county justice was not always successful in defending his reputation. Matthew Read, a Kent county magistrate, sued John Spurdance in an action of defamation. When, however, Read failed " to prove his suit," Spurdance asked that Read should be nonsuited. This request the county court granted and made Read pay the cost of the suit. (LIV, 206).

In one case the wife of a Kent county justice was successfully sued in an action of defamation. This was when Hannah Jenkins brought action against Isabella, the wife of William Head, a county magistrate. The Kent county court compelled Mrs. Head to ask Hannah's forgiveness " in open court," and to pay all costs of the suit. (LIV, 251).

[21] I, 286, 350. See also same volume, pp. 158, 159 and III, 117.

[22] X, 136 (Mee); LIV, 167 (Bradnox).

[23] IV, 395, 396.

[24] XLI, 315, 316, 333. Peter Sharpe, a chirurgeon, was accused of refusing to assist the sheriff of Anne Arundel county " in the apprehending of a delinquent." For this " contempt " Sharpe was fined five hundred pounds of tobacco. (XLI, 316). Sharpe and John Gary were also suspected of having broken open some official mail. (XLI, 573, 574).

[25] IV, 401, 402, 434-437. William Elliott was accused of erasing the broad arrow " in contempt of law and order " which a sheriff had placed on the door of his tobacco house. The broad arrow was used by the sheriff to mark tobacco seized by him for fines or fees due to the Lord Proprietor. (XLIX, xxvi, xxvii, 85, 86). In re Margaret Brent, see index CMS.

[26] XLI, 602, 603. It does not seem likely that Innis escaped scot free for his assault on the sheriff.

[27] LIV, 170, 171.

[28] XLI, 553.

[29] III, 61, 96, 97, 117; I, 450, 451; VII, 39, 68-70; XLI, 87.

[30] XVII, 42, 57, 58, 60-63, 96. Inglish had previously been employed as a servant by Colonel Nathaniel Utie. In re Peirce, see XVII, 283, 284.

[31] XLIX, 477, 478 (Sprigge); XLIX, 137, 138 (Evans). By the terms of an early law if the sheriff neglected to require sufficient bail of a party arrested, or otherwise " consent to be the cause of his escape," then the sheriff was liable to pay and satisfy the court judgment himself. (I, 492, 493). A sheriff was entitled to receive twenty pounds of tobacco a day for " tending upon a prisoner." (I, 308, 360; II, 222, 223; VII, 255).

[32] I, 410, 411; XLI, 91.

[33] XLI, 418 (Rigby); LIV, 220 (Elliott). See also case of John Everett in CMS, p. 224.

[34] XLI, 333 (Hall); XLI, 334 (Bagby). In re Edward Lloyd, see index, CMS.

[35] VII, 456, 457, 464, 487, 497-499, 547.

[36] X, 342 (Gerrard); X, 491 (Peake). In re Cornwallis, see index, CMS.

[37] X, 106-108 (Brent); IV, 119 (Robinson). For more in re Margaret Brent and Robinson, see index, CMS.

When Mary Willkins was on trial she left a session of the Kent county court without the knowledge or permission of the justices. For doing this she was ordered brought before the next meeting of the court for contempt. (LIV, 264).

[38] LIV, 493; XLIX, 317, 386 (defending suit); LIV, 462, 494, 502, 505, 522, 548 (prosecuting suit); LIV, 513 (action of debt).

[39] LI, xii, xiv, xxii. For an account of the early judges and attorneys, see Proceedings of the Maryland Court of Appeals, 1695-1729. Edited by Carroll T. Bond, pp. xviii-xxiv. As Judge Bond points out " in some instances the distinction between attorneys in fact and the trained attorneys at law is not easily applied; some of those who acted as agents developed a business of serving as attorneys in the courts and overstepped and obscured the dividing line. But there were, and continued to be, attorneys in fact only, or proxies, who were not, and did not pretend to be attorneys at law, even if set down as ' attorneys ' without distinction." In regard to attorneys in the colonies during the seventeenth century, see also Studies in the History of American Law, by Richard B. Morris, pp. 41-45, 132, 133.

Members of the council thought that it was necessary for " the aged, and impotent not able to travel . . . and those absent in parts beyond the seas " to have attorneys to represent them, or else they could not in court " seek their right or defend themselves from wrong." (II, 175).

[40] XLI, 71; LIII, 490, 491 (deputation under handwriting) ; LIII, 119 (Charles county cases) ; XLI, 130, 131 (typical power of attorney). Usually the power of attorney was witnessed by two men. For other examples of powers of attorney, see XLI, 242, 247, 327; LIII, 158, 407-409; XLIX, 47, 74, 262, 479; LIV, 479, 480, 495, 496, 504, 514, 519, 520, 614, 615, 680, 700, 701. For an example of the revocation of such a power, see LIV, 627.

[41] LIII, 356, 307, 580 (aptment of other attorneys) ; XLI, 256 (atty. of non-resident) ; XLI, 233; LIII, 319; X, 19 (women as attys.) ; LIII, 443, 444 (servant as atty). If one held the office of justice of the peace, sheriff, or clerk, he could not, during the time he held any of these offices, act as an attorney in the county court. (II, 132, 322, 323). In one case a man was allowed to appear as attorney only in what concerned himself, unless the court gave him permission to plead " in other men's causes." (XLI, 10, 11). The Statute of 3d James, 7 ch. discusses the qualifications of attorneys in England.

[42] IV, 465, 327, 328 (Gess) ; LIII, 482, 483 (Fouke).

[43] II, 409-411, 467, 468. This was the Act of 1674. In 1676 Charles Calvert authorized the provincial secretary to demand of every attorney practicing before the provincial court a yearly fee of twelve hundred pounds of tobacco. Failure to pay this amount would result in the attorney not being able to practice before that court. (XV, 79).

From the records of Talbot county it would appear that an attorney received from fifty to sixty pounds of tobacco for each day he was in court in behalf of his client. (LIV, 553, 562). See also XLIX, 270.

That tobacco was worth about two pence a pound is shown by the following valuations of that commodity. On Kent Island, during William Claiborne's administration of that island, it was worth about three or four pence a pound. (High Court of Admiralty Proceedings, Examinations, Bundle 181, case 1038. In Public Record Office, London, England). In Maryland, tobacco was valued at different times at one pence per pound (II, 407, 408; V, 268; XIII, 172), one and a half pence (I, 231; II,·220, 299), two pence (I, 444; IV, 92; II, 388; XIII, 171), and three pence per pound. (IV, 74). From these estimates, it is fair to assume that the average price of tobacco, during the proprietary period, was about two pence per pound.

[44] XVII, 36, 37.

[45] I, 496, 497; II, 222, 397; XIII, 122 (orphan's court) ; VII, 70 (court regulations).

[46] LIV, 486, 524, 566; XV, 71.

[47] LIII, 511 (Charles co.) ; LIV, 652 (Somerset co.). For a discussion of the administration of justice in the colony, see Maryland as a Proprietary Province, by Newton D. Mereness, p. 228 et seq.

CHAPTER II—BIBLIOGRAPHICAL NOTES

[1] XLIX, 538-545; LI, xvi, xvii, 121-130; II, 370, 377. By giving an account of the proceedings in the trial of Pope Alvey for larceny, it is hoped that the reader may be able to visualize the way in which a criminal trial was conducted during the seventeenth century. The reason for taking this particular case was that the proceedings are given in greater detail than in most cases. It should be noted that Colonel William Evans, whose cow Alvey was accused of stealing, was a member of the court which tried him for this offense. While the death penalty for stealing and killing a cow seems very severe, Alvey did deserve the death penalty in the case where he beat his maidservant to death. See text, pp. 108-110 for this case.

In re benefit of clergy, see article by Carroll T. Bond in LI, xxiii. Benefit of clergy was not abolished in England until 1827. This privilege is also discussed in History of the English Law, by Pollock and Maitland, vol. I, pp. 424-440; History of the Criminal Law in England, by Stephen. See also Benefit of Clergy in the American Criminal Law, by Arthur L. Cross, in Proceedings of the Massachusetts Historical Society, vol. LXI, 154-181.

Benefit of clergy was claimed in other cases in Maryland. When Thomas Smith was convicted of piracy, he claimed this privilege. The court, however, refused to allow this claim on the ground that it could not be granted " in this crime, and if it might, yet now it was demanded too late after judgment." (I, 16, 17). In another case of theft, John Oliver, after his conviction, claimed benefit of clergy, but it was found that he could not read. He was, however, pardoned by the governor on condition that he serve as hangman during the rest of his life. (LI, 214).

The distinction between grand and petty larceny is now abolished in England by statute. In most of the United States, a difference, similar in theory, being made in the punishment, based upon the amount stolen, or the circumstances of the theft.

For information regarding Charles, Philip and William Calvert, and Truman, Brooke, Evans and John Jarboe, see index CMS. The same index can be consulted for information about Thomas Smith, who is referred to in this note.

[2] I, 83, 184, 186, 210. In case of a judgment obtained against a servant, who had no property, the court could compel him to do some corporal labour. (I, 188). No one could appeal from the judgment of a lower court, such as a county court, unless he put up sufficient security. If they lost on their appeal to the higher court, such as the provincial court, then they were liable to pay double damages " to the party grieved." (I, 184; II, 562, 563; VII, 71-73). Query, did the successful litigant also recover all expenses incurred while attending court, including his attorney's fees, clerk's and sheriff's fees, and the expenses of his own witnesses? It would appear that he did. See XLIX, 343, 387, 415, 551.

[3] I, 411, 412 (500 lbs. tob. fine). If the offending party did not have sufficient property to pay this fine, he or she could be imprisoned for two months " without bail or mainprise." (ibid.). Mainprise, now obselete, was a form of bail. It implied a laxer degree of responsibility on the part of the sureties than did bail. For cases in which jurors were fined, see XLIX, 319. For those in which wit-

nesses were summoned, see XLIX, 423, 188, 260; XLI, 194, 198, or fined for not appearing, see XLIX, 193, 231; LI, 35, 36.

LIII, 212, 604; LIV, 541, 551, 562, 597 (jury and witness pay). Witnesses also received thirty pounds of tobacco for each day it took them to go and come from their homes to the place of trial. (LIII, 558, 559, 569, 570). By the terms of a law passed in 1666, jurors in civil cases were only allowed ten pounds of tobacco for each case in which they served on the jury. (II, 137, 138). For cases in which juries refused to render a verdict before being paid, see LIII, 515, 543, 603; I, 521. Grand juries were summoned twice a year in the counties and at St. Mary's "to enquire of offences and misdemeanors." At first the jurors had to pay their own expenses, but later they were allowed as a body 2500 lbs. tobacco for their services. (II, 392, 462).

[4] IV, 393 (Goneere); IV, 445 (Howell). The laws passed regarding perjury were equally severe. By the terms of an act passed in 1642, any person giving false witness upon oath in court, or persuading, or hiring another to give such false witness, might lose his or her right hand, or be burned in the hand, or " to any other corporal shame, or correction, not extending to life, or be fined, as the court shall think fit." (I, 159, 193). Eight years later the penalty for perjury was changed. Thereafter any one guilty of this offence was to be nailed to the pillory and loose both ears, or corporally punished in some other way, provided it did not involve the taking of life. (I, 286, 287, 350).

By the terms of a statute enacted in 1642, any one counterfeiting " the hand or sign manual," or any of the seals of the Lord Proprietor, or wilfully falsifying, corrupting or embezzling a record, was punishable in the same way as those guilty of giving false witness under oath. (I, 159, 193). See also Act of 1649 (I, 247). One woman was accused of embezzling and making away with the estate of her deceased husband. Although the sheriff was ordered to bring the woman before the provincial court to account for the estate, there is no record of what punishment, if any, she received. (X, 559). In another case, no action appears to have been taken when a man embezzled and made away with the estate of an orphan. (LIV, 314).

[5] II, 483, 484, 490, 491, 496, 499. The doorkeeper of the assembly was empowered " to press horse and man " to summon Sheriff James to appear for his trial. (Ibid.).

[6] X, 506, 556.

[7] XLIX, 258 (whore); LIV, 167, 168 (thief and liar). Captain Thomas Bradnox, commissioner of Kent county, alleged that a maidservant of his had taken " a false oath in court." The captain, however, failed to prove his contention. (LIV, 225, 226). One witness was disqualified from further testifying for saying while under oath that he saw a man " coming from the rum cask with a bottle full of rum, but whether he came from the rum cask, or no, he knows not." (LIII, 416, 447, 521, 568).

[8] XLIX, 44, 45, 53, 76, 77, 86, 87. This Elizabeth Greene should not be confused with the woman of the same name, unmarried, who was convicted and executed for murdering her bastard child. See text, pp. 127, 128.

In Charles county one man demanded a warrant against another in an action of forgery, but nothing seems to have come of his request. (LIII, 154). Richard Royston was, however, convicted of forgery, but upon his petition asking for pardon, and promising to behave himself in the future, he was pardoned and was not corporally punished as had been decreed. (V, 481, 482).

[9] VII, 77. Thomas Smith, of Calvert county, was accused of this offence but later pardoned. (V, 562, 563).

[10] IV, 258; I, 153 (1643); I, 159 (fines); X, 291 (1653); X, 365 (Fox). In regard to Fenwick and Packer, see index CMS.

[11] I, 455, 456 (Hannah Lee); I, 490, II, 67, 415 (Act of 1663, stocks, etc.); XLIX, 491; II, 470 (gallows). At about this time, the governor and council, in order to facilitate the building of the new prison, ordered that carpenters, with other workmen " to attend them," should be " pressed " into service. (III, 460). One man was punished by being compelled to build a pair of stocks. (X, 424).

[12] II, 139, 140. Despite the provision made for confining debtors in this prison apparently none were confined. (II, 373).

[13] II, 370, 371, 377-379 (need of new prison); II, 383, 386, 404-407, 487, 507, 509, 511; XV, 39, 196; III, 556 (new prison, state house, etc.). The new state house and prison had not been standing many years when both buildings were greatly in need of repairs. (VII, 294, 295, 299, 300; XIII, 23, 45, 85, 187, 188, 194, 197, 200-203, 207-209, 223-225). A replica of the state house was built in 1934, but not on the same site, unfortunately.

[14] VII, 116, 117.

[15] See citations under 13 supra.

[16] X, 291; (Kent county); I, 490-492; XLIX, 16, 17 (stocks, branding irons, etc.).

[17] LIII, 432, 459, 523, 634. There were stocks in Talbot county by 1669. (LIV, 445). In the printed court records of early Maryland no case could be found of any one being punished on the ducking stool. Possibly some of the unprinted records contain such cases.

[18] II, 413, 414, 507; VII, 220, 234; XIII, 57.

[19] VII, 70. Michael Dalton, The Country Justice.

[20] II, 542, 414. In re Baltimore county, see II, 430; V, 473. There was some complaint of the way in which the court house in Anne Arundel county had been built. (VII, 186). There were prisons in some of the counties by 1671, including Charles (LI, 322, 323), Calvert (LI, 363, 364), Talbot (LI, 346, 348), and Cecil (LI, 180). There is mention of a court house in Calvert county in 1684 (XIII, 83), and, in 1680, there was a court house and prison in Kent county. (XV, 350, 351).

[21] II, 542. What did it cost to execute a criminal? The trial and execution of one man cost over 7500 lbs. tobacco, or allowing two pence a pound for tobacco, this was over £62. Among those who exhibited their accounts " for satisfaction " were the sheriff and the clerk of the court. They asked to be reimbursed for " the imprisonment and the necessary and usual fees concerning the trial and execution " of the prisoner. The sheriff received 5000 lbs. tobacco for his services, the clerk over 600 lbs. tobacco. The innkeeper, who had furnished food and lodging to the grand and petty jurors during the trial, recovered nearly 800 lbs. of tobacco. The man who had acted as the prisoner's guard received 400 lbs. tobacco. Curiously enough, the smallest sum received was by the prosecuting attorney who recovered only 350 lbs. tobacco " for his pains and trouble." As the prisoner had escaped after his arrest and fled to Virginia, the expenses incurred in his capture and return to Maryland amounted to nearly 500 lbs. tobacco. All these items added up, as has been stated, to over 7500 lbs. tobacco. The provincial court allowed this amount to be recovered from the deceased's estate. This was in 1657 before the Act of 1662 was passed. (X, 547, 548, 557).

[22] II, 224-226. This was the act passed in 1669. In re Herrman, see index CMS.

[23] II, 542; XVII, 66 (irons); XLIX, 566 (woman's petition); XVII, 415 (man's petition). For cases of false imprisonment in early Maryland, see X, 423, 424; XLI, 369, 370, 424; LIII, 307-309.

[24] I, 164; II, 222, 223 (sheriff's fees); I, 445, 446 (Act of 1662). The Act of 1662 was reenacted in 1674 with a slight change in the wording. (II, 394, 395). As early as 1649, the sheriff of Kent county sought to recover from a man the cost of his imprisonment. (LIV, 3). Two cases occurred in St. Mary's county where the sheriff sought to recover the cost of imprisoning offenders. (XLI, 371, 368, 383, 398). This was in 1659. Another similar case occurred in 1662. (XLIX, 315, 439, 509). Except in the case of Indians, all others must pay the cost of their imprisonment. (XIII, 201; XVII, 79). For executing a negro and two Indians, a sheriff was paid out of the public account. (II, 31, 94, 95. See also LIII, 617). Some thought it "against common right and decency" to compel a sheriff to act as a hangman. (I, 175).

[25] I, 512, 546; II, 395. For cases which arose under Act of 1664, see XVII, 423, 424; LI, 214, 215. For one which arose before it was passed, see X, 292.

[26] XVII, 59 (Inglish); I, 184 (gentleman). It is interesting to note that Inglish himself had once been a servant of Colonel Nathaniel Utie. (XVII, 96).

[27] Drunkards were sometimes set in stocks. See text, p. 148. Nailing by the ears to a pillory has been discussed in this chapter. For the negro hanging see XLIX, 491.

[28] LIV, 103, 51, 407.

[29] X, 487, 488, 495, 519.

[30] V, 498, 499. For an account of other unusual punishments, see The Law of Adultery and Ignominious Punishments, by Andrew McF. Davis, in the Proceedings of the American Antiquarian Society, at the Semi-Annual Meeting, April 24, 1895, and also by the same author, Certain Additional Notes Touching upon the Subject of Ignominious Punishments, etc., in Proceedings of the American Antiquarian Society, n.s. XIII, pp. 67-71.

CHAPTER III—BIBLIOGRAPHICAL NOTES

[1] IV, 321 (Calvert); IV, 86-88 (Wintour). In re both these men, see index, CMS.

[2] IV, 99 (Adams); IV, 46 (Mottershead). In re Adams, Mottershead, and Leonard Calvert, see index CMS. For the inventories of other burgesses, see Richard Loe (IV, 74), Richard Lusthead (IV, 94), Justinian Snow (IV, 79-83), and Thomas Allen (IV, 405, 406). For scattered references to articles of gold or silver, such as rings, cups, buttons, etc., see IV, 86, 388; X, 315, 316; XLI, 80; II, 518.

[3] LI, 150, 151.

[4] LIII, 391, 392.

[5] A Relation of Maryland, in Narratives of Early Maryland. Edited by Clayton C. Hall, p. 93 (servant's clothing); I, 97 ("custom of the country"). In addition, the servant was allowed a few farming implements, three barrels of corn, and fifty acres of land. (Ibid.).

[6] IV, 51 (Lee). For other inventories of early colonists, see IV, 30, 43, 47, 48, 74-100, 103-107, 328. In re Lewger, see index CMS.

[7] X, 301, 302.

[8] The following references taken from the early records go to show that 25 by 20 feet was the average size of the 17th century dwelling. In Charles county mention is made of a house of these dimensions. (LIII, 26, 27). In the same county there is a reference to a house 20 ft. square. (LIII, 127). In Kent county a dwelling was built which was 30 ft. long and 18 ft. wide (LIV, 80), and in Somerset county a house of exactly the same dimensions was constructed. (LIV, 670).

In the provincial court records mention is made of a house ten feet in length (X, 197), another fifteen feet long (X, 476), another twenty feet long (X, 363, 364), and of one 20 ft. long and 15 ft. wide which had "a welch chimney." (XLI, 281, 282). One house was 30 ft. long (X, 405, 454), and the largest of which a record could be found was fifty feet square. (LI, 72). A dwelling house 25 ft. square "with a shed and an outside chimney" cost 1200 lbs. tobacco. This was about ten pounds sterling at the prevailing price of tobacco. (LI, 71).

In re Cornwallis' house, see Letter of Cornwallis, in Calvert Papers, Maryland Fund Publication, no. XXVIII, p. 174; X, 253, 254. Advice to colonists about locks and glass. (A Relation of Maryland, in Hall, p. 98. Also see Leah and Rachel, by Hammond, in Hall, pp. 284, 297, 298). There are but few references to glass for windows, lead, solder, etc. (IV, 110, 111; XLI, 510). Some houses were made of logs. (LIII, 232, 356, 357). In re Cornwallis, see index, CMS.

[9] In some instances two chimneys were built (LIV, 80; XLI, 526, 527), and in one case the chimneys were in the middle of the house. (IV, 189). Brickmaking in Maryland. (X, 213, 214, 267, 268). Cornelius Canneday was a brickmaker by trade. He lived in a house 25 ft. long by 16 ft. wide. The furnishings were not elaborate, consisting of a bed filled with feathers, one pillow and rug, two dishes, six spoons and one pot. (Ibid.). Brickmasons, (I, 3; IV, 110, 111) and brickmoulds (IV, 328). Advice to colonists. (A Rel. of Md. in Hall, pp. 98, 99); loam in Maryland (ibid. p. 81). A Jesuit priest suggested

269

to Lord Baltimore that he should have "a brikeman" in Maryland from whom all the settlers would have to buy bricks. (Letter of Father Andrew White, in Calvert Papers, Md. Fund Publication no. XXVIII, p. 207).

[10] XLI, 367.

[11] XLI, 526, 527. In re Neale, see index, CMS.

[12] V, 265, 266. As tobacco became important in the economic life of the colony, many tobacco houses were constructed. As a general rule, these were larger in size than the dwelling houses. They varied in size from 30 to 50 feet square. (LIV, 54, 382, 443, 444; LIII, 230; LI, 71, 72). In some instances the tobacco house was longer in length than in width. One planter had two tobacco houses, one 100 ft. long, the other 90 ft. long and both 32 ft. wide. (X, 278). By the terms of a law passed in 1676, every tobacco planter was obliged to build "a good tight house with a good door lock and key" for the storage of his crop. Failure to do so subjected the planter to a fine of 500 lbs. tobacco. (II, 519).

[13] Captain Thomas Cornwallis boasted that his house was furnished with "plate, linen hangings, bedding, brass, pewter and all manner of household stuff worth at least a thousand pounds." (X, 253). This was, of course, exceptional. Those interested in household, kitchen and tableware owned by the average colonists should consult III, 178; IV, 30, 43, 46, 48, 74-100, 103-107, 320, 321, 406, 526; X, 180, 226, 253; XLI, 60, 65, 100, 103, 114, 115, 124, 135, 136; XLIX, 285, 363-366, 422, 436, 581, 582; LIII, 398, 502; LIV, 80, 81, 92, 93, 97, 101, 107, 226, 227, 571, 572, 670. See also A Relation of Maryland, in Hall, pp. 94-96.

A complete set of farming implements and carpentry tools could be bought in England for a little over one pound sterling. Those interested in these articles should consult citations given above. Also see Hall, pp. 94, 95. Firearms were expensive, a good musket costing about a pound.

[14] II, 134, 135, 219, 220, 321, 322; VII, 246.

[15] LIV, 272, 273, 299, 323. In re bridge, see LIV, 51.

[16] LIV, 453, 481, 540, 544, 553, 648, 661, 678.

[17] II, 349, 369, 408, 409 (Wicomico road); XVII, 423 (St. Mary's roads).

[18] Dankers' and Sluyter's Journal, in Memoirs of the Long Island Historical Society, vol. I, pp. 211, 212.

In proprietary Maryland letters, that is, official ones were sent from house to house "till they be safely delivered as directed." Every householder receiving a letter was, within a half an hour of its receipt, compelled to relay it on to the next plantation. Failure to do this made the offending householder liable to pay a fine of one hundred pounds of tobacco. (I, 415, 416). Two prominent colonists were accused of not sending letters down to the governor and for "contemptuously nailing up a letter of the sheriff's directed to the governor." (III, 441, 442).

In the case of some emergency special messengers could be "pressed," or drafted, into service for the purpose of carrying dispatches. (III, 524). Towards the end of the proprietary period, a man known as a public post was appointed whose duty it was to carry letters. (VII, 584; XIII, 89, 130). For an example of the way letters were carried in early Maryland, see XLI, 573, 574. Letters were sometimes forwarded from one constable to another. (Journal of the Dutch Embassy to Maryland, by Augustine Herrman, in Hall's Narrs. p. 323).

[19] I, 534, 535; LIII, 7, 560. In 1658 an act was also passed providing that all counties, except Kent county, maintain one ferry. Apparently this law was

never put into effect. (I, 375, 376). For character of boats used on Chesapeake Bay, see CMS, chapter III.

²⁰ V, 118, 119; XV, 54-56, 93, 94, 288, 289. In re ferry over the Patuxent river, see also II, 315; VII, 38; XIII, 201; X, 470. For ferry over Severn river, see XVII, 123, 124. For reference to ferry on Eastern Shore, see LIV, 424, 445, 472, 481, 544. This ferry was run by Jonathan Hopkinson who also kept an inn or ordinary. The boat left from a landing near his house where he lodged travellers. It is not clear from the records on which river this ferry was maintained.

There was also a ferry maintained by Thomas Banks in Calvert county. (VII, 38, 168, 169; XVII, 375, 376). For other references to ferries, see X, 410; II, 318; LIII, 443.

²¹ II, 145, 146, 402, 403. See also I, 160, 194 and II, 225, 226 in re pass from province. II, 298-300 (pass from county); I, 451, 452 (right to examine strangers).

²² The Sizes of Plantations in Seventeenth-Century Maryland, by V. J. Wyckoff, in Md. Hist. Soc. Mag. vol. XXXII, ft. note, p. 339. See also V, 266.

²³ For the way in which trading was conducted in Maryland, see CMS, pp. 50, 51. For the penalties imposed on those harbouring runaway servants, or buying from any servant, see text, pp. 51, 111, 112.

²⁴ I, 35, 36 (Richardson); LIV, 297 (Guinn).

²⁵ X, 291, 292. Smith, who was a hired servant and not indentured, was also told that in addition to the corporal punishment which he received, he must pay all the expenses incurred by his imprisonment and punishment. If the wages which he received from his master were not sufficient to cover this, then he must do additional work. (Ibid.).

²⁶ LI, 214, 215. This is the only case found in the early records where the man could not read when he asked benefit of clergy. For a discussion of this privilege, see text pp. 27, 28.

There are other cases in early Maryland of where a servant was accused of stealing. See, for example, the following cases: Jane Palldin (XLI, 382, 432, 433), Frances Stockdell (XLI, 450, 451), Sarah Taylor (LIV, 213). All these were cases of maidservants. See also case of William Wake (LIV, 420).

²⁷ V, 498, 499.

²⁸ I, 500, 501; II, 526, 527; LIV, 511. In order that all might have notice of this act, the secretary of the province was required to give a copy of it to each master of a vessel which arrived in the colony. This notice must then be nailed to the mainmast of the ship so that all seamen on board might read it. (Ibid.). For another case where this statute was enforced, see XLIX, 495.

²⁹ Ibid.

³⁰ LIII, 28 (Wilson). For the Ellis case, see XLIX, 485, 504, 505, 508. In this case as the shirt was valued at three shillings, Ellis should have been charged with grand and not petty larceny. In the Wilson case nothing is said about the value of the articles stolen. At this time the theft of anything worth over twelve pence constituted grand larceny. The distinction between grand and petty larceny is now abolished in England by statute. See also Chapter II, bibliographical notes, citation i. In re Alvey, see text, pp. 21-29.

³¹ XLI, 207-213, 221-223, 255, 258, 259. One wonders why the prisoners did not claim benefit of clergy. In regard to the amnesty see text, p. 167. For other cases of theft in early Maryland, see X, 439 (Morley and Canneday), X, 477, 478 (Mrs. Turner), X, 478 (Tennis), LIV, 48 (Wickes), LIV, 71 (Salter),

LIV, 508-510 (Sullavant), XLIX, 45 (Macklin). See also XVII, 280, 281, 384, 385.

During most of the proprietary period, tobacco was the medium of exchange. It varied in value, sometimes worth a penny, sometimes two pence a pound, never more than three pence. It would hardly have been worth while to steal much of this commodity. If a thief had secured thirty pounds of tobacco he would only receive a few shillings for it if he succeeded in disposing of it.

[32] XLI, 245, 246, 289. The phrase now used is " No bill," " No true bill," " Not a true bill," or " Not found," although in some jurisdictions " Ignored " is still used. For another case of theftbote, see case of Dr. Lumbrozo against whom this charge was made (LIII, 609, 616).

[33] XLI, 457, 458 (Gibbons); XVII, 36 (Withrington). For two other cases in which pardons were granted in cases of theft, see LI, 213 (Price) and LI, 214 (Jones). In re peak and roanoke, see CMS, pp. 65-67.

[34] I, 158, 192, 193, 344; X, 424.

[35] VII, 201-203. As none of the county or provincial court records for 1681 and subsequent years have been printed, it is impossible to give examples of the application of the Act of 1681.

[36] XLIX, 230-235. In 1644 Edward Ward was accused of having committed a burglary " on the house of John Halfhead." Apparently the case never came up for trial. (IV, 281). Robbery and burglary were in 1642 made capital offences and punishable by death, loss of member, burning in the hand, etc., etc. (I, 158, 192, 193). No case of robbery was found in the early records, although in 1684-1685 the province was stirred up by the robberies committed by Roger Mackeele and his associates or confederates. As they acted more like pirates than robbers, a consideration of their depredations does not properly belong in this chapter. (XVII, 349-352, 372, 373, 389, 390). In re benefit of clergy, see text, pp. 27, 28. For types of firearms used in early Maryland, including matchlocks, etc., see CMS, chapter X.

[37] XLIX, 538-543. For cases of where Indians were accused of stealing from colonists and vice versa, see CMS, pp. 364-369.

[38] XLIX, 375, 376, 379, 380, 396, 498-503. In 1643 John Nevill complained against William Edwin and his wife for a forcible entry into his house. Nothing came of this case. (IV, 253).

CHAPTER IV—BIBLIOGRAPHICAL NOTES

[1] III, 142, 143 (colonial secretary). Giles Brent received two steers yearly out of his Lordship's livestock for his own " expense in regard of his care " of the proprietary cattle. (III, 142). Cecil Calvert, the Lord Proprietor, was at all times anxious about " the well ordering and disposal " of his livestock. He wanted a list made of the number of his cattle and to have them " marked duely with my mark." He also did not want them sold except when necessary and even then he wished that it might be done with the least prejudice to the increase and good of his stock. (III, 140-143). In order to protect the cattle of the Lord Proprietor, an act was passed forbidding any one to sell or buy them without his Lordship's permission. All who offended in this respect might be imprisoned. (I, 303, 304). In 1647 Leonard Calvert, governor of Maryland and brother of Cecil Calvert, took steps to recover the livestock belonging to his brother on Kent Island. (IV, 309).

I, 227 (soldier wages). For other cases where soldiers were paid for their services with cattle, see IV, 449; X, 6. Livestock is often written as two words.

[2] XLI, 54 (Cawsine); IV, 519 (Hebden); X, 392, 393 (Marshall).

[3] LIV, 110, 111, 125, 137 (schooling); IV, 310 (advance portion). For other examples of advance portions, see IV, 341, 374, 398; X, 14; XLIX, 34. IV, 317, 318, 338, 386, 401 (cattle by will). For other examples of where livestock was bequeathed by will, see LIV, 123, 208, 209, 669, 670; LIII, 208, 209.

[4] X, 447, 448; XLI, 30 (gifts); IV, 336, 337, 354, 355, 373 (services rendered); XLI, 97 (building house). In another case, a man claimed that he had failed to receive a cow for work which he had done. During the trial, it appeared that when the animal was offered to the man who had done the work he refused to take it as he did not wish " to carry the cow over upon the ice." (XLI, 317, 318). XLIX, 454, 496, 497 (Gerrard); LIII, 96 (cowkeeper). In another case, a cowkeeper for his services received a pair of shoes and some corn. (XLI, 253). William Claiborne rewarded a faithful servant with a cow. (IV, 208).

Cattle was given as security to pay a debt and would be forfeited if the debt was not paid by a stated time. (IV, 116, 178, 190, 242, 491, 516; X, 32, 168, 198, 264, 320, 497; LIV, 27, 62, 63).

[5] IV, 251 (beaver skins); IV, 272 (salt); IV, 272, 371 (corn); IV, 272, 371 (swine); X, 36; LIV, 226 (land); X, 486 (canoe); XLI, 83 (gun); X, 461 (boy servants); IV, 120, 121 (manservant); IV, 284, 371; X, 120 (tobacco).

From the early records we can obtain some idea of the value of cattle. In one case a bull was valued at 500 lbs. tobacco (LIV, 97), while in another instance a cow was sold for 700 lbs. tobacco. (X, 190). For other cases where cattle was valued, see LIV, 62; XLIX, 174, 205.

[6] IV, 132 (proprietor); X, 206, 223 (Cornwallis); IV, 152 (Fenwick); XLI, 240, 241, 161 (other cases). One planter was awarded 800 lbs. tobacco damages because his steer was killed by another colonist. (LIV, 473, 474). In another case, only 400 lbs. tobacco was allowed the owner of a castrated bull which had been killed. In awarding this amount the court said that the sum was " something mitigated in regard of the offensiveness of the beast," but that it was not intended to set a precedent that might encourage any one to kill another's cattle. (X, 117).

For other examples of litigation about cattle, see IV, 443, 444, 456, 457, 492-

495, 497, 549; X, 246, 340, 341, 433, 452, 484, 493, 552-554; XLI, 52, 53, 165, 251, 252, 416, 481, 482, 523, 585, 586; XLIX, 258, 260, 271-273, 444, 471, 480, 481, 549; LIII, 12, 34-37, 47, 54, 61, 254, 481, 482, 495, 496, 514, 515; LIV, 478, 558, 559, 577. In one case where the cow did not have enough milk to feed her calf, an attempt was made to feed the calf with mush, using a rag to force it down the throat. The calf, however, choked to death on the rag. (XLI, 71, 174, 175, 252, 253).

[7] III, 22 (King's request); A Relation of Maryland, in Hall, pp. 76, 77, 97, 98 (instructions to colonists); III, 30 (sentiments of Virginians); X, 253 (Cornwallis).

[8] I, 84, 97, 160, 161, 251, 252, 349, 350; III, 98 (corn planting). Steps were also taken to prohibit the export of corn. (I, 96, 161; III, 194). I, 96, 344, 413, 414, 446, 537 (fencing laws). The owner of swine was not compelled " to law " them, that is, he did not have to injure or maim their feet in order to prevent them running wild. (I, 96). I, 158, 192; II, 350, 398, 399 (fence burners). By the terms of the law which was passed in 1674, those guilty of willful fence burning were only liable for the damage caused to the fence, although the offender might also suffer such punishment as " the laws of England " provided against " such malicious persons and practices." (II, 398, 399).

[9] X, 48, 49, 209, 210 (wild cattle); I, 418, 419, 429, 486; II, 346, 347; XV, 155, 156; XVII, 241, 242; V, 568, 569; VIII, 36, 37 (rangers). At one time the lord proprietor claimed all wild unmarked cattle. (III, 295; X, 210, 324). It was for this reason that Mistress Mary Brent, sister of Margaret and brother of Giles Brent, became involved in difficulty. Mistress Mary maintained that all the cattle which had been killed were on the family manor land on Kent Island. (X, 149-152, 164, 186, 348, 349).

[10] I, 251, 295 (markings recorded). Some of the colonists had cowpens. (I, 419; III, 295). LIII, 65 (Fendall); LIII, 221 (Lewger); XLI, 160 (Philip Calvert); XLI, 463 (Wm. Calvert); IV, 509, 288; XLI, 61, 266, 268; LIII, 261 (other markings). Mistress Margaret Brent's marks were " the right ear cropped " (IV, 517), while Mistress Jane Cockshott had as her markings " cropped on the right ear and two slits in the crop and overkeeled on the left ear." (LIII, 3). In some cases a man's wife, or daughter, would record her markings which would be only slightly different from his. (LIV, 57, 208). In Somerset county from 1665 to 1722, over 650 people recorded their marks for livestock in the county court records. (LIV, 741-788).

[11] For sales of cattle, see X, 260, 261, 266, 301, 304, 318, 319, 351, 393, 394; XLI, 97; LIII, 57, 71, 72, 95, 120, 122 266, 346, 347, 477, 478, 525; LIV, 57, 58, 63, 66, 105, 106, 613, 614, 624, 631, 643, 644, 666-668, 675, 699. For gifts of cattle, see LIII, 78, 79, 90, 126, 155, 242, 295, 329, 330, 442, 443, 450, 451; LIV, 600, 617, 618, 628, 629, 645, 646, 650, 651, 653, 654, 676, 677; XLIX, 25, 26. Sometimes a planter would assign his livestock marks to another as when he disposed of all of his cattle. (LIV, 753, 766).

[12] IV, 480, 494; X, 119, 120, 131. Cows were known to swim from one island to another. (IV, 534).

[13] IV, 379-385, 416, 528; X, 3. Captain Thomas Cornwallis contended that one of his cows had been " mismarked." (X, 218, 219). Mrs. Fenwick brought a suit against Henry Pope because he laid claim to her mark and had killed a steer which belonged to her. (XLI, 70). XLI, 74, 75, 159, 160 (Gerrard v. Evans). Thomas Gerrard recorded a different mark for the cattle which he

owned at Bramley House and at Basford Manor. (XLI, 234). For other suits about livestock marking, or altering the marks on cattle and swine, see IV, 422-427, 429, 431, 433, 435, 447-449, 462, 486-489, 506-508, 528, 540, 541; X, 6, 7, 29, 167.

[14] IV, 308 (Harwood); LIV, 111 (Deare). There are several instances where sheep are spoken of in early Maryland. (XLIX, 363, 582; LIII, 482, 493, 494; LIV, 645, 646). In 1672 the export of sheep or lambs, "dead or alive," was forbidden. (V, 105, 106).

[15] IV, 142-149, 151, 163, 165, 182, 207.

[16] IV, 153 (Cornwallis); IV, 174, 176, 280 (Thorowgood and Hollis); X, 220, 239-245, 273-275, 329, 330 (Taylor and Brooke). For other cases instituted at about this same time for injury to swine, see IV, 216, 226, 387, 391, 392, 397, 412, 413, 441, 461, 468; X, 100, 101, 132, 167, 168, 232, 234. In re Brooke, Cornwallis, Fenwick, Hollis and Thorowgood, see index CMS. Regarding wolves in early Maryland, see same book, pp. 2-5.

[17] III, 255, 293; X, 232; LIV, 7 (revocation of licenses); I, 251, 444, 503, 504 (new law). This law, passed in April 1649, also contained a provision regarding the killing of swine in his Lordship's forests.

[18] XLI, 522, 523.

[19] XLIX, 477; I, 251. For other cases in the provincial court records regarding the unlawful killing of swine, see X, 478; XLI, 541, 552, 553; XLIX, 486.

[20] LIII, 544-548. Each of the four prisoners was compelled to pay over twelve pounds sterling apiece. These fines were imposed in accordance with the terms of the act regarding the unlawful killing of swine.

[21] LIII, 551-553.

[22] LIII, 205, 206, 237-239. William Robisson was involved in other cases where the charge of hog stealing was raised. (LIII, 220, 250, 251). For other cases of hog stealing in Charles county, see LIII, 84, 91-94.

[23] LIV, 42, 43, 49-51, 60. At another session of the Kent county court, Joseph Wickes, himself a member of the court, asked that the constable should be given a warrant to search for hogs "upon some suspicion of loss he hath received." (LIV, 141).

[24] LIV, 396 (Tilghman); LIV, 399 (Dell). In re Tilghman see text p. 250 and also CMS, pp. 240, 241. Dell was involved in another suit in which the jury found him guilty of hog stealing. (LIV, 404). For another case of hog stealing in Talbot county, see LIV, 582-587. For cases of hog stealing in Somerset county, see LIV, 678, 679, 712.

[25] LIII, 628; XLI, 480, 481.

[26] LIII, 628, 634, 636. In one case Thomas Gerrard sold land to another man with the understanding that the latter could take alive, or kill, any wild hogs upon the land, provided he gave him as lord of the manor half of all such swine. (XLIX, 586). For a discussion of manorial rights in early Maryland, see Old Maryland Manors, by John Johnson, in Johns Hopkins University Studies in Historical and Political Science, vol. I. For the provincial policy adopted towards Indians accused of stealing or killing swine, see CMS, pp. 420-425.

[27] I, 455 (Act of 1662); II, 140, 141, 277-279, 350 (Act of 1666).

[28] Ibid.

[29] Ibid. Under the Act of 1666 about the only swine which a man could kill with impunity were those on his own land which were unmarked and which were over three months old. (Ibid.).

[30] XIII, 184 (negro); LI, 203, 205, 206, 208-212 (chancery cases). In Kent

county in 1681 a man by the name of Ellis Humphryes appears to have caused the inhabitants of that county considerable consternation by his killing their hogs, disturbing their cattle, and threatening to burn their houses. (XV, 337-340, 344).

[31] I, 500, 501. As was the case with cattle there are records of the sale of swine. (LIV, 33, 631, 632). There are, however, fewer instances of such sales as swine were not as valuable as cattle. For other cases in the provincial and county courts in regard to swine, see X, 420, 432, 433; XLI, 20, 420; LIII, 369-371; LIV, 85, 86, 88, 89, 206, 207.

[32] XVII, 422, 423.

[33] IV, 314 (Calvert); XLI, 262, 263 (Fenwick); LIII, 504, 505 (Harris). For other examples of where horses are left under the provisions of a will, see LIV, 100, 669, 670; XLI, 582, 583.

[34] VIII, 6, 7, 24; XLIX, 518 (mill and carts); XV, 56 (expeditions against Susquehannocks). For use of horses by rangers, see CMS, chapter VIII. XV, 99, 277, 278, 377, 378, 380; II, 494; XIII, 165 (messengers); II, 484; XIII, 165, 226 (sheriff and doorkeeper); XV, 378 (supply at Mattapany); II, 469, 540; VII, 144; XIII, 103 (claims for horses); XIII, 196 (Col. Pye); LIII, 427 (chirurgeon); LIII, 601; LIV, 100, 101 (saddles, bridles, etc.). In re Pye, see index CMS.

[35] LIV, 594 (first horse race); LIV, 550, 551 (wager lost); XV, 392, 393 (race at Rumsey's plantation).

[36] IV, 543 (Brent); XLI, 226, 227; XLIX, 440, 441 (guarantee); LIV, 201, 220 (description); LIV, 201, 220, 623, 668, 669, 679, 727 (names, age); XLIX, 446; LIII, 448 (branding).

For other cases of the sale of horses in the provincial court records, see XLIX, 29, 252, 253. There are sales of horses recorded in Talbot county (LIV, 427, 513, 526, 527, 530, 531, 589), in Somerset county (LIV, 623, 668, 669, 679, 727), and in Kent county (LIV, 201, 220). Horses sold at from 800 to 3000 pounds of tobacco apiece, or from six to twenty-five pounds sterling. It does not seem to have made any difference whether the horse was a stallion or a mare. (X, 438, 468, 469, 471, 563; XLI, 226, 227; IV, 475; XLIX, 101, 102, 181, 182; LIII, 393, 448, 553-557).

For an example of a gift of a horse, see LIV, 594, 595. As security for a debt which he owed, one man promised to give his creditor several horses in case the debt was not paid. (XLI, 28, 29).

[37] XLI, 219 (Bigger); LIV, 469, 470, 488 (White).

[38] XLIX, 473 (unlawful riding); X, 339, 340 (Hammond).

[39] XLIX, 503-505, 507. Thomas Gerrard and Cuthbert Fenwick also were involved in litigation over a colt. (X, 238, 276).

[40] XLI, 575-578. After this decision, Fendall "having nothing to say," Kidson and his master were nonsuited.

[41] LIV, 452, 453 (Wickes); LIV, 367, 368 (Coursey). For other cases of litigation about horses, see IV, 300, 305, 312, 315, 347, 355; XLIX, 60, 61, 108, 158, 159, 172, 196-200, 436, 437, 480; XLI, 77, 141, 142, 167, 185, 217-220, 277-279; XVII, 112, 119, 120; I, 297 298.

[42] LIII, 634 (Tenison); LIII, 636 (Poulter); LIII, 637 (stray horse).

[43] III, 194; I, 218.

[44] II, 254, 255, 281, 333, 334.

[45] LIV, 597 (Robbins); LIV, 541, 542 (Sturdivant and Newnam).

[46] II, 358. At this time Captain Hugh Oneale was permitted to import horses

from New York provided there were not more than ninety-eight and that they were brought into the colony " in one gang." (II, 358, 383). At the same time a license was granted to William Thomas to import not more than six horses from Virginia, the reason for issuing this license being that Thomas' wife had been left these horses by her father. (II, 384).

⁴⁷ XV, 317 (Hawkins) ; XVII, 241, 242; V, 568, 569; VIII, 36, 37 (Darnall, Diggs and rangers). In 1683 Augustine Herrman complained that the people living on the Delaware river drove away the horses and livestock of the Maryland colonists. (XVII, 135, 136).

⁴⁸ VII, 275, 277, 292, 302-305, 338, 341. For an account of the wars with the Indians, see CMS, chapters XV-XXIV, XXVI.

⁴⁹ VII, 482-484, 555, 564, 568-570.

⁵⁰ A Relation of Maryland, in Hall, p. 98 (mastiff and spaniels) ; XLIX, 13 (attack on trespassers). One man by the terms of his will left another his bitch. (LIV, 197). There was a dog called Towser in early Maryland. (XLIX, 13).

⁵¹ XLIX, 351 (Armstrong) ; XLIX, 77 (Anketill).

⁵² LIII, 337, 338. In one case a woman obtained one hundred pounds of tobacco from a man who had killed her dog. (LIII, 347).

⁵³ X, 236 (hunting swine) ; LIV, 369, 370 (Wollman) ; LIII, 633, 636 (manor cases).

CHAPTER V—BIBLIOGRAPHICAL NOTES

[1] A Relation of Maryland, in Hall, pp. 95, 96, 98, 99. Robert Wintour, member of the governour's council, said that experience had shown that in most of the English plantations where servants worked in the tobacco fields his labour would " produce to his master, at the least, fifty pounds sterling of yearly clear profit into purse, most commonly far more." (Letter of Robert Wintour, unpublished manuscript in the collection of Dr. Hugh H. Young, of Baltimore, Md.).

[2] Ibid.

[3] I, 465; X, 253 (Cornwallis); III, 250, 251 (Mitchell); III, 256 (Brooke); IV, 85 (Wintour); X, 302 (gentleman's agreement); V, 144 (one to three servants). In re Brooke, Cornwallis and Wintour, see index, CMS.

Thomas Gerrard, councillor and lord of St. Clement's Manor, left his son, Justinian, by his will six men servants, together with other bequests of household goods, etc., as his " portion." (XLIX, 582, 571).

Joseph Wickes, a commissioner of Kent county, had eleven servants (LIV, 127, 330), Henry Morgan and Thomas Bradnox, commissioners of the same county, were also owners of servants. (LIV, 144, 228, 214).

[4] I, 463 (Cornwallis); I, 516, 517 (Barber). In re Act of 1676 prohibiting import of felons, etc., see II, 540, 541 and also XV, 136, 137. One of the annual Jesuit letters speaks of a man of noble birth who, having been reduced to poverty, had come to the colony as an indentured servant. (Annual Letters of the Jesuits, in Hall, p. 122). While not of noble birth, Cuthbert Fenwick is a case of where a man of a very good family came to the province as an indentured servant. In re Fenwick, see index CMS.

[5] Letter of Robert Wintour, unpublished manuscript. In re English and Irish servants, see V, 126, 129, 268.

[6] A Relation of Maryland, in Hall, p. 99; XLI, 385-388; XLIX, 82, 189; LIII, 401, 462, 463, 579, 580; LIV, 248, 543 (examples of indentures). LIV, 554, 555 (cordwainer); LIV, 503; LIII, 410 (learning); XLI, 19 (mortar). In his will Thomas Allen stated that he did not want his sons sold " for slaves or mortar boys." (IV, 404). By another indenture the servant agreed to work at joiner's work for four years. The master promised to allow the joiner each year a third part of what he should " by his labour gain." (XLIX, 103, 104). Another man entered into an indenture to saw plank for his master for one year. (XLIX, 106).

[7] IV, 173, 327, 328; XLI, 446, 447; LIII, 158, 180, 181, 461 (tobacco); IV, 210 (money); IV, 166; XLI, 468, 469 (cattle, corn, clothing, land); IV, 271 (governor). In consideration of his working for him, one master agreed to cure the servant of a disease with which he was afflicted and also to give the servant at the end of his employment tobacco, corn and clothing. (X, 496). Where, in one case, the servant was taken sick during the time of his employment, it was held that he could not recover his wages. (LIII, 312, 313). In another case, it was said that a master taking advantage of the " weak capacity " of a servant, had made the latter work for " inconsiderable wages." (LIV, 490, 491, 498, 505). In Kent county, a day's work was valued at twenty pounds of tobacco. (LIV, 41). In one case, a maidservant sued her master for wages. The master maintained that the woman had not completed the term for which

278

she had been hired. When the maidservant retorted that this was only because she had been abused by her master, the court held that the woman should recover her wages. (LIII, 14, 15).

There was other litigation about wage agreements. See IV, 201, 291, 356, 394, 397, 468, 469; X, 473, 522, 546; LIV, 73, 76, 77.

[8] III, 141, 142. The county courts were authorized " to moderate the bills, wages and rates of artificers and labourers." (I, 97). In re Cecil Calvert, see index, CMS.

[9] X, 353 (interpreter); LIII, 43 (sawyer); XVII, 167, 168 (joiner); X, 213, 214 (brickmaker); LIV, 544 (smith); XV, 199, 200; XLIX, 230, 232 (tailor); IV, 306 (attend forge). One man agreed to serve a master provided he taught him the trade of carpentry (X, 311); another, provided he was instructed in the trade of cooper. (LIII, 462).

[10] X, 252 (bayly); XLIX, 52 (corn, tobacco crop); LIV, 553 (cowkeeper); XLIX, 174; XV, 339 (women labourers); XV, 339 (housekeeper); IV, 224 (ladies' maid. See also I, 464); X, 389, 390(governess); XLI, 515, 516 (read and sew). In one case the servant complained that his master kept him at " other employments than the usual employment or trade of a servant." (XLIX, 164). Many indentures contained a statement that the master would employ the servant in such service and employment " according to the custom of the country in the like kind." (XLI, 385-387). See also A Relation of Maryland, in Hall, p. 99. Cecil Calvert gave instructions to those in charge of the first settlers in Maryland that they " cause all the planters to employ their servants in planting of sufficient quantity of corn and other provision of victual and that they do not suffer them to plant any other commodity whatsoever before that be done in a sufficient proportion which they are to observe yearly." (Lord Baltimore's Instructions, in Hall, p. 23).

[11] Dankers' and Sluyter's Journal, in Memoirs Long Island Historical Society, vol. I, pp. 191, 192, 216, 217.

[12] Ibid. And see also I, 21 (opinion of assembly); IV, 306 (discharge). In one case the servant made the master agree that he would not have to work on Saturday afternoon. (LIII, 462). George Alsop, who himself had served as an indentured servant in Maryland, gave a pleasing but not accurate picture of the duties required of servants. (A Character of the Province of Maryland, by Alsop, in Hall, pp. 357, 358, 378, 380). John Hammond gave a similar account of the life of a servant in Virginia. (Leah and Rachel, or, the Two Fruitfull Sisters, Virginia and Maryland, by Hammond, in Hall, pp. 284, 290-295).

In regard to servitude in Maryland, see White Servitude in Maryland, 1634-1820, by Eugene I. McCormac, in Johns Hopkins Univ. Studies in Hist. and Pol. Sci., vol. XXII, p. 119 et seq. See also White Servitude in the Colony of Virginia, by James C. Ballagh, same series, vol. XIII, p. 265 et seq.

[13] XLIX, 291, 325-329, 343, 385, 387, 403, 415, 461, 462, 525, 546; XLI, 554. In Talbot county, there are several instances of a man being employed as an overseer of servants. Chirurgeon Richard Tilghman claimed that John Madberry had undertaken to be his overseer and " to take the charge of his hands and build for him." (LIV, 359). Nathaniel Evett, who was employed as an overseer by Edward Webb, said that one day he set two servants " to succouring of tobacco " and another " to mauling of timber." (LIV, 372). One man who was employed as an overseer went outside the scope of his employment when he became the father of a bastard child. (LIV, 545).

[14] A Relation of Maryland, in Hall, pp. 99, 100; X, 494 (printed indentures). Cecil Calvert said that he had heard that a carpenter's services were worth two thousand pounds of tobacco a year. (III, 141).

[15] Leah and Rachel, by Hammond, in Hall, pp. 284, 289, 293, 337; Character of the Province of Maryland, by Alsop, in Hall, 356, 357; A Relation of Maryland, in Hall, p. 99.

[16] III, 169 (Cornwallis); V, 373, 374 (English law).

[17] II, 147, 148. For the earlier laws limiting a servant's period of indenture, see I, 352, 409, 428, 443, 444, 453, 454. The Act of 1666 was, with slight changes, in force during the proprietary period. (II, 335, 336, 351-353, 527). By the provisions of the by-laws of the village of St. Mary's, adopted in 1685, all persons must bring their servants to the mayor's court to have their ages determined. (XVII, 420-422). All servants brought to the colony from Virginia must complete their term of service in Maryland which " they ought to have served in Virginia and no more." (II, 147, 335, 527. See also LIV, 388).

In Talbot county, there are three instances in which certificates to show the servant's age were introduced in court. (LIV, 515, 529, 582). For a case involving extending time of first indenture, recording indenture within six months, etc., see Halfhead v. Nicollgutt, XLIX, 215, 220, 221, 237, 238, 261, 265, 315, 351, 380, 381. For a case where a servant's age is questioned, see X, 451. There were disputes as to the length of time a servant was bound by his indenture. (IV, 195, 205, 385; LIV, 293, 294, 518). See also IV, 207, 208, 396, 422. In regard to giving a servant clothing, food, etc., at the end of the period of indenture, see text, pp. 90-93.

Captain Thomas Harwood complained to the provincial court that two of the seamen on his vessel had sold two men servants to John Grammer, a Maryland colonist, before the seamen had paid him for the passage money of the servants, which amounted to six pounds, ten shillings each, and six shillings, six pence apiece, for " petty charges." In order to reimburse himself, Harwood attached some tobacco which was in Grammer's hands, the man to whom the servants had been sold. Grammer was compelled to pay the cost of the servant's sea passages to Maryland. (XLI, 514).

[18] XLI, 422 (Coursey); XLI, 478 (Gerrard); LIII, 158, 443, 456 (Adams). Other commissioners of Charles county brought their servants to court to have their ages determined, including Walter Beane (LIII, 368, 424, 452, 564), Zachary Wade (LIII, 451, 601), Thomas Baker (LIII, 452), William Marshall (LIII, 452, 541) and Francis Pope (LIII, 501).

Jacob Lumbrozo, Jew chirurgeon, also presented a young man to have his age judged. (LIII, 452). Robert Hundley presented a young girl by the name of Zara Tusan to have her age determined. Later Hundley was accused of having a bastard by this sixteen-year-old girl and then transporting her out of the colony. (LIII, 451, 599). One boy whose name was Peter Blackbeard was brought before the county justices to have them pass on his age. One wonders whether this youth lived up to his name. (LIII, 527). For other examples of servant boys and girls being presented to the Charles county commissioners to have their ages determined, see LIII, 178, 179, 204, 207, 294, 318, 353, 485, 501, 541, 542, 564, 585. See also XLI, 23. In re Coursey and Gerrard, see index, CMS.

[19] LIV, 228 (Foster); LIV, 249 (girl); LIV, 286 (Bringgergrass). Matthew Read, a commissioner of Kent county, was one man who seems to have been willing to accept fewer years of service than that to which he was entitled.

In the case of two servants he showed this generous spirit. (LIV, 127, 243). William Charlton, of Talbot county, showed an equally generous spirit. (LIV, 465). For other cases of where boys and girls were brought before the Kent county commissioners, see LIV, 246, 260, 329, 330, 343, 348, 349. In Talbot county also, a number of servants were brought before the county justices to have their ages determined. See LIV, 356, 357, 366, 376, 377, 386, 393, 416, 418, 419, 424, 426, 430, 431, 436, 439, 441, 448, 454, 458, 489, 496, 515, 524, 526, 534, 545, 555, 556, 564, 565, 567, 588, 590, 591.

[20] XLI, 476-478. See also I, 403.

[21] LIV, xxi, 28, 29.

[22] LIII, 182, 183. When Elizabeth Lippins, a child of six, was upon the death of her father, left a charge to Kent county, the court justices ordered William Granger to clothe and lodge the girl until she was eighteen. Elizabeth, on her part, must serve Granger and his wife. (LIV, 286). In a similar case, Joseph Edlow, an orphan, was placed in the custody of a guardian. (XLI, 599). In one instance, a young girl was assigned over to a Jesuit priest for ten years as an apprentice. The priest agreed not only to clothe and feed the child during this time, but also to educate her. (X, 305, 306). John Dabb, a commissioner of Kent County, bound over his daughter as a servant to another man and his wife. Dabb signed his name by a mark. (LIV, 246, 247). In one instance, the indenture stated that the agreement was made with the consent of the servant's parents. (LIV, 124, 125).

[23] IV, 464 (Mathewes); LIII, 62 (Gess). In re Greene and Trew, see index CMS.

[24] X, 127, 247; XLI, 67, 68 (descriptions); XLIX, 98, 99, 142, 143 (jury). One man agreed to sell another "three able menservants." (X, 145, 146). In one instance a manservant was said to be not only an able man, but also capable of carrying a heavy weight, such as four bushels of salt. (XLI, 5, 6).

[25] XLI, 275 (perfect health); LIII, 168-170 (Neale). In re Neale, see index, CMS.

[26] I, 352, 353 (Act of 1654); LIII, 6, 7 (corn and clothes); XLIX, 103, 104, 140, 141 (new master liable); XLI, 350, 371; IV, 519, 520 (several new masters). One maidservant complained that although she was bound in England to one master, she had been assigned over to another man, "whereupon she knew not well who was her master." (XLI, 370). It is doubtful if the servant's consent to an assignment to a new master was necessary. There was, of course, no reason why it could not be asked. Thus, in one case a man agreed to sell his boy servant to a new master provided the lad was willing. (IV, 538). In one case, however, the servant complained that his master had sold him to another man without his consent "contrary to law and equity." (XLI, 588). In another case, where a servant was assigned over to a new master, the servant refused to serve the new master and left him. (XLIX, 451, 486, 545). One man claimed that a servant that had been sold him ran away soon after the sale was consummated. (XLI, 601, 602). George Alsop, himself an indentured servant, said that servants were free to choose their masters, but Alsop is a questionable authority. (A Character of the Province of Maryland, by Alsop, in Hall, pp. 356, 357. See also Leah and Rachel, by Hammond, in Hall, pp. 284, 289, 293, 337). One maidservant was advised to make herself so disagreeable to her mistress that she would want to sell her to another person. (XLIX, 194, 195).

There are many instances of the sale or assignment of servants in the pro-

19

vincial and county court records. See, for examples, IV, 497, 533, 537; X, 115-119, 216, 217, 223-225, 235, 236, 325, 326, 379-381, 396, 397, 568; XLI, 581, 582; LI, 3; LIII, 179, 462, 463; LIV, 68, 156, 302, 377, 401, 402, 413, 600.

William Batten alleged that he had employed Richard Smith to procure servants for him in England to be transported to Maryland for his (Batten's) use. Batten also said that he was to pay for the servants' transport and had promised to give Smith " a salary " for his services. According to Batten, when Smith shipped the servants he took out the indentures in his own name, and, later, " contrary to the faith reposed in him, disposed of the said servants." Batten asked that Smith should be compelled to make some satisfaction to him for this disposal of his servants. Smith, denying what Batten said was true, asked him to submit proof. When the latter failed to do so, the jury, before whom the case was being tried, returned a verdict: " No cause of action." (XLI, 363).

In another case, Hugh Stanley attempted to hold Captain Samuel Tilghman responsible for having allowed three servants to perish. It seems that the three servants were drowned at St. Mary's as they went ashore upon the ice. This was rather hard luck after having made the dangerous Atlantic crossing. Stanley asked the provincial court to take the matter " into their serious and most tender consideration " and to make Captain Tilghman pay for their loss on the ground that he had exposed them to the danger by which they perished. Tilghman, in defence, said that the servants who were drowned went ashore " without his privity or consent, he being then ashore, and not able to go aboard his own ship, neither doth he conceive himself liable to make satisfaction for all sudden accidents, which lieth not in his power to hinder." The judges of the provincial court agreeing with Captain Tilghman, decided there was no cause of action and nonsuited Stanley. (XLI, 358).

[27] X, 553; XLI, 22, 23, 77, 78; XLIX, 300, 452, 51, 166, 205, 210; LIV, 11, 561; (servants sold for tobacco); LIII, 360, 361 (for horse); IV, 120, 121; X, 461 (for livestock); X, 210; LIV, 406 (house and land); LIII, 84 (boat); XLI, 7; LI, 547 (servant for servant). Two colonists entered into an agreement to exchange maidservants. One of them wanted something " to boot," as his servant had a longer period to serve, but the other colonist refused to grant this request and said that they must exchange " at even hand." (LIII, 392).

Although frequent mention is made in the early records of a servant being sold for so many pounds of tobacco, it is not always possible to learn from these sales what was the yearly value of the man or woman's services. This must, of course, depend upon the sex, age, and qualifications of the servant.

Cecil Calvert thought that if his indentured or apprenticed servants were sold, it would be reasonable to value the labour of an ordinary servant at fifteen hundred pounds of tobacco a year and that if the servant knew some trade, such as carpentry, his worth might be appraised at two thousand pounds of tobacco a year. (III, 141, 142). From the provincial court records we can obtain other estimates of a servant's yearly worth, although it must be admitted that some of them are rather inconsistent. Thus one manservant with six years to serve was valued at two thousand pounds of tobacco, while another with not quite two years to serve was valued at almost as much, or fifteen hundred pounds of tobacco. The two men who had made this appraisement were informed by the justices that they had made " a very just one." But had they? Unless the servant who had six years to serve was very young, inexperienced, or physically unfit, this difference in the value of the two servants does not seem justified. (XLIX, 240, 241; LIII, 455, 456).

In the estate of John Bateman, two menservants "having two years and upwards to serve" were appraised at thirty-two hundred pounds of tobacco, while two boy servants "having almost two years to serve" were appraised at the same amount. In this case apparently the ages of the servants had little to do with the determination of the value of their services. (XLIX, 363). In one instance, we learn that the services of a boy servant, who had five years to serve, were valued at one thousand pounds of tobacco a year although the youth was not seasoned. "Old hands" were worth fifteen hundred pounds of tobacco a year. (XLI, 259). In another case, a man was offered one thousand pounds of tobacco and three barrels of corn for the hire of his servant "the time of the crop," notwithstanding the servant was a new hand and therefore expected to be sick. (XLI, 9).

A maidservant having two years to serve, whose "worth" was appraised by two sworn appraisers, was valued at five hundred pounds of tobacco for each year's services. (XLIX, 319). Among the items listed in the estate of Thomas Hawkins were three servants. One of these, a girl who had six months and a half to serve, was valued at six hundred pounds of tobacco, while a boy servant who had about the same period to serve was valued at one thousand pounds of tobacco. A man who had a year and a half to serve was valued at fifteen hundred pounds tobacco. (LIV, 102). It is apparent that the services of a woman were not worth as much as those of a man.

In one case, a servant was attached at the request of a woman who claimed that the servant's master was indebted to her children. (XLI, 195, 256, 270, 338, 339, 348). In another case, a party litigant desired an attachment against a servant for "the securing of his debt," (LIV, 389) while, in still another instance, a manservant was seized by a sheriff "towards the satisfaction of a judgment." (LIV, 378, 379). The master of one servant maintained that his servant was kept prisoner by a sheriff "to the damage and loss of your petitioner in respect of the want of his said service." The provincial court ordered the sheriff to return the servant to his master to serve out the remainder of his time. After that, the servant must serve additional time in order to pay for the sheriff's fees and other expenses incidental to his imprisonment, "unless otherwise agreed." (XLIX, 510). For "the true performance" of his obligation to pay a tobacco debt, one man assigned to his creditor a manservant and two cows. (LIII, 498).

[28] IV, 51 (will); XLI, 6, 7, 26, 277 (Mackey). For cases where servant was freed because of brutal treatment by master, see text, pp. 95, 96.

[29] IV, 47; LIII, 74; XLI, 499, 500; LIV, 28, 376 (oral or written statement); LIII, 295 (open court); LIV, 418 (death); XLIX, 7, 81-84; LIII, 22, 23 (friends or relatives); IV, 243 (skins); LIV, 248 (cure).

In one case, where a Jesuit priest promised some of his servants their freedom on condition that they were "to make a crop" and pay him specified amounts of tobacco, the provincial court refused to uphold the agreement and decided that the servants must complete the terms of their indenture. (X, 186, 187). The Jesuits would buy the freedom of servants in Virginia, who were of the Roman Catholic faith, and bring them to Maryland. (Annual Letters of the Jesuits, in Hall, p. 123).

Many suits were instituted by servants to win their freedom and in almost all of them they were successful. The reason that the servant was freed in most of the following cases was that it was shown the servant had completed the period stated in the indenture. (IV, 411; X, 218, 247, 254, 255, 277,

527, 528; XLI, 7, 8, 11, 19, 67, 179, 180, 435, 436, 471, 493, 494, 496, 497, 515, 516; XLIX, 331, 332, 387, 388, 565, 566; LIII, 443, 444, 585, 592, 599, 600, 624, 625; LIV, 9, 10, 191, 192, 194, 239, 363, 375, 485, 502, 503, 582, 660, 661, 684, 692, 710.

[30] XLI, 456, 457. At a session of the Charles county court, one Mary Hews declared under oath that John Waltom not only bought her freedom from Captain Josias Fendall, but also had promised to marry her as soon as his wife was dead. Later, however, Waltom seems to have had change of heart for, on one occasion, he threatened to stick a knife into Mary's " guts." The latter said that she did not know why he said this unless it was because she had not washed one of his shirts. (LIII, 224, 225).

[31] V, 268 (desire to be planter); I, 97 (Act of 1640).

[32] I, 352, 353 (Act of 1654); X, 213, 214 (Cannaday); I, 353 (agreements upheld). Despite this favorable agreement, Cannaday ran away to Virginia. Curiously enough, Cannaday later had the audacity to sue Gerrard for the house and " divers other goods," which, Cannaday said, were to be allowed him for his service in Maryland. Gerrard, of course, defended the suit on the ground that Cannaday by running away had not completed the conditions of his servitude. (X, 248, 271). Cannaday was sold by Gerrard to a Virginia master. (X, 214, 215).

[33] I, 496; III, 47, 48, 99, 221, 223, 231, 233. See also II, 525 and A Relation of Maryland, in Hall, pp. 91, 99.

[34] LIII, 185 (Hayes); X, 41 (runaway). For suits instituted by servants to recover their allowance of corn, clothing, etc., see IV, 361, 447, 456, 470, 471, 539; X, 48, 52, 218, 238, 247, 254, 255, 261, 382, 406, 415, 436, 505, 506; XLI, 67, 417, 418; LIV, 416, 442, 448, 468, 498, 575.

[35] In re suicides among servants, see text, pp. 138-139, and in regard to the two masters hanged for murdering their servants, see text, pp. 122-126. The case about the servant preferring to live with Indians, can be found in X, 474, 482, 484, 485. In this case, the servant complained of "the hard and cruel usage " received at his master's hands, who was John Little. It appears that the servant had been compelled to beat at the mortar and also to bake bread on Sunday.

Of course, there must have been a great number of masters who were kind to their servants. Such masters would not be called before the courts. One master, who was fair in his treatment of servants, was Robert Brooke. John Clifford, one of his servants, was found drowned in the Patuxent river not far from De la Brooke, as the plantation of Robert Brooke was called. Those who went to investigate found no " bruise or hurt " on the naked body which would indicate that Brooke had beaten this servant. Servants in Brooke's employ were called upon to testify as to the treatment Clifford had received from his master. One day, it appears, that Clifford had been found asleep in the garden when he was supposed to be working. Brooke, instead of whipping or disciplining Clifford, told him that if he or any of the other servants "disliked of his service," they were free to choose another master. While it was true that Clifford had " to beat at the mortar," always an arduous job, he was only employed at this task about three times a week and then only compelled to beat one bushel of corn at a time. According to the servants in Brooke's employ, Clifford really had better usage " and more favour and respect from Robert Brooke, their master, than the rest of the servants." (X, 157-159). In re Robert Brooke, see also, CMS, pp. 169, 170.

Ralph Beane was another master who was exonerated of any blame in having caused his servant's death. One day, it seems, Beane told Ralph Loe, his servant, to go out into the woods to maul some timber. At that time Paul Simpson, a neighbour, happened to be in Beane's house. Simpson said that Beane went out to see how Loe was coming along, but returned soon afterwards saying that Loe had been taken suddenly ill. Simpson then returned with Beane to see what was the matter. They found Loe " groveling on the timber, rattling in the throat . . . drabling at the mouth." They at once carried Loe back to Beane's house where the servant "fetched a great groan and died." Simpson was convinced that Loe had died probably as the result of apoplexy. He had not seen, he said, any blood about Loe which would indicate that his master had struck or beaten him. Beane, however, was charged with having caused Loe's death. A jury, after hearing Simpson's testimony, was not long in bringing in a verdict which completely exonerated Beane. Thereupon, the court discharged him, " as not anyways guilty of the death of his servant." (X, 52, 73, 74). In re Ralph Beane, see CMS, pp. 73, 316.

[36] X, 439, 440 (Stevens); XLI, 316, 317 (Morgan). In Kent county, when one master succeeded in proving to the satisfaction of the county commissioners that he had a just complaint to make against his servant for some act of insubordination, the justices ordered that the servant have twenty lashes on his bare back " well laid on." (LIV, 248).

[37] X, 401, 402 (Goulson); XLI, 451 (Woosey); X, 403 (Taylor); LIV, 120 (Bradnox).

[38] XLIX, 8-10. In another case, however, where a servant complained that his master did not allow him sufficient clothing, the provincial court ordered the master to properly clothe the servant. (X, 401). It is interesting to note that when Leonard Calvert, first governor of Maryland, died, he left some of his clothes to two of his servants. (IV, 314). In re Richard Preston, see CMS, pp. 77, 116, 572.

[39] XLI, 513, 532, 554-556. In re Robert Kedger, see CMS, pp. 67, 72.

[40] X, 416 (Frizell); X, 191 (Jones). Frances Shambrooke was a maidservant employed by two masters, William Bishop and Robert Palmer. Frances complained to the Kent county court of the abuses and ill usage she had received at the hands of these two men. To prove her contention, she not only had three witnesses testify in her behalf, but also showed the court " the marks and tokens " of her beatings. The county justices ordered Bishop and Palmer to pay one hundred pounds of tobacco for the girl's maintenance, until she was able to get work elsewhere. (LIV, 292).

[41] XLIX, 318, 319. The court also ordered that whoever bought Sarah Hall must agree to give her at the expiration of the two years " her corn and clothes," this being in accordance with the custom of the colony.

Margaret Roberts, another maidservant, said that her master John Hambleton was detaining her after her period of servitude had ended and that he sometimes beat her. The provincial court referred this case to the next session. In the meantime, Hambleton was instructed " to use her well and not strike her." Should she later be adjudged free, then her master must pay for the extra time she had served. This case never seems to have been finally settled. (XLI, 68).

[42] LIV, 167-169, 171, 173, 176.

[43] LIV, 167, 178-181, 213. During all this time, Sarah Taylor had a friend in Captain Joseph Wickes, also one of the county commissioners. He lent a

willing ear to all her complaints. On one occasion, Captain Wickes informed his fellow commissioners that Mrs. Bradnox "broke the peace in striking her servant before him, being a magistrate." It seems that Sarah had come with her mistress to complain of the treatment she had received at the hands of the latter. This Mrs. Bradnox resented.

[44] LIV, 224-226, 228.

[45] LIV, 234; XLI, 482, 506, 525.

[46] LIII, 410, 411.

[47] X, 505.

[48] X, 521.

[49] LIV, 9, 10, 125, 126.

[50] XLI, 467, 471, 475, 478-480. Mrs. Nevill was accused of manslaughter, not murder. For the distinction between these two crimes, see text, p. 119.

[51] XLI, 268, 269.

[52] XLI, 482, 500-506; LIV, 230. After the successful outcome of this case, Mrs. Bradnox attempted to hold Sarah Taylor and John White responsible for what she maintained were false complaints made against her late husband. (XLI, 525).

[53] LIV, 8, 9.

[54] IV, 391, 396, 444 (steer); III, 455 (conversion of funds). The charge of killing the steer was later dismissed by a grand jury. (IV, 447, 448). Francis Brooke also accused Bradnox of driving away his cattle, but failed to show that he had suffered any real damage. (IV, 390, 391, 393). In regard to Bradnox's ignorance, see LIV, 5, 11, and regarding his drunkenness and profanity, see text, pp. 146, 147, 162. On one occasion, Captain Bradnox was so drunk that he could not stand up. See text, pp. 178, 179, where will also be found an account of his wife's adulterous relations with John Salter.

[55] IV, 436 (Brent); IV, 460 (Vaughan); IV, 444 (accusation). In regard to the Ingle Rebellion, see index, CMS, under this heading and also under Richard Ingle.

[56] XLIX, 307-312. Grammer was accused of manslaughter, not murder. For the distinction between these two felonies, see text, p. 119.

[57] XLIX, 351, 401. In Talbot county, Francis Carpenter was suspected of murdering his servant, Samuel Youngman. A witness, named Edward Fullward, testified before the Talbot county commissioners how, on one day when he was at Carpenter's house, he saw Youngman's head bleeding and that he bled "from sunset till two hours within night." When he asked Carpenter how it had happened, the master replied that he had struck the servant's head with a stick. Fullward said that when he dressed the wound he found the skull bare. It seems that about three weeks after this Fullward returned to Carpenter's house. This time he found Youngman sitting "by the mortar with his face all bloody." Again Fullward asked his master what had happened and again Carpenter admitted that he had struck his servant with a stick. The very next day, according to Fullward, Youngman died. After young Fullward had finished testifying another youth, by the name of Henry Wharton, was called to the witness stand. Wharton also testified that Carpenter had admitted striking his servant.

Immediately after Youngman died, a coroner's inquest was held to determine the cause of his death. They found that the servant's head and body was bruised and that there was a depression in the skull in one place. Death was due to lack of care in attending to these wounds, the jurors declared. In view of this finding of the coroner's jury and the testimony of the two young men, the Talbot county justices decided that there was "sufficient cause" to refer this case to the provincial court for trial. The final outcome of this case can probably be found in the unprinted provincial court records. The case came before the Talbot court in March, 1665/6. (LIV, 390, 391).

[58] XLIX, 166-168, 230, 233-235; LI, 121-123, 129, 130. Alvey was accused of manslaughter, not murder. For distinction between these two offenses, see text, p. 119. In regard to benefit of clergy, see text, pp. 27, 28, where an explanatory bibliographical note will be found. Alvey tried to plead benefit of clergy again when, later, he was accused of theft. See text, pp. 27, 28.

[59] XLIX, 453, 555.

[60] See text, pp. 21-29.

[61] XV, 21; LIII, 132; III, 126 (hue and cry); I, 410 (constable); II, 299, 300, 524, 525 (reward). In one case, where a man was asked to return a servant whom he had borrowed, the borrower said that it was impossible for him to do so as the servant had run away. Although he had raised "hue and cry," he could never find him. (X, 335). In another case, a man authorized two of his friends "to apprehend and take into their custody" four runaway servants and promised that he would "save harmless" and indemnify his friends "in what one or both shall do, in or concerning the said servants." (LIV, 28).

At first any one who captured a runaway servant could turn him over either to his master or the nearest justice of the peace. (II, 146, 147). Later, however, a law provided that he must in all cases turn him over to the justice of the peace. (II, 299, 300, 325).

[62] LIV, 502, 517, 527, 554. In re Kinemant, see LIV, 443.

[63] II, 26, 27, 299, 524, 525. For agreements with various Indian tribes in regard to runaways, see V, 30 (Nanticokes); XV, 171, 214 (Assateagues); XV, 291 (Piscattaways); III, 277, 421 (Susquehannocks); III, 433 (Passayonckes); III, 486 (Delawares). A matchcoat was a wrap, or mantle, of fur worn by the Indians. The English would make them of a coarse cloth called match-cloth and sell or give them to the savages.

[64] II, 528 (400 lbs. tobacco reward); III, 134 (letter); III, 367, 368, 372 (Stuyvesant); III, 430, 431 (New Amstell). In re Stuyvesant, see index, CMS.

[65] II, 224-226, 258, 261. During this same year, the colonies to the north were asked to return all runaways. (II, 226). In re Herrman, see index, CMS.

[66] II, 495; X, 442, 515, 516 (sheriffs); IV, 319, 327; X, 15, 27 (suits). In 1643, Upkin Powell, who acted in behalf of Captain Henry Fleet, then in Virginia, was authorized to take into his custody several servants who had fled to Maryland and to take them back to Virginia there "to answer the allegations of Captain Henry Fleet in point of service." (III, 133). A constable was, in one instance, instructed to seize a runaway from Virginia. (XLI, 91). The Lords Commissioners for Foreign Plantations gave directions that neither the Virginians, nor the Marylanders, should receive "any fugitive persons belonging to the other." (III, 22).

[67] I, 107, 124 (Act of 1641); I, 249 (Act of 1649); I, 348, 349 (Act of 1654); I, 249, 349; II, 403, 526 (transport). The damages which the master was to receive, when his servant was transported out of the province, varied from time to time. The Act of 1649 allowed the master double damages, the Act of 1674 such damages "as shall appear justly due unto such master for want of such servant," while the Act of 1676 allowed the master treble damages. (Ibid.).

[68] IV, 165. Jane Cockshott complained not only against White, but also

against wife of David Whitcliff for "unlawful dealing" with one of her maidservants because she had kept "a red base wastcoat lined with silk galon," which she had received of the girl. (IV, 154).

In 1644 a commission was issued to James Neale to apprehend Edward Robins, Daniel Duffill and others, "to answer to their crime of open rebellion in arms to commit felony in carrying servants out of the province, and in case of resistance, to shoot them." (IV, 280).

[69] IV, 268, (Clarke); IV, 268 (Pope); LIV, 477, 478 (Wickes). Lt. William Lewis was accused of helping runaway servants to escape. (X, 20, 21, 24).

[70] XLI, 232, 233. For another case in which the plaintiff failed to prove that the defendant had enticed his servant away, see Hosley v. Wilde, XLIX, 173, 193.

[71] LIV, 433-435. Thomas South, another commissioner of Talbot county, claimed that William Bagley took his servant away from his work "to go to drinking." (LIV, 406, 407). Obadiah Judkins attempted in vain to hold Thomas Norris responsible for the escape of his servant named William Bunch. The Talbot court declared that they saw no reason why Norris should have been compelled to apprehend Bunch. (LIV, 491-493).

[72] I, 250, 349 (Acts of 1649 and 1654); I, 451 (Act of 1662); II, 146, 147 (Act of 1666). In 1671 the penalty for "wittingly or willingly" detaining, or entertaining, a servant who did not have his master's or mistress' permission to be absent was made five hundred pounds of tobacco "for every night or four and twenty hours." (II, 298, 299, 524).

[73] IV, 219, 220 (Price); XVII, 188, 189 (King). In re Price and Slye, see index, CMS.

[74] X, 132-135. About a year before this suit, Father Fisher himself had brought an action against John Hallowes for having influenced a servant of his to leave his (Fisher's) services. The priest asked three thousand pounds of tobacco damages. (IV, 406). For information in regard to Father Fisher, see index, CMS.

[75] LIV, 169, 171, 176.

[76] LIV, 168. In re Captain Bradnox's treatment of Sarah Taylor, see text, pp. 96-98.

[77] LIV, 350, 351 (Andrews); LIII, 592, 602, 603 (Edmonds). In the Andrews case, the defendant at once appealed to the provincial court. In another Charles county action, Vincent Young sued Mathias Obrian for harboring his runaway maidservant. When, however, Young's attorney could not prove his letter of attorney "by witnesses or any other legal attestation," his client was nonsuited. (LIII, 604).

For other instances of where a colonist detained, or harbored another's servant, see Posey v. Hatch (IV, 496) and Pakes v. Smoot (IV, 500). These two cases came before the provincial court. There are also several instances in the court of chancery, where the master or mistress appears in court to complain that some other colonist had taken away and "unjustly detains" their servant or servants. (LI, 81, 103, 104, 224-226).

In the case of Dorrington v. Holman, we find a runaway servant testifying against the man who had befriended him. (X, 500, 523, 524).

[78] I, 107, 108, 124 (Act of 1641); I, 249, 250, 348, 349 (Acts of 1649 and 1654); II, 146, 524 (Act of 1666). During the fall of 1663, complaint was made that many servants ran away in company with negro slaves. As the latter were under servitude for life and therefore could not be made to serve additional

time for their running away, it was made the law that the servants running way with them must "pay singly or proportionably," if more than one servant, all the damages incurred by their masters. (I, 489). In regard to negroes in early Maryland see bibliographical note on p. 290.

[79] I, 451, 452 (Act of 1662). See also II, 298, 523. Other steps were taken in 1671 to prevent the escape of runaway servants. All persons, it was decreed, must have a pass with the county seal on it, if they wished to leave the county in which they lived. For this pass they must pay ten pounds of tobacco or one shilling in money. (II, 299, 524).

[80] X, 516, 517.

[81] X, 504, 505 (persuasion); X, 322 (abuse); X, 396 (correction).

[82] LIV, 416, 420, 426, 428, 429, 431, 501, 511, 542 (ten days); LIV, 540 (Vaughan); LIV, 398 (Armstrong). For other cases of runaway servants in Talbot county, see LIV, 401, 402, 564, 570. In Kent county, a servant by the name of Thomas Guinn had to serve his master ten days for each day he was absent. (LIV, 297). In a case which came before the provincial court Owen Morgan, a servant, was only required to double the time of his unauthorized absence. This was in accordance with the provisions of the Act of 1649, then still in force. (XLI, 316, 317). For other cases that came before the provincial court where the servant was compelled to serve additional time, see Cauther v. Eason (IV, 126, 162), and Perry v. Smith (IV, 289, 290, 302, 303).

If a man bought a servant in Virginia and then brought him to Maryland where he ran away and left his service, the servant, if captured, must serve the additional time required by the Maryland law. (LIV, 400).

In not every case was the master successful in his suit to compel a servant to serve additional time for his unauthorized absence. See, for example, Cornwallis v. Cage (IV, 213, 244), and Brent v. Langworth (IV, 290).

[83] LIII, 560.

[84] LIII, 538.

[85] LIV, 184. Robert Chessick, servant of Thomas Thomas, was accused of plotting with other servants to run away. Chessick was given thirty lashes on his bare back. It appears that he had run away on a previous occasion. (X, 511-514).

[86] LIV, 443.

CHAPTER VI—BIBLIOGRAPHICAL NOTES

[1] I, 158, 192 (Act of 1642). For the Treason Act, see 25 Edward III, c. 2, 1351. For the Jacob case, see XLIX, 486, 489-491. Spesutie Island is near the head of the Chesapeake Bay. In re Nathaniel Utie, see index, CMS.

[2] XLI, 190, 191, 204-206. Antonio is also referred to as a slave and servant. There are several references in the early records to male and female negroes who were servants. (XLI, 262; XLIX, 179, 205). For other references to negroes in Maryland records, see II, 164, 165, 327-329, 517, 518; VII, 197, 199; XIII, 179, 181; IV, 189, 304; X, 398; XLI, 587; XLIX, 342, 363; LIII, 174, 238; LIV, 396, 520-522. See also Letter of Secretary Kemp of Virginia, in Calvert Papers, no. XXVIII, pp. 149-151 and Letter of Lewger, in same volume, p. 199. The Negro in Maryland, by Jeffrey R. Brackett, in Johns Hopkins Univ. Studies in Hist. and Polit. Sci., extra vol. VI (1889), should also be consulted.

John Baptista, a Moor of Barbary, was also a servant. (XLI, xii, 460, 485, 499, 500).

[3] XLIX, 290, 303-307, 311-314, 351, 393. Lawrence Organ was another witness who testified that he saw Haggman being severely beaten by his master. For a case in which a master escaped hanging for beating his servant to death by pleading benefit of clergy, see text, p. 109. One cannot help but wonder why Fincher did not claim the same privilege. Possibly he could not read.

[4] X, 522, 534-545.

[5] XLI, 385.

[6] XLI, 504. In re Watson, see text, pp. 102-104.

[7] XLIX, 212, 217, 218, 220, 231-236. This Elizabeth Greene should not be confused with another woman of the same name accused of forgery. See text, pp. 31, 32. As in the case of Fincher it is surprising that Elizabeth Greene did not claim benefit of clergy. A woman named Mary Axell appears to have been hanged for the murder of her child. (XVII, 439). For cases of where Indians were accused of murdering colonists, and vice versa, see CMS, chapters XIII, XIV.

[8] XLI, 430-432.

[9] X, 456-458.

[10] LIV, 250. Jane Crips was brought before the Talbot county court, in June 1666, "upon suspicion" of murdering a child to which had given birth. Charles Harbert, a youth, told how—

> Jane Crips was delivered of a child without doors on the plantation and she would not let it be known that she had a child, but that he went and fetched a midwife, and two women more, for to examine her. Then she confessed that she had a child and the hogs had eaten it.

In view of Harbert's testimony and Jane's confession that she had given birth to an infant, the Talbot justices referred the case to the provincial court for trial. An investigation of the unprinted records of this court for this date would probably reveal the final decision. (LIV, 394, 395).

[11] X, 464, 465, 488.

[12] IV, 403, 404 (Allen); XLI, 429, 430 (Rogers). In Somerset county Macum Thomas and his wife were murdered while asleep in their bed by Henry Parratt, their manservant. (LI, 257).

[13] I, 158, 192. The Act of 1642 said that what constituted murder should be determined by the court " as near as may be to the law of England."

[14] III, 472; XLIX, 10-16. James Lewis was another man, who, when convicted of manslaughter, had his hand burnt. (VIII, 23, 24). The Act of 1642 provided that any one convicted of homicide might be burnt in the hand. (I, 158).

[15] IV, 254, 255, 260; III, 146, 187, 188. Alexander D'hynoyossa was accused of stabbing a man in the belly. (V, 494). For other cases of manslaughter, see chapter V.

[16] LI, 321-324.

[17] LI, 346-348. Thomas Folkes, though indicted for murder, was found by a trial jury guilty only of homicide " by misadventure." (XVII, 37).

[18] LI, 210, 211 (Floyd); LI, 202, 203, 210 (Curre).

[19] X, 141-144. In 1661 William Gylls was suspected of murdering Nathaniel Proctor. A grand jury, after considering all the facts, refused to indict Gylls for murder, as they were of the opinion that Proctor was killed accidentally. (XLI, 471, 475).

[20] IV, 9, 10. The jury in this case also found that by the fall of the tree Bryant " had his blood bulk broken." In re deodand, see Studies in the History of American Law, by Richard B. Morris, pp. 225-230.

[21] LIII, 626.

[22] LIV, 382 (servant); LIV, 389 (child).

[23] X, 154, 155 (Lisle); XLIX, 457, 458 (mill case).

[24] LIII, 362, 363 (maidservant); LIII, 401, 402 (Wood). Philip Saloman and Thomas Elston were also " accidentally drowned." (XLI, 451, 452). For another case of accidental drowning, see XLIX, 259. One woman was thought to have been " casually drowned." (XLI, 452).

[25] LIV, 7, 10.

[26] LIV, 21 (forfeiture of property); LIV, 360-362 (Short). During the winter of 1663 John Jerome was found drowned. Members of a coroner's inquest declared that in view of the reports they had received of Jerome's " melancholy discontent of late " and considering that his footprints were discovered leading into the water, they were inclined to believe that he had willfully drowned himself, although they could not say " so much upon their oaths." (XLIX, 114).

[27] XLIX, 88 (Vaughan); XLIX, 113 (Teedsteed). In Charles county, a coroner's jury found that Elizabeth Johnson, another servant, " hath hanged and murdered herself with a bridle rein." (LIII, 501, 502). In 1661 the body of Jane Copley, a maidservant of Thomas Turner, was found lying at the foot of a tree. The coroner's inquest summoned to determine the cause of her death were of the opinion that after running away from her master, Jane had starved to death. (XLI, 452). For cases of brutal treatment of servants, see chapter V.

[28] XLIX, 215, 216. Francis Holland, the foreman of this jury of inquest, himself had a servant named James Cowill who hanged himself. Another inquest was ordered and again the finding of the jury was that Cowill had willfully and feloniously murdered himself. (XLIX, 216).

[29] LIII, 140, 141.

[30] LIII, 502. In Talbot county Robert Hawkins was declared a suicide. There was no evidence that Hawkins had been maltreated by his master. Hawkins hanged himself. (LIV, 372, 373).

[31] XLIX, 314, 351, 374, 375, 394. A grand jury in 1664 also presented Edward Skidmore, of Anne Arundel county, for not having arranged with a coroner's inquest to investigate the circumstances surrounding the drowning of

one of his servants. There is no record of the final disposition of this case. (XLIX, 314, 351).

[32] I, 158, 192. For the time within which an action of assault and battery must be brought, see II, 201, 202.

[33] IV, 317, 318 (Hudson); XLI, 474, 475 (Allanson).

[34] XLIX, 433, 434, 439.

[35] XLI, 164. The provincial court ordered that this case should be sent down to the county court of St. Mary's " to be there heard and determined."

[36] X,31, 32, 163. When, in another case, Morris Murfee maintained that Captain Robert Troope had wounded and lamed him, the jury awarded Murfee eight thousand pounds of tobacco damages. (XLI, 588, 591).

[37] LIII, 367, 382, 383. Not long after this case was settled, Joan Nevill swore in open court that as she went in fear of her life of Thomas Baker, she humbly begged the commissioners of Charles county that Baker " may be bound to the peace." Baker testified against Joan not only in the suit under discussion, but also in an action which Richard and Mary Dodd had brought against her and her husband for slandering them. Whatever the reason for his attitude towards Joan, the county justices ordered Baker to put up sufficient security for his " good abearance (behaviour)." Before the next meeting of the same court, however, something must have occurred to make Joan change her attitude towards Baker because at that session she said that she wanted to release him " from the oath she swore against him the last court by virtue whereof he was bound to the peace." (LIII, 377, 378, 381-383, 432, 443).

[38] LIV, 384, 385. Seth Foster was a commissioner of Talbot county.

[39] LIV, 466.

[40] LI, 332, 333. Suits for assault and battery were also instituted in Kent county (LIV, 184, 307, 310, 311), and in Somerset county. (LIV, 636, 707, 708).

[41] LIII, 627, 628, 636 (manor court cases); XLI, 484 (Gerrard).

CHAPTER VII—BIBLIOGRAPHICAL NOTES

[1] III, 354-358, 366, 375, 383, 384, 387, 396, 407-409. The St. George's river was later known as the St. Mary's river. (see footnote p. 73, in Hall's Narratives of Maryland). For Cecil Calvert's letter in re drunken councillors, etc., see I, 333, 334. For further information about Coursey, Fendall and Gerrard, see index of CMS.

[2] LIV, 116, 119-121, 173, 178 (Bradnox); LIV, 27 (Read). Both Read and Bradnox were so ignorant that they could not write their own names. (LIV, 28, 201, 220). For Read as Kent commissioner, see LIV, 260, 261.

[3] I, 333, 334 (Calvert); XIII, 148-150 (Joseph). See also index, CMS, in regard to these men.

[4] Dankers' and Sluyter's Journal, in Memoirs of the Long Island Historical Society, vol. I, pp. 218, 219 (Dutchmen); XLI, 161, 181 (drinking and dancing); XLIX, 29, 30 (Furnifield). Regarding the Indian's use of liquor, see CMS, p. 373 et seq.

[5] XLIX, 113, 114 (Styles); LIV, 412 (Anderson). For cases of suicide, see text, pp. 138, 139.

[6] I, 159, 193, 286, 342, 343. Early laws against drunkenness were the Acts of 1642, 1650 and 1654.

[7] I, 375. This was the Act of 1658.

[8] X, 558 (Scott); LIII, 429 (Charles co.); LIV, 352 (Kent co.); XLI, 525 (Gaylourd). For other cases of drunkenness, see I, 404; X, 202, 414; XLI, 323; LIII, 310, 415-418; LIV, 170, 172.

[9] A Relation of Md., in Hall, pp. 97, (stock for ship and land); X, 302 (supply for gentleman).

[10] Letter of Robert Wintour, unpublished manuscript; Letter of Father Andrew White, in Calvert Papers, Md. Hist. Soc. Fund Publication, no. XXVIII, p. 208. For other references to the possibility of producing wine in Maryland, see Hall's Narratives of Early Maryland, pp. 9, 82. In re White, see index, CMS.

[11] Dankers' and Sluyter's Journal, vol. i, 197, 199, 218 (use of cider); XLIX, 517 (Gerrard). References to beer (X, 485, 488, 495; LIII, 274; LIV, 231). One planter told another that as he had no beer at his house he might go to the nearest creek for a drink of water (XLIX, 494). References to cider and perry (XLIX, 197; LIII, 258; LIV, 119, 218, 322).

[12] Narrative of a Voyage to Maryland, in Amer. Hist. Review, XII, p. 329 (use of brandy); XLIX, 502 (morning draft); LIV, 445 (delegates); XLIX, 330, 562; XLI, 589; LIII, 264, 265, 601 (mention of brandy); XLI, 538 (metheglin); XLI, 574 (" white coin "); LIII, 416, 258, 259, 301; XLIX, 421 (rum); IV, 402, 410, 414, 417, 420; X, 66, 302, 331; LIII, 601, (sack). There are many references to ordinary wine. This was probably imported (XLI, 226; LIII, 232, 233; XLIX, 499; LIV, 39). IV, 432; X, 302, 315, 316, 331, 391; XLI, 443; LIII, 12; LIV, 87, 88, 218, 231, 233; XLIX, 210 (drams). Drams given to sick man (XLIX, 305) and to jurymen (LIV, 231). Cordials (LIII, 190). For other references to alcoholic beverages, see LIII, 430, 336; LIV, 415, 431, 463, 464, 483, 517, 547, 573, 594. Burnt brandy, punch, canary wine and canary sack are among the beverages mentioned in these last citations. See also I, 532; XV, 391; VII, 592, 593. In re Virginia colonists, see Lord Baltimore's Instrucs. in Hall, p. 18.

While he was in command of Kent Island, it was said that Captain William

Claiborne received " several Spanish and French wines and vinegars " which he sold on the island at a profit of at least one hundred pounds sterling. (Claiborne v. Cloberry, Md. Hist. Mag., vol. XXVI, p. 396). During one year (1635) there was spent on the island " in wine and drink " one thousand pounds of tobacco, or £16, 3s, 4d. During Claiborne's administration of the island, mention was made of drinking jugs, sack and " hot waters," beer and a malt mill. (High Ct. Admlty, Libels 100, no. 63, in Md. Hist. Mag., vol. XXVIII, p. 30 et seq.).

[13] For the use of boats on the Chesapeake and its tributary rivers, see CMS, chapter III.

[14] II, 34, 96 (accommodations lacking) ; I, 447, 448; II, 432-435 (need of ordinaries) ; XV, 78-80 (county ordinaries).

[15] III, 493, 494; XLI, 47, 48; VII, 65-68. A man generally had to deposit about one to two thousand pounds of tobacco before he could obtain a license to maintain an ordinary. Any one was subject to a fine if he ran an ordinary without obtaining a license. (Ibid. See also XV, 21, 22; V, 123; XIII, 215). While running an ordinary, either in the counties or at St. Mary's, the ordinary keeper could not at the same time be a delegate, county commissioner, or hold any office at St. Mary's. (VII, 63, 67).

[16] II, 295-298. See also II, 148, 149, 214, 215 where the prices for liquor, etc., are given in pounds of tobacco.

[17] Ibid. See also II, 346, 347 (beds). If a man kept an ordinary near a county court house, he would have to maintain more than four extra beds. II, 407, 408, 560, 561 (horses). The Kent county commissioners promulgated an order which gives some idea of the cost of living in early Maryland. The justices declared that if any housekeeper would " entertain Robert Bartrum and let him have lodging and necessary food," they would pay the housekeeper one hundred pounds of tobacco a month. At the prevailing price of tobacco this was a little over sixteen shillings a month. (LIV, 322).

[18] II, 407, 408, 560, 561. These were the Acts of 1674 and 1676. By the terms of the latter act no ordinary keeper could credit any freeman (not a freeholder but depending upon his labour) for more than four hundred pounds of tobacco in any one year.

[19] VI, 66, 67. This was the Act of 1678.

[20] I, 447, 448 (Act of 1662) ; II, 297, 298 (Act of 1671). In other words, the ordinary keeper could not recover more than three shillings, six pence, from any one guest no matter how much he had furnished him.

[21] XIII, 213-215. This was the Act of 1688. At first every county was obliged to allow their burgesses enough to pay for the cost of their " meat, drink and lodging " while attending a session of the assembly. (I, 440, 456, 505). Later, in 1666, every county paid direct to William Smith, an ordinary keeper at St. Mary's, for all necessary expenses incurred by their respective burgesses. (II, 151). Still later, in 1676, expenses of both burgesses and councillors, while in St. Mary's " about the country's business," was borne and defrayed " by the public." (II, 554, 555).

Near the end of the proprietary period, or in 1688, complaints were raised about the expense involved in holding assemblies. This was due, it was said, to ordinary keepers charging more to their accounts than was actually consumed by the delegates, or burgesses. To rectify this abuse, it was proposed to limit each assemblyman to an allowance of so much per day. Apparently nothing came of this suggestion. (XIII, 180, 194, 197, 202). Indeed, rather inconsistently, not long after this it was proposed to allow innkeepers a twenty per cent increase in the rates, or prices, which they could charge and the plan was actually carried out. The increase was defended on the ground that ordinary keepers had been compelled to purchase their liquors and provisions " for the entertainment " of the recent assembly when money and tobacco was " at extraordinary dear rates." (XIII, 205, 208).

Members of grand juries, who were summoned and chosen from among all the colonists, to attend the provincial court meetings, were allowed for every session they attended as jurors 2500 pounds of tobacco for their expenses. Every county had to pay their proportion of this expense. While in St. Mary's, the jurors apparently stayed with the ordinary keeper of " the country house." (II, 462; VII, 430).

[22] X, 219 (fine) ; A Narrative of a Voyage to Maryland, in Amer. Hist. Rev., vol. XII, pp. 336, 337; V, 470, 502 (Sweatnam).

[23] LIII, 636.

[24] XLI, 138, 139.

[25] A Relation of Maryland, in Hall, p. 100 (establishment of ordinaries); VII, 65, 66 (licenses); II, 431 (accommodations); VII, 66, 67; XIII, 55; XVII, 422 (rates). Tobacco at this time was worth about a penny a pound (V, 268; XIII, 171). In view of this, compare rates given on text p. 153. At one time the secretary of the colony was authorized to issue licenses to ordinary keepers. (XV, 78-80). No ordinary keeper at St. Mary's was allowed " to charge the public faith " with any provisions which he sent out of his tavern to " any private man's house or chamber." (VII, 269).

[26] I, 455, 456, 538; III, 459. It was necessary for Mrs. Lee to go to court in an attempt to recover the 12,000 pounds of tobacco which she had been promised. (XLIX, 73). She also asked for and received a license to trade with the Indians for corn in order to replenish her stock of provisions at the ordinary which she kept. (III, 447, 448). Mrs. Lee was so illiterate that she could not even sign her own name. (XLIX, 203).

[27] XLIX, 344, 368, 395, 396.

[28] XLIX, 377, 378, 449, 550, 553-555 (Hollingsworth); XLIX, 221-223, 242, 273-276, 315, 399-401, 525 (orphan's estate).

[29] XLIX, 566. The records of the provincial court that have not been printed might diclose the final outcome of this litigation. In regard to the cost of Hannah's imprisonment, see II, 51, 129.

[30] I, 505, 482.

[31] III, 465, 490-494; XLIX, 180, 296, 297; LIV, 661. Jolly was involved in much litigation. (X, 510; XLIX, 181, 182, 290). See citation 26 above in re Mrs. Lee's ignorance. See CMS, chapters III and IV in re trading with the Indians.

[32] III, 492, 532, 557; I, 538, 539; II, 27-29, 50, 51, 58, 87, 95, 115, 123-127, 138; LIII, 522. Smith sued Jolly for 6,000 pounds of tobacco. (XLIX, 287). Members of both the lower and upper houses of the assembly stayed at Smith's ordinary, as well as Indians on their mission to St. Mary's. (II, 151). See also XLIX, 285, 428, 429, 478 in re Smith.

[33] V, 494, 498.

[34] XV, 345. John Baker and Daniel Clocker were also told to supply sufficient accommodations.

[35] XV, 349. Others who kept ordinaries at St. Mary's were William Howkins. He was one of the first to be licensed to keep " an ordinary or victualling house." (III, 99). A man by the name of Garrett Van Sweeringen also kept an ordinary in St. Mary's, where he accommodated burgesses, jurors, etc. He was also sheriff of St. Mary's county and built the stocks and whipping post at St. Mary's. (V,

117, 118, 469, 470; II, 339, 415, 416, 469, 554; VII, 139, 140, 230, 239, 269, 429; XIII, 208, 227). See also V, 411-417 in re Van Sweeringen.

John Nuttall (or Nutthall) supplied accommodations at St. Mary's to proprietary officials and to Indians. (II, 32, 33, 57, 58, 151, 233, 234. See also III, 449, 450 and index, CMS, in re Nuttall). Other ordinary keepers in St. Mary's who entertained delegates, jurors, etc., included Daniel Jenifer (II, 233, 234, 268, 303, 304), Charles and Elizabeth Delaroche (II, 469, 554) and Robert Ridgely (II, 554). Richard Moy (XV, 35; II, 339, 340, 416), Thomas Beale (V, 491; XIII, 205, 208, 227), and John Deery (LI, 510) conclude the list of ordinary keepers at St. Mary's.

[36] XIII, 204-206, 208 (Underwood) ; II, 432, 454, 455 (Calvert).

[37] III, 303, 304. Belcher at one time sought to recover seven hundred pounds of tobacco from the estate of John Dandy, who had been hanged for murder. (See text, pp. 125-127). This amount was asked for the " diet " of Dandy and his friends, and the grand and petit jurymen. (X, 547). Dandy must have been living in the vicinity of the Patuxent when this trial took place. Captain Thomas Bradnox, on the Eastern Shore, had Belcher send him a rundlet of drams. (LIV, 120) Thomas Chandler and William Russell were also innkeepers in Anne Arundel county. (II, 432).

[38] XVII, 389. Richard Keen also kept an ordinary on the Patuxent river. (LI, 155), and Richard Bally was an innkeeper in Calvert county. (LI, 351, 352).

[39] VIII, 6-9, 24-26, 39; XIII, 227, 178, 179. In Charles county, Nicholas Emerson had a license to keep an ordinary (LIII, 601), as did the Jew doctor, Jacob Lumbrozo, who kept an ordinary, or inn, at his house on Nanjemoy creek. (XLIX, 455).

Ordinary keepers in St. Mary's county, included Hugh Lee and Thomas Innis. Their licenses were good for five years instead of the more usual one year period. Hugh Lee was the husband of Hannah Lee. (See text, p. 33 and XLI, 398, 399, 412, 413, 442; XLIX, 6). Francis Van Enden was also an ordinary keeper in St. Mary's county. He seems to have spent most of his time in court suing different people. (IV, 429, 457, 467; X, 14, 43, 74, 99, 122, 123, 146, 153, 154, 160, 161, 205, 211, 215, 258, 259, 269). In view of the fact that John Hammond agreed, in 1654, to build the county court house in " the county of Mary's and Potomac," he was granted a license to sell liquors and wines in his house near the court house. (X, 410).

There were several ordinary keepers in Talbot county, including Richard Deaver (LIV, 404; XLIX, 528), Henry Hawkins (LIV, 461, 462), Jonathan Hopkinson (LIV, 464, 486, 487). Hopkinson had previously maintained an ordinary on the South river, in Anne Arundel county. (XLIX, 440). John Scott, who acted as crier of the Talbot county court, also ran an ordinary. (LIV, 544). Francis Armstrong had such a large ordinary that he employed a bookkeeper. (LIV, 435).

There were ordinaries in Kent county. (LIV, 43, 50). Among the ordinary keepers was Christopher Andrews. (LIV, 328).

[40] Dankers' and Sluyter's Journal, in Memoirs Long Is. Hist. Soc., vol. I, p. 218 (opinion of visitors) ; I, 159, 193, 343 (laws against swearing). At first the penalty was five pounds of tobacco. The offense of swearing was usually proved by the testimony of two sworn witnesses.

[41] I, 404 (Hills) ; LIV, 51 (Price and Salter). When John Cherman took the name of God in vain before a meeting of the Charles county court, he was fined ten pounds of tobacco. (LIII, 84).

[42] X, 421-423, 435.

[43] LIV, 173, 174, 178. Thomas Bradnox was a picturesque but brutal character. For his cruel treatment of his servants, see text, pp. 96-98. He lived at Craford plantation or fort (LIV, 120), although he also seems to have owned land near the northern end of Kent Island at Love Point. (LIV, 11). As early as 1641, he had become a resident of Kent Island. He held the rank of " commander's mate " under Giles Brent. (III, 97). He was involved in the Ingle Rebellion, but was later pardoned on taking the oath of fidelity to the Lord Proprietor. (III, 181, 182). Mistress Margaret Brent said that during the Ingle Rebellion, Bradnox " carried himself as captain of a certain crew of rebels in the Isle and made her house his garrison for some time." Mistress Brent added that the captain also burnt down one of her houses and killed some of her cattle. (IV, 436, 444, 391, 396, 447-449). Notwithstanding his record, Bradnox was one of the commissioners of the Kent county court, a position which he held at the time of his death in 1662. (III, 183, 424, 455. See also X, 291; XLI, 89). Captain Bradnox was also made sheriff of Kent county in 1653. (LIV, 21). For Bradnox's captain's commission, see III, 349, 351.

[44] LIV, 158, 161, 163-166. Apparently this penalty was in accordance with the provisions of the Act of 1649. (I, 245, 246, 286). It was Cecil Calvert's wish that any one holding an office in the colony who was twice convicted of being " an usual swearer or curser," should be suspended from office. (I, 333).

[45] I, 245, 246, 286. See also Virginia and Maryland, in Hall, footnote p. 216.

[46] I, 343, 344 (Act of 1654 in re gunning, etc.) ; VII, 51, 52 (Act of 1678). Apparently it was the custom of some people both on week days and on Sundays to assemble together " to drink, tipple and game." Sometimes they met at the house of some planter for their sports and carousal. (XV, 154, 155).

[47] LIII, 250, 251 (Charles co.) ; LIV, 193, 195 (Clay) ; LIV, 78 (Russell).

[48] LIV, 41, 59 (Rogers) ; LIV, 27 (Read).

[49] III, 493, 494; XLI, 47, 48 (regulation of ordinaries) ; II, 414, 415 (Act of 1674). The Act of 1674 also prohibited the private sale of liquor on Sundays. See also VII, 53.

[50] XVII, 419, 420, 422.

[51] I, 244, 245, 286. Any one thrice convicted of making reproachful remarks about the Virgin Mary, or any of the apostles or evangelists, would have to forfeit all his possessions and was banished from the province.

By the provisions of a law passed in 1642, any one guilty of sacrilege might be put to death, or burnt in the hand, or imprisoned for life. Other punishments could also be imposed. (I, 158, 192, 193).

During the spring of 1676, Articles of War were adopted by the governor and council of Maryland for the regulation of the provincial militia. Among these articles was one which read:

> If any officer, or private soldier, shall presume to blaspheme the Holy and undivided Trinity, or the persons of God the Father, God the Son, or God the Holy Ghost, or shall presume to speak against any known article of the Christian faith, he shall have his tongue bored through with a red-hot iron. (XV, 80).

In re Articles of War in England and Maryland, see CMS, Appendix H.

Cecil Calvert did not wish any one to hold office in Maryland who professed himself " of no religion." (I, 333, 334).

[52] XLI, 203, 258, 259. For additional information in re Lumbrozo, see text, p. 226.

[53] Atlantic Voyage in the Seventeenth Century, in Md. Hist. Mag., II, p. 322

(ship Johanna) ; Father White's Brief Relation, in Hall, p. 31 (his belief). In re White, see index, CMS.

[54] I, 158, 192 (Act of 1642) ; III, 422, 534, 535, 554; V, 53; XV, 66, 397; XLI, 87, 88; LIII, 129; LIV, 198, 261, 354, 634; LI, 75, 78, 82, 342, 349, 366 (county commissioners) ; XLIX, 476, 486, 508 (grand jury).

[55] Witchcraft in Maryland, by Francis Neal Parke, in Md. Hist. Mag., vol. XXXI, pp. 282-284. See also A Report of English Statutes, by William Kilty. Annapolis. 1811. p. 190.

There are other cases in Maryland records where persons were indicted for witchcraft although in none of them was the accused hanged. In the year following the execution of Rebecca Fowler, a grand jury indicted Hannah Edwards for bewitching Ruth Hutchinson. The form and words of the indictment were practically the same as those used in the indictment of Rebecca Fowler. In the case of Hannah Edwards, however, a trial jury found her not guilty. (Witchcraft in Maryland, by Parke, in Md. Hist. Mag., vol. XXXI, pp. 284-286).

Another case of witchcraft came before the provincial court in 1712 after Maryland had become a royal colony. In this case the indictment charged Virtue Violl, of Talbot county, with not only causing the body of Elinor Moore " to waste, consume and pine," but that she had also by the use of witchcraft rendered speechless the tongue of Elinor Moore. (Parke, pp. 287-289). In Anne Arundel county, during the year 1702, Charles Killburn, in a petition to the justices of that county, said that he was in " a very languishing condition" because of witchcraft practiced on him by Katherine Prout. This woman, however, was not convicted of witchcraft. (Parke, pp. 295, 296).

[56] II, 425, 426.

[57] LIII, 54, 55. Hatch was one of the members of the council of Josias Fendall.

[58] LIII, 139, 142, 156. This was not the only case of witchcraft in which Francis Doughty was involved. Before coming to Maryland to settle along the banks of the Patuxent river, Doughty was pastor of Hungars parish on the Eastern Shore of Virginia. While there, he had Barbara Winbrow brought before the justices of Northhampton county on the ground that she led a wicked life and was supposed to be guilty of " witchery."

Doughty, the witch hunter, had an interesting career. When it was proposed that Doughty should be given " a benefice in Patuxent river," one man objected, saying that the minister was " given to many vices and especially to drinking."

Francis Doughty served in many pastorates. Born in England, he was for a time vicar of Old Sodbury, in Gloucester county. Here he lived with his wife and three children, whose names were Mary, Francis, Elias and Enoch. Because of his Puritan tendencies, the Anglican Church authorities forced him out of Old Sodbury. Doughty and his family then emigrated to New England where they lived at Dorchester, near Boston, and later at Taunton.

From Taunton Doughty went to Rhode Island and from there to Manhattan. As there were many English residents in this Dutch colony, Doughty organized them into a separate congregation of which he became the pastor. While in Manhattan, his daughter, Mary, married Adrian Vanderdonck. Later Francis Doughty moved to Flushing where he was chosen pastor of that recently founded town. From Flushing the minister went to Virginia where, as has already been stated, he became pastor of Hungars parish on the Accomac peninsula. His daughter Mary Vanderdonck, now a widow, and his son, Enoch, went with him to Virginia. His two other sons, Francis and Elias, remained on Long Island where they later became prominent citizens.

In Virginia, his first wife having died, Doughty took unto himself a second wife. She was Ann Eaton, widowed sister of William Stone, one time governor of Maryland. From Virginia the Doughtys, including Mary and Enoch, moved to Maryland. In 1659 we find them living along the banks of the Patuxent in what was then a part of Charles county. In this county Francis Doughty organized a church and officiated as its first pastor. This, probably the first Protestant church in that county, was founded in 1660 or 1661.

As Francis Doughty never remained very long in one place, we are not surprised to find him making another move. This time he returned to Virginia, but not to the Accomac peninsula. Instead, he settled in Rappahannock county where he served as minister. In 1668 a controversy arose between Doughty and his vestry because of non-conformity on part of the minister. Doughty was finally compelled to leave Virginia.

His son, Enoch, and his daughter, Mary, did not go with Doughty upon his return to Virginia. They remained in Charles county where Enoch became a planter. Mary married again, this time to Hugh Oneale. She acquired quite a reputation as a healer of the sick. (The First Church in Charles county, by Louis D. Scisco, in Md. Hist. Mag., vol. XXIII, pp. 155-162, 349; Institutional History of Virginia in the Seventeenth Century, by Philip A. Bruce, vol. I, pp. 218, 219, 280; Early Presbyterianism in Maryland, by J. W. McIlwain, in Johns Hopkins University Studies in Hist. and Polit. Sci., 8th series, at pp. 6-9; Narratives of New Netherland, edited by J. F. Jameson, pp. 334, 366, 401. See also XLI, 84; LIII, 228, 396).

[59] LIII, 139, 142, 143.

[60] LIII, 139, 144, 145. "Goodie," or goody, was an appelation formerly used in addressing a married woman of lowly station. "Goodman" was used in addressing men under the rank of gentlemen.

[61] LIII, 139, 145.

[62] X, 399.

[63] III, 306-308; Jesuit Letter of 1654, in Hall, pp. 140, 141.

[64] XLI, 327-329. In re John Washington, see index, CMS.

CHAPTER VIII—BIBLIOGRAPHICAL NOTES

[1] These remarks were made by William Joseph, who, during the absence of Charles Calvert in England, acted as governor of Maryland. (XIII, 149, 167, 179-181). For other information regarding Joseph, see index, CMS. For the law against adultery and fornication, see I, 286, 344, 345. For cases of intercourse between Indians and white women, see CMS, pp. 376-378.

[2] III, 250-252. Captain Mitchell must have been a man of some means as he not only paid for the transportation of himself and his wife to Maryland, but also brought over " divers artificers, workmen, and other useful persons, in all to the number of twenty persons." Mitchell came to Maryland in 1650.

[3] X, 80, 81, 148, 149, 161, 170, 171, 173-186. On the voyage to Maryland, Mitchell's wife died under circumstances which made it appear that he himself " had brought his wife to an untimely end." After he had separated from Mrs. Warren, Captain Mitchell went and lived with another woman by the name of Joan Toast. The provincial court ordered them to separate " till they be joined together in matrimony in the usual manner."

[4] I, 332-334.

[5] LIV, 366, 371, 387. Thomas Hynson, Jr. was also a sheriff in Talbot county. (III, 541). In re Thomas Hynson, Sr., see LIV, xvii.

[6] LIV, 116-121. Thomas Bradnox and his wife lived in the old Craford fort which had been built by William Claiborne, when he was in command of the island, as a protection against the Indians. At this time (1657) the fort must have been in a delapidated condition.

[7] X, 188-191, 203. One man testified before the provincial court that Henry Carlien and Elizabeth Garnier " did lie in bed together and went under the notion of man and wife." (X, 494). A man named John Stowell was presented to the Kent county court for living with Jane Fox " as his wife." Stowell, by way of defence, said that while he " did lie with Jane Fox . . . and that they have lived together as man and wife he can prove that they are man and wife by Mr. Ladamore." The county commissioners at once ordered that Stowell should prove what he said was true to the satisfaction of the court " at the present coming up of Mr. Ladamore." (LIV, 172, 173).

[8] X, 499, 500, 531-533.

[9] LIII, 560.

[10] LIII, 225.

[11] LIII, 131, 225, 226.

[12] LIV, 9, 10.

[13] X, 506-509, 521, 558, 560. In re John Nevill, see CMS, pp. 108, 109.

[14] LIV, 379, 382, 385.

[15] X, 109-112.

[16] X, 515, 516.

[17] LIV, 55, 56. Possibly Captain Fleet was Henry Fleet.

[18] X, 272, 280-290, 339, 340, 366. The Maryland court, it would seem, might have saved much time and trouble had they in the beginning refused to consider this case on the ground of want of jurisdiction. Possibly, however, they could not make this decision until after all the facts in the case had been considered.

[19] LIV, 386, 407. David Farebanck was accused of going " naked to bed " with one of the maidservants of Ralph Elston. The Talbot commissioners ordered

the sheriff to take Farebanck into his custody and hold him until he gave sufficient bond for his good behaviour in the future. (LIV, 571).

John Foster and Elizabeth Stuckey were brought before the Talbot justices "for living together in incontinence." The commissioners ordered that Elizabeth return to her master's service until the next session of the court, at which time her master was instructed to bring her before the court for trial. Foster was ordered to remain in prison until the next meeting of the court, unless he could give sufficient security for his appearance at that session. There is no record of the final outcome of this case. (LIV, 429, 430).

Thomas White said that he and Margaret Brent, a maidservant of William Marshall, "had passed their faith and troth together." White said that he was determined "to buy her off" from her master, and, if necessary, to substitute another maidservant in her place so as not to cause her master, Marshall, any loss by the lack of her services. White was also alleged to have remarked that if he should die before he married Margaret Brent, his wish was that she should have "all he had." When, soon after this, White died, Margaret asked the provincial court to give her possession of all of White's property. The court would not, however, allow her claim. (XLI, 26, 27, 91, 92, 175, 176). This Margaret Brent should not be confused with the woman of the same name who was the brother of Giles Brent. See index, CMS.

Colonel Henry Coursey had a maidservant by the name of Mary Cole who seems to have had rather promiscuous sex relations. See X, 549-551.

[20] LIII, 133, 134.

[21] X, 337, 365, 366.

[22] X, 525, 526.

[23] I, 373, 374, 441, 442; II, 396, 397, 528; XIII, 212. In order to show that a particular man was the father of the child, this must be proved either by the testimony of witnesses, confession of the guilty man, or "pregnant circumstance agreeing with her declaration in the extremity of her pains and throes of travail, etc." If the girl servant claimed that the man promised to marry her, this, too, must be proved by the testimony of witnesses, or confession of the man.

The Act of 1658 was, with minor modifications, in force during the entire proprietary period.

[24] XLI, 522, 528, 603, 604.

[25] LIV, 540. Other women servants were indicted for having bastards. Mary Brown, a servant of Cornelius Comegys, was presented by the grand jury of Kent county court for having a bastard child which died soon after birth. When Mary was summoned to make her personal appearance in court to answer the charge against her, she failed to appear, but her master, Comegys, promised to bring her to the next session of the court. For some unknown reason, however, Mary did not appear at the next meeting of the court, or at any subsequent session. (LIV, 324, 325, 329).

At a meeting of the provincial court, in 1661, the attorney general appeared before the grand jury to inform them of the case of a woman servant, who, while in the employment of Colonel Henry Coursey, gave birth to a bastard. Although an investigation was ordered, apparently none took place. (XLI, 524).

In the Talbot county records, there were two women servants presented for having bastards, but there is no record of the final disposition of either case. (LIV, 495).

[26] See citation twenty-three supra.

[27] LIV, 291, 292.

[28] LIV, 527, 533, 545, 595.

[29] LIV, 513, 518. A jury of inquest in Talbot county also presented two other servants for having a bastard child, but the case seems to have been dropped. (LIV, 514).

[30] LIV, 610, 615, 621, 622, 642, 643, 656, 659, 671, 686, 691.

[31] LIV, 513, 519.

[32] LIV, 495, 496 (Somers); LIV, 578 (Chapling). William Bagley and Elizabeth Crookshanks, a former servant, were presented to the Talbot county commissioners for committing fornication, but no action seems to have been taken in this case. Elizabeth charged Bagley with having "bodily and carnal knowledge of her several times . . . and that she is with child by him." (LIV, 376, 387, 388).

[33] LIV, 233.

[34] LIV, 486, 488.

[35] LIV, 127, 186, 205, 206, 211, 212. When Elizabeth Lockett was asked whether Mathew Read, her master, had ever had relations with her, she replied in the negative and maintained that her master was "at Severn" when the child was begotten. She only knew Read "by his face and hands," she said, and that one night, when they were in the tobacco house together, her master "did but tickle her."

[36] X, 516; XLI, 14-18. The proprietary government did not approve of the marriage of English women and negro slaves. White women who married a negro slave were considered forgetful of their station in life and a disgrace to their race and nation. In order to break up this practice, a law was passed in 1664 which provided that if in the future any white woman married a negro slave, she must serve the master of the slave during the life of her husband. Furthermore, all children of such a marriage were to be slaves "as their fathers were." (I, 533, 534).

This Act of 1664 referred only to "freeborn" English women. Nothing was said about white woman servants. About twenty years later, or in 1681, a new law was passed which applied to white women servants who married negro slaves. If the master or mistress of a white maidservant either arranged or permitted her marriage to a slave, the new statute declared that the owner of the servant at once lost all claim to the services of the woman who was made free by her marriage to a slave. The children of this marriage were also to be free. For having permitted or arranged her marriage to a slave, the owner of a maidservant must pay a fine of ten thousand pounds of tobacco. Under this law, it would be to the interest of owners of women servants to do all they could to prevent the marriage of their servants to negroes. If the owner of a maidservant did not do all he could to prevent her marriage to a negro, he not only lost all claim to further service from the woman, but he also had to pay a very large fine. (VII, 203-205).

[37] LI, 146, 147, 460-462.

[38] VIII, 34.

[39] LIII, 560 (Cooper); LIV, 524 (Johnson); LIV, 391 (Chessill). A grand jury, or jury of inquest, as it was called in Talbot county, presented Walter Rowles and Joyce Cox for having an illegitimate child. Rowles was later compelled to pay one thousand pounds of tobacco "for himself and Joyce having a bastard child." (LIV, 513, 518).

There is one case in the Talbot county court records which it is rather hard to explain. This was when the commissioners of that court ordered Robert Euan to maintain the bastard son of Elizabeth Kent, of whom a man named John Bewly was the reputed father. This decision seems unfair to Euan. (LIV, 597).

In a similar case, Robert Lamden was ordered by the Talbot justices to maintain William, the bastard son of Judith Hadway, until he came of age. In this case, however, Lamden might have been the father of the child. (LIV, 597).

Hannah Moore, under oath before the Talbot commissioners, told them that John Cooper was the father of her illegitimate child. Thereupon a man by the name of William Cary came forward and offered to pay five hundred pounds of tobacco in order to save Hannah from punishment. It is not clear just why Cary did this. (LIV, 545).

At a session of the Charles county court held during the fall of 1665, there were several bastardy cases. Elizabeth Smallridge was presented for having a bastard. Constable Thomas Gibson accused Robert Hendly of having transported out of the colony a young woman "that hath had a bastard, whose name was Sarah." The same constable alleged that at the house of John Douglas there was a woman servant " illegitimately got with child." Gibson said that the same thing had occurred at the house of John Morris. There is no record of the final decision in these cases. (LIII, 599). At the 1662 session of the Charles county court, Robert Robins and Elizabeth Weekes were presented for " having a bastard." (LIII, 250, 251).

At a meeting of the Kent county court, in 1676, although the grand jury presented Mary Howten for having an illegitimate child, the case does not seem to have come up for trial. (LIV, 331).

[40] XLI, 329-331.

[41] LIII, 2, 6, 28, 30-33, 37, 38; XLI, 291-294.

[42] LIV, 661, 671, 672, 686, 691.

[43] LIV, 383, 386, 387. It is interesting to note that it was Constable Purse who presented Thomas Hynson, Jr. a justice of the peace in Talbot county, for " committing fornication " with Ann Gaine. (LIV, 366).

[44] LIII, 78.

[45] I, 97.

[46] IV, 66. That some law, similar to the Act of 1640, was passed soon after the founding of the colony in 1634 is indicated by the entry of several other marriages in the provincial court records. Thus we find that in 1638 William Edwin, a planter, married Mary Whitehead, and that a short time afterwards John Hollis, a trader with the Indians, married Restituta Tue, and that James Courtney wed Mary Lawne. Edwin, Hollis and Courtney took oaths similar in wording to that taken by Richard Thompson. (IV, 24, 25, 52). In re John Hollis and Thompson, see index, CMS.

[47] I, 374, 442, 443; II, 148, 522, 523. The same " form of words " was used whether a magistrate, or clergyman, was performing the ceremony.

[48] I, 443; II, 522 (witnesses); I, 450; II, 523 (minister's fee); I, 454, 476 (secretary's fee). The governor of the colony could also issue " a particular license," or certificate, to persons wishing to marry. (I, 443; II, 522). In the Somerset county court records, there are many entries stating that so and so had been married after " setting up their names at the court house, etc." in compliance with law. (LIV, 638, 649, 657, 666, 671, 679, 689, 703, 708, 712, 729).

The clerk of every county court was required to keep a list, or register, of all births, marriages and deaths. A fine was imposed on those who failed to report either births, marriages, or deaths. (I, 345, 373; VII, 76, 77). For examples of how a list of births, marriages and deaths was kept in Kent county, see LIV, 38, and in Talbot county, see LIV, 600-609.

[49] LIII, 147, 148.

[50] XLIX, 43, 85.

[51] XLIX, 42, 84, 85.

[52] VIII, 32-35. After having Burnet put in jail, Colonel Goursey took the little girl to live with him. He then wrote to the members of the provincial court asking them what he should do with the child. In reply they told him to send her to the next meeting of the provincial court, when some action would be taken in the matter. Possibly the unprinted records of this court, after 1688, will disclose what disposition was made of the little girl. In re Colonel Henry Coursey, see index, CMS.

[53] XLI, 336-338.

[54] LIII, 599.

[55] XVII, 399, 404, 405.

[56] X, 471.

[57] XV, 319-322. When her husband later died, Mrs. Tennisson brought suit to compel the executors of her husband's estate to abide by the terms of this judgment. (XVII, 131-133).

In Kent county, Elizabeth Martin "for divers reasons and causes," which are not stated, released and discharged her husband, Robert Martin, from all claims she had on him, or his estate, "at present or for the future," for such "good considerations" as she had received. (LIV, 81, 82).

[58] X, 501-504, 555; XLI, 20, 50, 51, 85; LIII, 4, 33, 34. Several years after this, Robert Robins and Elizabeth Weekes were presented by a grand jury of Charles county for having a bastard child. This makes one believe that what Elizabeth Robins had said of her husband during the trial was true. Elizabeth had declared that her husband "spent his means upon whores." (LIII, 250, 251).

[59] XIII, 165, 166 (Clarke); XV, 206, 207 (Leshley). William Sewick, of St. Mary's county, was hanged "for buggery." (VII, 393). By the provisions of a law passed in 1642, sodomy was made a capital offence for which the death penalty could be imposed. (I, 158, 192).

[60] I, 158 (law against polygamy); XVII, 417, 418 (Thompson).

[61] XLI, 149-151, 228-230, 243, 244, 258, 259, 528. In regard to the amnesty, see text, p. 167. For the difficulty which arose between Christian Bonnefeild and David Holt, son of Robert Holt, after the latter's death, see III, 463; XLI, 591-593. A jury of inquest in Charles county presented James Lee for "having two wives," but nothing seems to have come of the case. (LIII, 250, 251).

Rape was a capital offence in early Maryland for which the death penalty could be imposed. (I, 158, 192). Joseph Spernon, of Cecil county, although found guilty of rape was later pardoned. (VIII, 50). By the provisions of the Articles of War adopted in 1676 any officer or soldier who "shall force a woman to abuse her" could be put to death. (XV, 83). See Appendix H, in CMS, in re Articles of War.

CHAPTER IX—BIBLIOGRAPHICAL NOTES

[1] Defamation may be oral (slander) or written (libel). All the cases considered in this chapter are actions of slander. LIV, 90, 91, 103 (knave). Instead of having to pay a lump sum of tobacco to Read for having injured his reputation, South was required to keep a bridge in condition for one year. South also had to pay the cost of Read's suit. Both South and Read were commissioners of the Kent county court.

Although, as has been stated in the text, there were many suits for defamation in early Maryland, there was only one law passed which referred to slander. This was the statute of 1654 which stated:

> That all such person, or persons, who, by slandering, tale bearing, or backbiting, shall scandalize the good name of any person, or persons, directly or indirectly, in such words and expressions as in the common acceptation of the English tongue, or such language as is understood, shall be counted slander, being lawfully convicted, shall be censured both by way of satisfaction to the party injured thereby, and also for the breach of the peace. (I, 343).

[2] LIII, 168, 198, 204, 205. Henry Carline was presented for having made slanderous remarks which were of a similar character. What Carline said and did was told by two witnesses, Hassadia Hill, and the young man's mother. According to Hill, Carline had called him " jackanapes and rascal, and many other abusive speeches." On one occasion, when he went to Carline's house to secure something which belonged to his father, the youth said that Carline had come towards him exclaiming: " Sure, I'll teach you to come a thieving to my house." When Hill replied that the article he was taking belonged to his father, Carline told him that he lied and thereupon kicked him " under the short ribs." Hill's mother confirmed what her son had said, how Carline had abused the boy and that if she had not been present Carline might have seriously injured the youth. Unfortunately, the record of this case is so torn and defaced that we are unable to learn whether Carline was held responsible for his abusive language and conduct. (LIV, 74, 75).

[3] LIII, 256. James Rumsey, of Calvert county, was accused of having abused a minister by calling him " a base, pitiful fellow." (XVII, 117). James Clofton (or Cloughton) sued Anthony Cotton " in an action of defamation " for having said of him that he should have been whipped and hanged. (IV, 17-20).

[4] In re perjury see text p. 30.

[5] IV, 434, 438 (Commins); LIV, 531, 532 (Finney). Roger Baxter also sued Commins for calling him " a perjured rogue." (IV, 392).

[6] X, 481-483, 502. Although Colonel Henry Coursey brought an action of defamation against Thomas Manning for saying of him that he had forsworn himself, that is, sworn falsely, the case never came up for trial. (XLI, 526, 530).

[7] LIV, 63, 64, 69, 70, 74.

[8] LIV, 78, 84.

[9] XLIX, 208, 209, 243, 255, 261, 265-271. Benjamin Hammond attempted to show that William Gother had maliciously defamed him by saying that he had cheated him out of his tobacco crop. When he failed to prove his case, Hammond was nonsuited. (LIII, 386, 387).

[10] IV, 430, 450, 451. The provincial court judges had the jury's verdict

305

entered "for the judgment." If John Smith, in the presence of a third party, calls John Jones a thief, or accuses him of any offence for which the party, if guilty, might be indicted and punished by a criminal court, it is not necessary for Jones to prove "special damages." Such accusations are actionable per se.

[11] LIV, 26. Although Brown appealed to the provincial court from this decision of the Kent county court, the case never came up for trial in the higher court. Perhaps this was because Brown was warned by the county justices that if he lost on his appeal he would have to pay "treble damages." Brown also had to give security that in the meantime he would remain "in the sheriff's hands."

[12] X, 473, 477, 478. Henry Pennington complained against Elizabeth Greene for slandering his wife, Rachel, by saying that she had received some goods which a servant in Mrs. Greene's employ had stolen. From the evidence which was produced, it appears that although Mrs. Pennington did receive some goods from Mrs. Greene's servant, she may not have known at the time that they had been stolen. It was for this reason perhaps that Mrs. Greene was not compelled to pay any damages for her slanderous remarks, only the costs of the suit. (XLI, 445).

[13] LIII, 406, 415-418.

[14] LIII, 21, 22. During a later session of the Charles county court, William Empson brought an action of defamation against William Robisson and his wife, Susannah. When, however, Empson did not appear to prosecute this case, the Robissons asked that Empson should be nonsuited. This request the court granted. (LIII, 91, 94). In the case discussed in the text, Courts is spoken of as "Goodman" Courts. For the meaning of this word and also "goody" see citation sixty, chapter seven. See also LIII, 234 for examples of the use of these words.

[15] XLIX, 550, 551. In the two cases which follow the plaintiff failed to prove his case. Thomas Burdett sued Captain Robert Morris in an action of defamation for calling him "thief and rogue." After hearing the evidence in the case, the court decided that no slander was involved and nonsuited Burdett. (XLI, 433, 434). In the other case, William Gaskin sued Ralph Elston in an action of defamation for having called him a thief and for saying that he had stolen two hoes from Elston's plantation. When Gaskin failed to prove his charge, the county commissioners dismissed the case. (LIV, 505).

During the early years of the colony, a mariner by the name of John Snow sued Michael Duggins, another mariner, for accusing him of theft "in the presence of divers others." This case never came to trial. (IV, 300).

Although not labelled as an action of defamation, James Neale alleged facts in regard to Joseph Edmonds which amounted to the same thing. Neale in his petition to the commissioners of Charles county said that Edmonds had circulated a report that Neale and some of the men with him had assaulted him and stolen goods out of his boat. In defence, Edmonds said that he did not remember having ever made such remarks which he knew were not true. Edmonds said that if Neale would withdraw the suit, he would agree to pay the court charges to date. As this suggestion satisfied Neale, his suit was withdrawn. (LIII, 321).

In another case in the Charles county court George Thompson sued William Robisson, a carpenter of Port Tobacco, alleging that he had been defamed by Robisson. Thompson went on to state that the reason for his suit was that Robisson had retained at his house a court roll or draft of the orders of the court. How the retention of a court order by Robisson could be considered as amounting to defaming Thompson, it is impossible to understand, but the county court so held, and required Robisson to ask Thompson's forgiveness and also pay the cost of the suit. (LIII, 47).

[16] LIV, 369, 370.

[17] LIV, 383-385. Knapp did not prosecute his action of assault and battery.

[18] LIV, 293.

[19] LIV, 644, 649. For some of the defamation cases instituted, see LIV, 624, 625, 627, 636.

[20] X, 167. Ballance also had to pay the cost of the suit. In re livestock markings, see text, p. 61.

[21] LIII, 374-376.

[22] LIV, 83, 85, 104; X, 493. Ringgold also had some litigation with Wickes about land. (XLI, 319-321). For additional information regarding Ringgold and Wickes, see LIV, xvi, xvii. Also see index, CMS, in re Wickes.

[23] LIV, 38, 78, 84, 85, 113, 121.

[24] LIV, 558. This is an action of slander, as the singing of a libellous song was slanderous.

[25] X, 114, 115. Apparently this case never came up for trial.

[26] LIV, 122. From what we know of Bradnox it is possible that Margaret may have been telling the truth about the treatment which she received at his hands. See text, pp. 96-98.

[27] LIV, 264 (Willkins); LIV, 391, 392 (Smith); LIII, 563, 570 (Grub). The grand jury of Kent county presented Ann Tumees, a servant, "for scandalizing" William Jones by saying that he had "got her with child." (LIV, 324).

[28] LIV, 576, 577.

[29] X, 487, 488, 495, 519.

[30] IV, 181, 183, 190.

[31] XLI, 550, 551.

[32] X, 587, 495. In a similar case, Mary Edwin complained against another woman for slandering her by saying that she had "lain with an Indian." This case, however, did not come up for trial. (IV, 258).

[33] LIV, 575, 576.

[34] LIV, 534-539.

[35] LIV, 478.

[36] XLIX, 56, 58, 72. This Elizabeth Greene was the wife of William Greene and therefore not the woman of the same name who was hanged for murder. The murderess was a spinster. See text, pp. 127, 128.

The wife of Peter Godson, a chirurgeon, got into trouble for saying of Michael Baisey's wife that her eldest son by a former husband was not really her son and that she knew the man who was the father of the boy. Indeed, Mrs. Godson said she considered Mrs. Baisey nothing but "a whore and a strumpet" who practiced her trade "up and down the country." In view of these slanderous remarks, Mrs. Godson was committed to the sheriff's custody until she put up security for her good behaviour. As Mrs. Godson was later released from this bond, the case may have been settled out of court. (X, 402, 403, 409).

Mary, the wife of Martin Kirke, was another woman who was very loose with her tongue. She said that Mrs. Bonyfield was a whore. She also declared that a man by the name of Henry Potter was a rogue and that she knew enough about him to hang him. In the same category she placed Mark Pheypo and described how he had flung her on a bed. Apparently Mrs. Kirke was not punished for any of these slanderous remarks. (X, 407, 408, 411).

[87] X, 234, 235. Mary Empson was said to have "a pocky whore's back." (LIII, 378). Although William Hampstead brought an action of defamation against William Browne for calling his wife "a whore and a bawd," this case never seems to have come up for trial. (XLI, 596, 601). The same was true of a case which Thomas Hollis brought against Thomas Boys for referring to his wife as a whore. (IV, 149, 150).

[88] I, 509-511, 514-522. The same facts can also be found in the provincial court records, XLIX, 78-80, 115-118, 145, 146. Spinke himself had previously been a servant. (IV, 470). Among the other grounds which Dr. Barber gave for appealing his case to the upper house of the assembly were that (1) the writ had "nothing in it of the declaration when it ought to have all, but only time and place and this is error," (2) that Spinke had not stated that Barber had made statements about his wife which were either false or malicious, (3) that the words spoken by him, Barber, were not "actionable," (4) that the jury should not have allowed damages for injuring Spinke more than in his "estate, credit or otherwise, he was worth, his birth, education and estate, being sufficiently obvious," (5) that the provincial court jury should not have awarded damages to Spinke when he never named or proved any definite amount.

[39] LIII, 319, 320.

[40] LIII, 355-357.

[41] XLI, 589-591.

[42] LIII, 609, 616 (receiving stolen goods); LIII, 424, 425, 427 (dress); III, 488 (naturalization); LIII, 619 (wolves); III, 526 (Indian trader); XLIX, 455 (ordinary keeper); XLIX, 156 (attorney for Brent); XLIX, 53, 76, 142, 145 (service on juries). For other cases in which Lumbrozo acted as an attorney, see LIII, 357, 368, 386. Lumbrozo also acted as "a commissionated appraiser" of an estate. (LIII, 502). In 1661 he sued James Jolly, another ordinary keeper. (XLI, 467). For an account of the charges of blasphemy and abortion made against this Jew, see text, pp. 166, 167, 240, 241.

[43] LIII, 367, 376-380. Possibly the unprinted records of the provincial court might disclose the final decision in this case. Mrs. Belaine is spoken of as "Gammer" Belaine. This meant an old wife, or woman. Mrs. Dodd and Mrs. Nevill are frequently spoken of as "Goodie" Dodd and "Goodie" Nevill. For explanation of this term, see citation sixty, chapter seven. See introduction LIII, p. lv in re explanation of sign of the horns.

This was not the only action of defamation which the Dodds brought against the Nevills. In a separate suit, Richard Dodd alleged that Joan Nevill had slandered him by calling him "a perjured fellow" who had taken a false oath "upon record." Mrs. Nevill, no doubt, was referring to the defamation suit discussed in the text in which she was accused by Dodd of slandering his wife. This case was also appealed to the provincial court. (LIII, 380-382).

[44] LIII, 168, 220, 231-234. In re John Nevill, see CMS, pp. 108, 109.

[45] LIII, 220, 234-237.

[46] LIII, 9, 10, 13, 235. In addition to the cases of defamation discussed in the text, a careful search of the provincial and county court records revealed that four other cases of defamation were instituted in the provincial court, thirty in the Charles county court, thirteen in the Kent county court, four in the Somerset county court, and two in the Talbot county court, or a total of fifty-three cases. Although most of these cases never came to trial, they are evidence none the less of how zealously the Maryland colonist guarded his reputation.

For the procedure where damages in action of slander were less than forty shillings, see II, 201, 202.

In 1651, Cecil Calvert advocated the suppression of all false rumours and reports and advised the passage of law to punish those who spread them. (I, 334, 335). In 1654, a law was passed which made any one liable to pay a fine of one thousand pounds of tobacco if he circulated "false news and reports." (I, 343). By the provisions of a new act passed in 1671, the penalty was doubled. Under the new act, all "idle and buss-headed persons," who invented false reports about any of the provincial or county court justices, could be fined or even corporally punished. (II, 258-260, 273, 274). In 1681, Philip Calvert issued a proclamation calling for the apprehension and punishment of all divulgers of false news. (XV, 391, 392). There is only one instance of where a man was prosecuted under these laws against spreading false rumours and that was when John Sollers, of Calvert county, was summoned before the council "for divulging false news." After a hearing Sollers was discharged. (VIII, 47).

CHAPTER X—BIBLIOGRAPHICAL NOTES

[1] I, 9, 106 (councillors sick); I, 216, 223, 275, 284 (delegates sick).

[2] III, 358; XLIX, 303 (judges sick or lame); III, 456; XLIX, 267; LIV, 495, 506, 550 (witnesses and litigants sick); XV, 338 (woman sick). For other cases of where sickness, or lameness, was given as an excuse, see X, 131, 367; XLIX, 129, 265; LIV, 319. Some litigants before the court leet of St. Clement's Manor were "essoined," that is, excused from coming to court because they were sick and could not attend "to do their suit." (LIII, 633).

In one case, a woman who was ready "to be brought to bed, being great with child," was unable to appear in court until after the delivery (XLIX, 505), while in another case one of the parties to the suit could not come to court "by reason of his wife being newly delivered of a child." (LIII, 61). A member of the assembly received permission to leave and go home because of his wife's sickness. (I, 281).

[3] A Character of the Province of Maryland, by George Alsop, in Hall, p. 385.

[4] XLIX, 94 (distemper); Annual Letters of Jesuits, in Hall, p. 119 (common sickness); LIV, 10 (ague).

[5] X, 97; XLI, 479; V, 65, 350; XV, 292 (ague and fever); V, 375 (tertian quartan ague); Annual Letters of Jesuits, in Hall, p. 131, (Jesuit).

[6] Letter of Father Andrew White, in Calvert Papers, no. XXVIII, p. 202. During the winter of 1636, White had an illness which impaired his hearing. (Ibid.).

[7] X, 466; XLI, 9; XLIX, 303.

[8] LIV, 9 (Wilson); XLI, 503, 504 (Watson).

[9] XLI, 501; XLIX, 303; Annual Letters of Jesuits, in Hall, p. 132 (apostemes or abscesses); Jesuit Letters, in Hall, p. 121 (ulcers); LIII, 145 (cankers); LIII, 410 (servant); XLI, 205, 206 (negro); XIII, 21, 80 (Young). In re Young, see index, CMS.

[10] X, 73; XLI, 385 (apoplexy); LIII, 401, 402 (heart attack).

[11] XLI, 9 (sickness); LIII, 312, 313 (lame); XLI, 5, 6 (stone). On Kent Island Henry Pinke, reader of prayers, broke his leg and was "unserviceable." (High Court of Admiralty, Libels 100, no. 63, in Md. Hist. Mag. vol. XXVIII, p. 32 et seq.).

[12] IV, 56; LIII, 84 (non compos mentis); XLIX, 402 (Armstrong).

[13] XLIX, 441, 442, 493 (Hasling); LIII, 93 (Grace).

[14] LIII, 385. Roger Evans, a servant, was said to have "the sleepy disease." Was this sleeping sickness? The same man was also said to have had a spell of "deep taciturnity." Was this melancholia? (LIII, 140, 141).

[15] V, 411; X, 15, 16. In re venereal diseases among the Indians, see CMS, p. 378.

[16] XLI, 270-275. One woman said that "the custom of women was upon her." Probably this was a reference to the menstrual period. The same girl claimed to have caught a venereal disease from the man with whom she had had intercourse. (LIV, 69).

Walter Peake (or Pakes) was a chirurgeon who considered himself capable of curing "the country duties." (X, 15, 16).

[17] IV, 255. The coroner was quite an important figure in early Maryland. Whenever the circumstances surrounding the death of a colonist made it neces-

sary, a coroner's inquest would be held. When, for example, John Dyatt died a coroner's jury was impanelled to investigate the cause of his death. Among the items charged against his estate were a fee to the coroner, another fee for impanelling the jury and sixty pounds of tobacco " to the jury in drink." (XLIX, 510). A coroner could ask two hundred and fifty pounds of tobacco for " viewing the bodies of any person, or persons, murthered, slain, drowned, or otherwise dead by misadventure." (II, 292-294, 536. See also III, 91).

Often a very prominent man held the position of coroner. On the Eastern Shore of Maryland, Colonel William Stevens held this position. In the order appointing him he is described as " a gentleman." (LIV, 638. See also index, CMS). Coroners in each county were appointed by the governor. (II, 130, 131). Any one who was appointed to this position and who refused " to execute the office of coroner," or refused to take the required oath, could be fined one thousand pounds of tobacco. (I, 411). The coroner must take the following oath:

The Coroner's Oath.

> You shall swear that you will well and truly serve the Right Honourable the Lord Proprietor of this province and the people of the said province as one of his Lordship's coroners of this county of . . . and therein you shall diligently and truly do and accomplish all and every thing and things appertaining to your office after the best of your skill and power both for his Lordship's profit and the good of the inhabitants within the said county and be content with the fees allowed by act of assembly in this province.
>
> So help me God, etc. (V, 96, 97).

See also LIII, xl-xlii.

[18] XLIX, 11, 12.

[19] XLIX, 307, 308.

[20] LIV, 391. Goddard appears to have received one thousand pounds of tobacco for opening the skull of the dead man. (LIV, 410).

The following reports by chirurgeons are also of interest. Robert Lloyd, a chirurgeon, after examining the body of Jeffrey Haggman, a servant, who it was thought had been beaten to death by his master, was asked after he had viewed the corpse whether he had found blue spots " upon the forepart and hinder parts." Lloyd replied, " Yes, I saw two strokes and a sore on his side that was formerly under my cure." Later Lloyd asked to be paid for his services for having made an examination of the dead body. (XLIX, 307, 332, 333).

When the body of John Dyatt was found at St. Mary's, a jury of inquest was impanelled by John Lawson, sheriff and coroner of St. Mary's county. The report of this coroner's inquest follows:

> We do find the cause of his death to be, he being sick and weak of body and nature decayed in him by reason of his age and for want of convenient sustenance and looking after and his lying under a cold bank; all was the instrumental cause of his death.
>
> Witness our hands, the 15th of October, 1665. (XLIX, 510).

When Henry Gouge was found dead and his master, John Dandy, was suspected of the murder, James Veitch, a sheriff, was ordered by the provincial court to visit the spot where Gouge was buried, taking with him some neighbours and also Richard Maddokes and Emperor Smith, two chirurgeons. These men were to open the grave and continued the court order:

> After the neighbours with the two chirurgeons have taken a diligent view of the corpse, then the chirurgeons in the view of those that shall be then

present, are to take off the head of the said corpse and after diligent view
and search to signify under their hands, how they find the said head and
corpse, and are to cause the head to be carefully wrapped up and warily
brought to the Court, with what convenient and possible speed as may be.

The chirurgeons did as they had been instructed and later petitioned the court
" for satisfaction for their trouble and pains in viewing the body of Henry
Gouge, lately murdered, and dissecting the head from the corpse." The court
decided to allow the chirurgeons one hogshead of tobacco " equally to be divided
betwixt them."

The neighbours who went with the two chirurgeons tendered a report of the
condition in which they found the body of the murdered youth. They said:

> We here detest under our hands, that we can see, nor find nothing about
> the said head, but only two pieces of the skin and flesh broke on the right
> side of the head and the skull perfect and sound, and not anything doth
> or can appear to us to be any cause of the death of the said Gouge.
>
> And also we do detest that we did endeavour what possible in us lay to
> search the body of the said corpse and could not possibly do it, it being so
> noisome to us all and being put at first in the ground without anything
> about it, as the chirurgeons and the sheriff can satisfy you. This is the
> truth and nothing but the truth as witness our hands.

These facts, while gruesome, throw interesting light on the steps taken in a
murder case. (X, 524, 525, 546).

[21] I, 97 (Act of 1640) ; I, 4; IV, 67, 107, 119, 139, 140, 153, 171 (Robinson).

[22] IV, 72 (licentiate in physick) ; IV, 149, 155, 158, 163, 169, 186 (Binx'
suits) ; IV, 221, 228, 229 (physic to servants) ; X, 432 (Sharpe) ; XLI, 272
(physic and blood-letting).

[23] Annual Letters of Jesuits, in Hall, p. 126 (cure of Indian) ; A Relation of
Maryland, in Hall, p. 79 (Indian use of plants).

Captain John Smith during his voyages of exploration up the Chesapeake
Bay, in 1608, made some interesting comments on the Indian's knowledge of
drugs. Said Smith:

> Every spring they make themselves sick with drinking the juice of a root
> they call Wighsacan, and water, whereof they pour so great a quantity,
> that it purgeth them in a very violent manner, so that in three or four days
> after they scarce recover their former health.
>
> Sometimes they are troubled with dropsies, swellings, aches, and such
> like diseases; for cure whereof they build a stove in the form of a dove-
> house with mats so close that a few coals therein covered with a pot will
> make the patient sweat extremely.
>
> For swellings also they use small pieces of touchwood, in the form of
> cloves, which pricking on the grief they burn close to the flesh, and from
> thence draw the corruption with their mouth.
>
> With this root Wighsacan they ordinarily heal green wounds. But to
> scarrify a swelling, or make incision, their best instruments are some
> splinted stone. Old ulcers, or putrified hurts, are seldom seen cured
> amongst them.
>
> They have many professed physicians, who, with their charms and rattles,
> with an infernal rout of words and actions, will seem to suck their inward
> grief from their navels, or their grieved places.

Captain Smith said that the savages had great respect for the English chirurgeons believing that any of their plasters would heal a wound, but added the captain:

> But 'tis not always in physicians skill,
> To heal the patient that is sick and ill;
> For sometimes sickness on the patient's part,
> Proves stronger far than all physicians art.

(The Generall Historie of Virginia, New England and the Summer Isles, by Captain John Smith. Vol. I, p. 71).

For an account of the Jesuit mission among the Maryland Indians, see chapter seventeen in CMS.

[24] A Relation of Maryland, in Hall, p. 79; Lord Baltimore's Colony, in Hall, p. 10; Letter of Robert Wintour, unpublished manuscript.

[25] X, 162, 163. George Alsop speaks of using an antimonial cup or powder as an emetic or purgative. (A Character of the Province of Maryland, by Alsop, in Hall, p. 382).

[26] IV, 30, 77, 78, 388; LIII, 85 (salves, mithridate, etc.); X, 175 (opium); X, 171, 178 (strong pills); XLI, 17, 20 (savin). A blue stone was thought to be good for sore eyes. (X, 389). Instruments, probably surgical instruments are mentioned in one place. (IV, 30). While William Claiborne was in command of Kent Island, fruit, sugar and spices, and physic were given sick men. (High Court of Admiralty, Libels 100, no. 63, in Md. Hist. Mag. vol. XXVIII, p. 32 et seq.).

[27] LIII, xxii, xxiii, 387-391, 496, 497. Ratsbane was given this girl.

[28] LIII, 503.

[29] VII, 42, 49, 104, 105, 160.

[30] LIII, 362, 424-427. The cost of the purging pills, blood-letting, plasters, ointment and other drugs furnished Mrs. Haggate was four hundred and sixty pounds of tobacco, or in English currency about three pounds, twelve shillings. The remedies given Mr. Haggate cost six hundred and eighty pounds of tobacco, or about five pounds, ten shillings. For the ten days time he took in "going and coming back afoot" to visit Mrs. Haggate, Chirurgeon Meekes asked four hundred pounds of tobacco, or about three pounds sterling. Meekes asked three hundred pounds of tobacco, or about two pounds ten shillings, for "boat and hands and time for four days and visit" to Mr. Haggate. All the estimates in English money are based on the assumption that tobacco was worth two pence a pound. Whether Meekes made a reasonable charge for his services, or whether he administered the right drugs, cannot be determined by applying present-day standards.

After the death of her husband, Mrs. Haggate married Richard Fouke, a planter. (LIII, 518).

[31] LIII, 431.

[32] XLI, 161-164. On another occasion, Peter Sharpe sued the executors of the estate of Bassill Little, merchant, for over two thousand pounds of tobacco for "physic and attendance expended and administered unto Little in the time of his sickness." At the same time William Dorrington asked to be recompensed for the expenses he had incurred in connection with Little's sickness and death, including "the diet" of three servants, and the funeral charges, for which he asked three thousand pounds of tobacco. (XLI, 22, 23).

[33] X, 396 (ignorance); X, 434, 439 (Iger).

[34] X, 399, 400. Peter Godson recovered nearly two hundred pounds of tobacco from the estate of Thomas Trumpeter for the physic which he had given

Trumpeter during his illness. John Ashcomb, although not a chirurgeon, also sued Trumpeter's estate for " attendance and provision in his sickness and for his winding sheet and the burying of his corpse." This took an additional two hundred and fifty pounds of tobacco from the assets of the estate. (X, 398).

[35] IV, 215, 229-231, 240.

[36] IV, 294 (Hervey); IV, 215, 216, 220, 301 (Plowden). George Binx, another chirurgeon, also sued Plowden for his " pains and physic " for the cure of one of his servants. (IV, 229). Ellyson brought suits against Henry Brooks " for pains and charge of chirurgery." (IV, 267, 294).

[37] IV, 249, 250. There were many other chirurgeons in early Maryland. Even before the colony was founded, Captain John Smith on his voyages of discovery up the Chesapeake Bay, in 1608, was accompanied by chirurgeons. On the first voyage Walter Russell went along and on the second voyage Anthony Bagnall was the chirurgeon. (The General History of Virginia, etc., by Captain John Smith, vol. I, pp. 115, 124). While Captain William Claiborne was in command of Kent Island, a Dr. Potts was the chirurgeon. (H. C. A. Libels, 100, no. 63, in Md. Hist. Mag., vol. XXVIII, p. 32 et seq.).

After the colony was founded, many besides those who are mentioned in the text were chirurgeons by profession. Dr. Luke Barber was one. He was a member of the governor's council and at one time served as deputy governor of Maryland during the absence of the governor. (I, 382; III, 331, 332. See also XLI, 353, 354; XLIX, 147).

Other prominent men who were chirurgeons included Thomas Gerrard, Lord of St. Clement's Manor (IV, 51, 56), William Brainthwaite, one time commander of Kent Island (IV, 137), and John Brooke, a member of the lower house of the assembly. (XIII, 106). Others who practiced chirurgery in the province, included Emperor Smith (X, 467, 468, 476, 477, 482), John Stanesby (XLIX, 552), George Hack (XLI, 492; III, 459), Gaspar Guerin (XLIX, 297, 376, 377, 401), Joseph Sempile (XIII, 227), Paul Scurfield (XIII, 178), John Stansley and John Peerce (V, 65) and a " Dr. Maddox." (XV, 97).

In the different counties there were many chirurgeons. Some of these were:

Calvert county. Thomas Wild. (XLIX, 466).

Kent county. Bartholomew Glenin. He was also a county justice (III, 532). Thomas Marsh (LIV, 3), William Hemsley (LIV, xxvi, 172), Thomas Ward (LIV, 10, 32, 36), and Mitchell de Conty (LIV, 272, 302).

Charles county. John Stone (LIII, 604, 605).

Somerset county. George Hoorsford (LIV, 618, 724, 769), Abraham Gale (V, 464) and Thomas Walley (LIV, 699, 704).

Talbot county. John Dolby (LIV, 461, 473), Humphrey Davenport (LIV, 597) and George Soley (LIV, 498).

Thomas Markeyne complained to the provincial court against William Chaplane on the ground that when he was lame Chaplane had without the approval of a chirurgeon cut a joint in one of his toes. (XLI, 81).

[38] IV, 197 (Trafford); X, 97 (ague and fever); IV, 256, 257 (Ottoway). Other lay practitioners in early Maryland included James Benson (XIII, 97), John Gay (LIII, 45), and Owen Griffith. (LIV, 500, 501).

[39] X, 301 (Stevens); LIV, 35 (Morgan). For similar cases, see X, 478 (John Crabtree), LIV, 437 (Jacob Brimington), LIV, 503 (William Pitt). In re Stevens see index, CMS.

[40] LIV, 466, 467. This case was decided in 1670. This was not the same Thomas Watson who was thought to have been beaten to death by Captain Thomas

Bradnox in 1661. See text, pp. 102-104. An agreement, similar to the one discussed in the text, was made with John Wade, chirurgeon. (X, 491, 496; XLI, 71, 72). In Kent county there was an agreement made between a master and servant by the terms of which the servant was to be set free if he cured his master's leg. (LIV, 248).

⁴¹ XV, 207, 208.

⁴² XLIX, 149.

⁴³ II, 14 (Howell); LIV, 690 (Waerum). Ellinor Martin in a petition to the provincial court alleged that for fifteen years she had been afflicted " with a great lameness in her legs, together with several other distempers, which does altogether make her incapable of providing for herself." For this reason she asked the court to let her have possession of a third part of " all the lands and edifices," which had belonged to her deceased husband. (XLI, 513, 514, 518, 521).

⁴⁴ LIII, 384, 385 (Watson); LIV, 425 (Smith).

⁴⁵ III, 411 (Stockett); XVII, 28 (Lemaire); XV, 403 (plasters, etc.); XV, 102 (public levy); I, 407 (pay). We also find Lemaire suing a planter for medicines administered to his family. (XVII, 79).

⁴⁶ I, 363, 397 (Sprye); I, 364 (Wallcott); XIII, 179 (bandages, etc.).

⁴⁷ IV, 379, 383 (Brent); X, 96 (agreement). Henry Hooper also sued two planters, one for chirurgery which he had " performed," and the other for " physic." (IV, 264, 285). Hooper was dead by 1653. (X, 321). By the terms of his will, after all his debts were paid the remainder of his estate was to be employed in " pious uses." (X, 11).

Jaques Peon, a chirurgeon, who came to the colony from France, was asked by the governor and council to dress the wounds of a man " until it be made a perfect cure." (III, 450).

It is interesting to note that mention is made in the early records of ship chirurgeons. (XVII, 92, 93). William Jennings was the chirurgeon on a ship sailing from Bristol to Maryland. (XLIX, 103). Samuel Pratt was chirurgeon on the ship " Figtree " (LIV, 448), while Francis French was chirurgeon on the ship " Globe." (VII, 281).

⁴⁸ LIV, 544 (attendance); II, 247 (Brooke). John Hinson made a complaint which was similar to Brooke's. (II, 248). In re Tilghman, see also II, 252, 253, 267, 269, 357, 358, 360.

⁴⁹ LIV, 324, 325. Tilghman brought several suits, in one case recovering one thousand pounds of tobacco " for physic and attendance " (LIV, 573), while in another case he recovered over three hundred pounds of tobacco for physic alone. (LIV, 393).

⁵⁰ See CMS, pp. 240, 241.

⁵¹ LIV, 415, 419, 481 and introduction pp. xxi, xxiv. The court also allowed John Edmundson five hundred pounds of tobacco for Smith's " accommodations " during the time he stayed at his house. A Mrs. Carpender was allowed twelve hundred pounds of tobacco by the Talbot county court " for entertaining a wounded person." (LIV, 481).

For information regarding soldier pensions in early Maryland, see CMS, pp. 241-243.

⁵² XIII, 179 (nurse's wage); LIII, 145-147 (Vanderdonck). For an account of Francis Doughty, see text, pp. 170, and citation fifty-eight, chapter seven.

⁵³ LIII, 148, 149, 215.

⁵⁴ LIII, 229-231.

⁵⁵ LIII, 248, 261, 262.

[56] LIII, 262, 263.

[57] LIII, 240, 241. Mary Oneale appears once more in the Charles county court records when she attempted " the cure " of William Bowles. (LIII, 329).

[58] IV, 446, 478, 479.

[59] LIV, 194, 222, 223, 236.

[60] XLI, 332, 333. Robert Long's wife, Jemima, attended Bridget Heard " in the time of her sickness." For his wife's " administering means unto the said Bridget," and for her attendance, charge and trouble, Long asked four hundred pounds of tobacco. The commissioners of Charles county ordered the administrator of Bridget's estate to pay this amount. (LIII, 605, 606).

In Talbot county, Christopher Denny, upon the death of Thomas Hynson, a commissioner of that county, asked three barrels of Indian corn from Hynson's estate for his wife's " attendance of Hynson and his wife in their sickness." Denny recovered the corn. (LIV, 417). John Nunne gave a cow calf to Mary Shertcliffe for the care that her mother had taken of him when he was sick. (IV, 485).

[61] LIV, 230 (Sprye); LIII, 85, 92, 93 (Mrs. Cherman); LIII, 190, 191 (John Cherman); X, 415; IV, 244, 268 (Hebden). John Cherman had to call on Mathew Gage, a chirurgeon, to heal his own finger. (LIII, 84).

[62] XLI, 327, 335.

[63] LIV, 503. A " Doctor Waldron " assisted at childbirth. In one instance, when the child was born dead, Waldron said that it was his opinion that the mother " had gone out of her full time, and that the child had been dead, as he did suppose, three weeks in its mother's womb." (X, 171).

There is rather an unusual agreement in the Charles county records. This was when Elenor Empson, wife of William Empson, deceased, bargained and sold to Richard Dodd two heifers " for the nursing, keeping and relieving " of Mary Empson, her daughter for the period of two years. Elenor stated in the agreement disposing of the child that she was compelled to do so, as otherwise the child might perish. (LIII, 136, 137, 193, 196).

[64] V, 131.

[65] I, 495; II, 330 (estate and quality of person); I, 156, 190 (first defrayed). The following citations contain references to burials and the customs observed. See IV, 101, 108, 113; LIV, 98; XLI, 254. One hundred pounds of tobacco was the usual cost of a coffin. (XLI, 327, 335). The same amount was asked for building " a pall fence " around a grave. (LIII, 372). Richard Trew, a boatwright, seems to have employed his spare time in making coffins. (LIII, 503; CMS, pp. 72, 77). Four yards and a half of some cotton or linen fabric was enough for a winding sheet. (LIII, 253). Sometimes wax-lights and even a hearse-cloth were provided. (IV, 388).

In his will a colonist would sometimes declare that he wished to be buried " in a Christian-like manner." (LIV, 13). John Lloyd, who described himself as " a gentleman," gave instruction for his body to be decently buried " in the ordinary burying-place in St. Mary's chapel yard." To those persons who carried his corpse to the chapel he bequeathed to each a black mourning ribbon and a pair of gloves. (XLI, 116).

Several cases came up for trial which throw light on the expenses incidental to burying a colonist. When John Leaven died, it cost over three thousand pounds of tobacco to bury him. His estate was held liable for this sum. (LIV, 572, 573). Yet in another case when Hopkin Davis died the court would only allow only four hundred pounds of tobacco for burial expenses instead of one thousand which had been asked. (LIV, 465). In still another case when

William Snaggs died, only one hundred and sixty pounds of tobacco was allowed for funeral expenses. (LIV, 429). When John Stringer died, William Wilkinson, a clergyman, sought to recover not only for the expense he had incurred by having Stringer in his house while he was sick, but also for " the charges of burial," which included one hundred pounds of tobacco for a funeral sermon and three hundred pounds of tobacco for a funeral dinner. The provincial court finally allowed Wilkinson over eight hundred pounds of tobacco out of Stringer's estate. Wilkinson seems to have eked out an existence by performing other duties besides ministerial ones. He was also clerk of the court and acted sometimes as an attorney. (X,173, 309, 311, 314, 330, 422, 475). Without knowing the amount of the estate, or the " quality of the person," it is impossible to decide whether the amounts allowed by the court in the cases discussed above were fair and reasonable.

[66] LIII, 193-195, 207-210. Thirty-six yards of black ribbon was distributed among the mourners and the three barrels of beer were consumed by them in two days.

NOTE

As almost all of the material on which this book is based has been taken directly from the Archives of Maryland, it is unnecessary to give a bibliography of sources, books, etc. The Archives of Maryland, published by the Maryland Historical Society, contain the proceedings of the assembly, council, provincial court, court of chancery, and the county courts.

INDEXES

INDEX TO NAMES OF PERSONS

NOTE. Additional information regarding many of the persons listed in this index may be found in Captains and Mariners of Early Maryland, by the same author. See index of that book.

A

Adams, Henry, 89.
Adams, Thomas, 41, 87.
Alexander, Charles, 108, 109.
Allanson, Thomas, 141.
Allen, James, 246.
Allen, Thomas, 130.
Allen, William, 65.
Alvey, Pope, 21-29, 52, 108, 109.
Anderson, Christopher, 105-107.
Anderson, David, 148.
Anderton, John, 192.
Andrews, Christopher, 115.
Anketill, Francis, 78.
Anther, Phillip, 135, 136.
Antonio, 121.
Arnald, Teressa, 218.
Armstrong, Francis, 77, 117, 235.
Aspinall, Henry, 23, 26.
Atcheson, Susan, 182, 183.
Avery, Anne, 218, 219.
Austin, Richard, 221.

B

Bagby, John, 15.
Bagley, William, 217, 218.
Baisey, Michael, 162.
Baker, Thomas, 66, 67, 89, 139, 140, 143, 207, 228-231.
Ballance, John, 215.
Barber, Luke, 81, 222, 223.
Barbary, Susan, 100.
Barbery, Anne, 195.
Barnes, Isabella, 254, 255.
Barnett, Mary, 190.
Barre, Phillip, 214, 215.
Bassett, Thomas, 12.
Batchelder, 231.
Batten, William, 226, 228.
Baxter, Mary, 5.
Bayley, John, 57.
Beane, Walter, 215.
Beetle, Ann, 139.

Belaine, 227, 228.
Belcher, Thomas, 160.
Bennett, Elisabeth, 168.
Bennett, Disborough, 8.
Bennett, Richard, 12.
Berry, William, 218, 219.
Bessick, John, 108.
Bigger, John, 72.
Bish, Ursula, 198.
Bishop, Abraham, 190.
Binx, George, 236, 238.
Black, William, 94, 95.
Blackwood, John, 230.
Blakiston, John, 74.
Bland, Thomas, 16.
Blunt, Richard, 114, 115.
Bonnefield, Christian, 205, 206.
Bosworth, John, 172.
Bowen, Hannah, 194.
Bowles, John, 8, 186.
Bowling, James, 215.
Boyden, John, 65.
Bradnox, Mary, 9, 93, 96-98, 101-104, 115, 178, 179, 254.
Bradnox, Thomas, 9, 96-98, 101-105, 115, 127, 146, 147, 162, 178, 179, 217, 234, 254.
Brasse, Alice, 101.
Brayfeelds, Susanna, 191.
Brent, Giles, 2, 226, 246.
Brent, Margaret, 10, 15, 16, 71, 72, 104, 249.
Bretton, William, 2.
Bright, Francis, 50.
Bright, Thomas, 193.
Briggergrass, James, 88.
Brispo, Anthony, 119.
Brooke, Baker, 21, 62.
Brooke (or Brooks), Francis, 130, 208.
Brooke, Mrs. Francis, 130.
Brooke, John, 237, 238.
Brooke, Michael, 202.
Brooke, Nicholas, 250.

Brooke, Robert, 63, 64, 81.
Brookes, William, 141.
Broomfield, John, 34.
Brown, Mathew, 117.
Brown, Nicholas, 211.
Bryan, Robert, 218.
Bryant, John, 136.
Budd, Katherine, 181.
Burgess, William, 125.
Burnet, Archibald, 200.
Burrowes, Nathaniel, 31.
Burton, Edward, 214.
Bushnell, Thomas, 4.
Bussey, George, 208.
Butler, George, 1.
Butler, John, 185.
Butler, Mary, 184, 185.

C

Cabell, John, 65.
Cage, John, 218.
Calloway, Peter, 196.
Calvert, Cecil, 3, 58, 82, 145-147, 150, 175-177.
Calvert, Charles, 1-3, 21, 29, 70, 160.
Calvert, Leonard, 41, 58, 71, 82.
Calvert, Philip, 21, 54, 61.
Calvert, William, 21, 22, 61, 62.
Cannaday (or Canneday), Cornelius, 92, 201, 202.
Cannaday, Susan, 201, 202.
Carline, 185.
Carline, Henry, 4.
Cartwright, Demetrius, 213.
Catchmey, George, 185, 186.
Catchpole, Judith, 129.
Catterson, Francis, 159, 160.
Cawsine, Nicholas, 58.
Chandler, Job, 230, 231.
Chaplin, William, 99, 200.
Chapling, Richard, 192.
Cherman, John, 255.
Chessill, Martha, 194.
Chew, Samuel, 124, 130.
Clark (or Clarke), Robert, 112, 135.
Clark, Thomas, 73.
Clarke, Daniel, 204.
Clay, Henry, 164, 222.
Clifton, Stephen, 237, 238.
Clocker, Daniel, 53, 54, 147, 148, 256.
Clocker, Mrs. Daniel, 256.

Clocker, Mary, 53, 54.
Clymer, John, 220, 221.
Cocher, Thomas, 134.
Cockerill, Robert, 143.
Cockshott, Jane, 112.
Codwell, Margaret, 207.
Cole, Josias, 166.
Cole, Robert, 69.
Commins, Edward, 2, 10, 11, 208, 211.
Connery, Edward, 69.
Constable, John, 140.
Cooper, Ann, 194.
Cooper, John, 190, 191.
Copley, Thomas, 114, 115.
Coppage (or Coppedge), Edward, 10, 182.
Cornelius, John, 4.
Cornwallis, Thomas, 15, 44, 59, 63, 80, 81, 84.
Cotton, Edward, 72.
Counyer, Quintin, 91.
Coursey, Henry, 74, 87, 98, 145, 190, 200.
Courtney, Thomas, 53, 64, 78, 219.
Courtney, Mrs. Thomas, 78.
Courts, John, 212, 213.
Cowman, John, 168, 169.
Cox, Frances, 254.
Cox, Richard, 219.
Cox, William, 254.
Crab, Martha, 248.
Cromwell, Oliver, 167.
Cromwell, Richard, 167.
Cully, John, 52.
Cullin, John, 34.
Curre, Thomas, 135.

D

Dabb, John, 234.
Dandy, John, 125, 126, 132, 133, 142.
Dandy, Mrs. John, 142.
Daniel, Jane, 219, 220.
Daniel, Leonard, 219, 220.
Darnall, Henry, 76.
Davis, John, 5, 113.
Dawson, William, 68.
Daynes, William, 217.
Deare, John, 63, 115, 211.
Deare, Mrs. John, 211.
Dell, William, 68.
Dennis, Robert, 57.

Denny, Christopher, 68.
Dent, Thomas, 128.
Dickes, Thomas, 12.
Digges, William, 76.
Dodd, Mary, 226-228.
Dodd, Richard, 143, 226-228.
Dollovan, Derby, 144.
Dorrington, Sarah, 144.
Dorrington, William, 144, 194.
Dorrosell, Joseph, 224, 225.
Doughty, Enoch, 170.
Doughty, Francis, 169, 170, 200, 251.
Doughty, Mary, 251.
Douglas, James, 51.
Dowland, William, 8.
Due, Patrick, 131, 132.

E

Edloe (or Edlow), Joseph, 160, 161, 195.
Edmonds, Joseph, 115, 116.
Edmundson, John, 247.
Edward, 132, 133.
Elliott, William, 14, 67, 163.
Ellis, John, 52.
Ellyson, Robert, 245, 246.
Emerson, Thomas, 219, 220.
Emerson, Mrs. Thomas, 219, 220.
Empson, Elenor, 199, 213.
Empson, William, 4, 229, 230.
Erbery, Edward, 2, 3.
Ereckson, John, 192.
Evans, Roger, 139, 140.
Evans, William, 14, 21-23, 26, 62.
Exon, Henry, 159.

F

Farmer, Michael, 100.
Fendall, Josias, 61, 73, 74, 146, 172, 251.
Fenwick, Cuthbert, 13, 32, 59, 64, 71, 73.
Fenwick, Cuthbert, Jr., 71.
Fenwick, Ignatius, 71.
Fenwick, Teresa, 71.
Finch, Francis, 144.
Fincher, Joseph, 122-125.
Finney, William, 208.
Fisher, Philip, 114, 115.
Fitzherbert, Francis, 187.

Fleet, 185.
Floyd, Thomas, 135.
Ford, Robert, 12.
Fossett, John, 166.
Foster, Richard, 79.
Foster, Seth, 87, 143, 214.
Fouke, Richard, 18.
Fowler, Rebecca, 168, 169.
Fox, Henry, 32.
Fox, James, 78, 230.
Frizell, Susan, 95.
Fuller, Edward, 190.
Fuller, William, 236.
Furnifield, John, 148.

G

Gaine, Anne, 177.
Gardiner, Luke, 1, 2, 69.
Gary, Elizabeth, 180, 181.
Gary, William, 192.
Gaylourd, James, 149.
Gee, John, 129.
Gellie, Robert, 159.
Gerrard, Thomas, 15, 35, 58, 62, 63, 68, 74, 79, 87, 88, 92, 144-146, 150, 156.
Gess (or Guest), Anne, 89, 252.
Gess (or Guest), Walter, 18, 89.
Gibbons, Thomas, 54.
Gibson, John, 178.
Gillett, German, 14.
Gillford, Mary, 183, 255.
Glover, Elizabeth, 224.
Glover, Giles, 224.
Goddard, Thomas, 238.
Godson, Peter, 171, 172, 245.
Goneere, John, 30.
Goodale, Elizabeth, 169.
Gordian, Mary, 241.
Gott, Henry, 12.
Gouge, Henry, 125, 126.
Gould, Anne, 235, 236.
Gould, John, 223-225.
Gould, Margery, 223-225.
Goulson, Sarah, 93.
Grace, Henry, 235.
Grammer, John, 105-108.
Greene, Mrs. Elizabeth, 31, 32, 221, 222.
Greene, Elizabeth, spinster, 127, 128.

Greene, John, 172, 173.
Greene, Leonard, 58.
Greene, Thomas, 58, 75, 89.
Greene, William, 31.
Greenhill, Thomas, 136.
Greenway, John, 135.
Greenway, Mary, 135.
Griffith, John, 191.
Grigg, John, 152.
Grosse, Roger, 188.
Grub, Mary, 218.
Guinn, Thomas, 50.
Gundry, 194.
Gunnell, William, 122, 123.
Guyther, Nicholas, 145.

H

Haggate, Anne, 241-243.
Haggate, Humphrey, 18, 241-243.
Haggman, Jeffery, 122-124.
Hailings, Thomas, 112.
Hall, John, 14.
Hall, Penelope, 189.
Hall, Sarah, 95, 96.
Hambleton, John, 187, 188.
Hammond, John, 23, 73, 225, 226.
Hammond, Mrs. John, 226.
Hanson, Andrew, 88.
Hanson, Mrs. Andrew, 88, 89.
Hanson, Hance, 89.
Harris, Elizabeth, 128, 129.
Harris, Samuel, 144.
Harris, Thomas, 71.
Harrison, Jane, 191.
Hart, Robert, 248.
Hartwell, Mary, 216.
Harwood, Edward, 221.
Harwood, John, 62, 63.
Harwood, Olive, 221.
Harwood, Robert, 180, 181.
Harvey (or Hervey), Nicholas, 72, 245, 246.
Hasling, Jeremiah, 235.
Hatch, 231.
Hatch, John, 73, 169.
Hatch, Mrs. John, 169.
Hawkins, John, 76.
Hayes, George, 250.
Hayes, Hercules, 92.
Heard, William, 140, 252, 253.

Hebden, Katherine, 255, 256.
Hebden, Thomas, 58, 63, 246, 255.
Helme, John, 244.
Hemsley, William, 234.
Herde, William, 203.
Herring, Bartholomew, 245.
Herring, Mrs. Bartholomew, 245.
Herrman, Augustine, 37, 111.
Hews, Mary, 181.
Hide, Henry, 42.
Higges, Henry, 5.
Hill, Richard, 162.
Hills, Thomas, 162.
Hinfield, Robert, 88.
Hobbs, Robert, 132.
Hodges, Charles, 140, 141.
Holland, Francis, 139.
Holland, Richard, 247.
Holliday, John, 10.
Hollingsworth, Charles, 103.
Hollingsworth, William, 158.
Hollis, John, 63.
Holmwood, John, 140, 141.
Holt, Dorothy, 184, 205.
Holt, Robert, 205, 206.
Hood, Joan, 8.
Hooper, Henry, 245, 249.
Hooton, Thomas, 213.
Hopkins, William, 93.
Howell, Alexander, 248.
Howell, Blanche, 30.
Howes, William, 112, 113.
Hudson, Edward, 141, 184.
Hughes, Tom, 222, 223.
Humes, Patrick, 115.
Humphrey, Ellis, 115.
Hunniford, 203.
Hunt, John, 56.
Hunt, Sarah, 106.
Hunt, William, 139.
Hunt, Mrs. William, 139.
Hunton, Benjamin, 222.
Husbands, Edward, 2, 241.
Hyde, Philip, 245.
Hynson, Thomas, Sr., 5, 177, 209.
Hynson, Thomas, Jr., 177, 178, 186.

I

Iger, Thomas, 245.
Ingle, Richard, 104.

Inglish, Edward, 13, 39, 194.
Innis, Thomas, 11.
Ireland, William, 99.

J

Jacob, 119-122.
James, Charles, 31.
Jarboe, John, 22.
Jenkins, Hannah, 129, 130.
Johnson, Bridgett, 220, 221.
Johnson, Daniel, 66, 225.
Johnson, David, 220.
Johnson, Elizabeth, 196.
Johnson, Mrs., 186.
Johnson, Sissilly, 194.
Jolly, Elizabeth, 211.
Jolly, James, 14, 112, 113, 158, 159.
Jones, Daniel, 250.
Jones, Mary, 95.
Joseph, William, 147.
Joyner, Robert, 128, 129.

K

Kedger (or Cager), Robert, 94, 95.
Keeting, Nicholas, 9.
Kidson, William, 73, 74.
Kinemant, John, 110, 118.
King, Alexander, 184, 185.
King, Eleanor, 214.
King, John, 114.
King, Samuel, 214.
Knapp, Robert, 143, 144, 163, 214.

L

Ladd, Edward, 122, 125.
Ladds, William, 144.
Ladds, Mrs. William, 144.
Lake, Catherine, 126.
Lamb, Richard, 118.
Langworth, James, 135, 136.
Laut, William, 7, 8.
Lee, Hannah, 14, 33, 157, 159.
Lee, Hugh, 33, 157.
Lee, James, 65, 66, 211, 212, 257, 258.
Lee, Joshua, 79.
Lee, Mary, 172.
Lee, Richard, 43.
Leeth, Susanna, 123.
Legatt, John, 200.
Lemaire, John, 249.

Lennin, Edward, 2.
Lenton, Joseph, 257, 258.
Lenton, Ursula, 257, 258.
Leo, Valerus, 247.
Leshley, Elizabeth, 204, 205.
Leshley, Robert, 204, 205.
Lewger, John, 43, 61, 212.
Lewger, Mrs. John, 43.
Lewis, James, 2.
Lindsey, Edmond, 230, 231.
Lindsey, James, 240.
Lisle, Thomas, 137.
Little, Bassill, 90.
Little, John, 39, 40, 218, 219.
Lloyd, Edward, 14, 98.
Lloyd, Robert, 123.
Lockett, Elizabeth, 192, 193.
Lomax, Thomas, 258.
Long, Mrs., 171.
Lovely, Deliverance, 147.
Lucus, Frances, 248.
Lumbrozo, Jacob, 166, 167, 200, 224-
    226, 240, 241.
Lutt, Alse, 100.
Lynes, Philip, 161.

M

Mackey, Robert, 90, 91.
Madberry, Elizabeth, 220, 221.
Madberry, John, 220.
Maidwell, Thomas, 142.
Mannering, Margaret, 217.
Manners, George, 11, 211.
Manning, Thomas, 24, 144, 200.
Mansell, John, 144.
Manship, Richard, 171.
Manship, Mrs. Richard, 171, 172.
Mardin, Anne, 187.
Markeen, Thomas, 99.
Marsh, Paul, 109.
Marsh, Thomas, 194.
Marshguy, Margaret, 129.
Marshall, William, 58, 90, 117, 203,
    254.
Martin, Elionar, 142.
Martin, Elizabeth, 208, 209.
Martin, Robert, 208, 209.
Mathewes, Hannah, 89.
Mathews, Thomas, 208.
Mawman, Thomas, 190.
Mayle, Anthony, 20.

Mecane, Ricckett, 88.
Medwell, Thomas, 239, 240.
Mee, George, 9.
Meekes, John, 18, 211, 212, 241-244.
Meredith, John, 217.
Mertine, Thomas, 126, 127.
Miles, Thomas, 123.
Mitchell, Henry, 200, 201.
Mitchell, Joan, 169, 170, 171.
Mitchell, Thomas, 169, 171.
Mitchell, William, 81, 174-177.
Mithridates IV, 240.
Molden, Grace, 200, 201.
Morgan, Frances, 68.
Morgan, Henry, 10, 67, 74, 246, 247.
Morgan, Owen, 93.
Morgan, Philip, 99, 100.
Morton, Richard, 131, 132, 237.
Mottershead, Zachary, 41.
Mouse, William, 118.
Mullins, William, 183, 184.
Munday, Thomas, 10, 141.
Mungummory, Anne, 196, 197.
Muns, John, 65.
Murell, Gregory, 162, 163.

N

Neale, Henry, 69.
Neale, James, 45, 90.
Nelson, Bridgett, 91.
Nevill, Anne, 100, 101.
Nevill, Joan, 143, 187, 226-230.
Nevill, John, 100, 143, 182, 183, 226-230.
Newman, Joseph, 98.
Newnam, John, 75.
Newell, Richard, 13.
Norman, John, 114.
Norton, John, 193.
Norton, Mrs. John, 193, 194.
Nottool, Arthur, 56, 57.

O

Oliver, John, 50, 51.
Oneale, Hugh, 251, 252, 254.
Oneale, Mary, 251-254.
Osborne, William, 74.
Ottoway, Francis, 246.
Overzee, Symon, 54, 121, 122.
Overzee, Mrs. Symon, 52, 53, 256.
Owens, Richard, 235, 236.

P

Packer, Edward, 32.
Paggett, Thomas, 209-211.
Pake (or Peake), Walter, 15, 23, 179, 180.
Palldin, Jane, 193, 194.
Parker, Grace, 127.
Parker, Joan, 252, 253.
Parker, Samuel, 252.
Parker, William, 200.
Parrott, William, 101.
Patee, Hubert, 208.
Pattison, James, 22, 26.
Pecheco, Anthony, 118.
Peirce, William, 13.
Phillips, James, 189, 190.
Pickard, Nicholas, 97.
Pinner, Richard, 115.
Pippett, Temperance, 71.
Plott, Thomas, 182, 183.
Plowden, Edmund, 246.
Pope, 231.
Pope, Francis, 257, 258.
Pope, Nathaniel, 112.
Potts, Elizabeth, 39, 40, 219.
Poulter, Henry, 74.
Powell, Howell, 251.
Prescott, Edward, 172, 173.
Preston, Richard, 93, 94, 166, 201.
Price, Hannah, 158.
Price, John, 1, 62, 114.
Price, William, 67, 157, 158, 162, 208, 209.
Purse, Anthony, 196, 197.
Pye, Edward, 72.

Q

Quigley, John, 34.

R

Randall, Benjamin, 19.
Rawlings, Nicholas, 95.
Rawlings, Ralph, 5.
Rawlins, Anthony, 18.
Read, Matthew, 68, 118, 147, 164, 192, 193, 207, 209.
Redfearne, Margaret, 100, 101.
Revell, Randall, 191, 214.
Rice, John, 194.
Richardson, Elizabeth, 172, 173.

Richardson, John, 50, 134, 135.
Richardson, Mary, 134, 135, 200.
Rigby, John, 14, 182.
Ringgold, James, 8, 97, 177, 178.
Ringgold, John, 67.
Ringgold, Thomas, 162, 163, 215, 216.
Risby, Elizabeth, 182.
Robbins, George, 75.
Robins, Elizabeth, 203, 204.
Robins, Robert, 203, 204.
Robinson, John, 15, 16, 238.
Robisson (or Robinson), William, 35, 66, 67, 186, 187, 207, 229-231.
Roe, Mary, 142, 143, 227.
Roe, Richard, 142, 143, 231.
Rogers, Edward, 164.
Rogers, Hannah, 130, 131.
Rogers, Sissly, 189.
Rowse, Abraham, 22.
Rowse, Gregory, 23.
Rumsey, James, 72.
Russell, Christopher, 251.
Russell, John, 164.

S

Salter, Jane, 68, 162.
Salter, John, 67, 68, 93, 162, 178, 179.
Salway, Anthony, 139.
Sandford, Alice, 108, 109.
Sandsbury, Francis, 168.
Saxon, John, 2.
Scott, Roger, 149, 188.
Scurfield, Paul, 161.
Shambrooke, Frances, 189.
Sharpe, Peter, 180, 239, 244, 245.
Shelton, Thomas, 181.
Short, John, 138.
Sibery, Jonathan, 5, 113.
Simmons, Thomas, 105-107, 237, 238.
Simpson, Paul, 15, 43, 179, 180.
Slye, Gerard, 114.
Slye, Robert, 112.
Smith, Elizabeth, 218.
Smith, Frances, 188.
Smith, Frances, 190.
Smith, James, 249.
Smith, John, 115.
Smith, John, Jr., 50.
Smith, Richard, 145, 187, 197, 200.
Smith, Susan, 175.

Smith, Thomas, 142.
Smith, Walter, 8.
Smith, William, 8, 159, 175, 250, 251, 256.
Smoot, William, 251.
Somers, Roger, 191, 192.
South, Thomas, 20, 67, 138, 207, 217.
Southern, Thomas, 102.
Spernon, Joseph, 15.
Spicer, Elizabeth, 212, 213.
Spicer, Hannibal, 212, 213.
Spim, Henry, 245.
Spinke, Ellinor, 222, 223.
Spinke, Henry, 222, 223.
Spurdance, Sarah, 183, 184.
Sprigge, Thomas, 13.
Sprye, Oliver, 249, 255.
Sprye, Mrs. Oliver, 255.
Standbridge, Thomas, 66.
Stanley, Anne, 93.
Stanley, Hugh, 209-211.
Stanley, John, 69.
Staplefort, Raymond, 33, 57.
Stavely, Adam, 244.
Stedhed, Mary, 192.
Stevens, David, 93.
Stevens, Richard, 130, 131.
Stevens, William, 2, 246.
Stockett, Francis, 249.
Stone, William, 62, 177.
Stratton, Lucy, 195, 196.
Sturdivant, William, 75.
Stuyvesant, Peter, 111.
Styles, William, 148.
Sweatnam, Edward, 155, 156.

T

Tarline, Mary, 78.
Taylor, Agnes, 181.
Taylor, John, 156.
Taylor, Mary, 93, 185, 186.
Taylor, Robert, 63, 185-188.
Taylor, Sarah, 9, 96-103, 115, 219.
Teedsteed, Thomas, 139.
Tenison, Elizabeth, 202.
Tenison (or Tennisson), John, 74, 202.
Theobalds, Clement, 134.
Thirle, George, 190.
Thomas, Trustrum, 189.
Thompson, Anne, 205.

Thompson, George, 229, 230, 257, 258.
Thompson, Richard, 198.
Thompson, Robert, 205.
Thorowgood, Cyprian, 63.
Thurston, Thomas, 10, 14, 15.
Tilghman, Richard, 68, 136, 137, 247, 250.
Tilghman, Samuel, 1.
Tompkinson, Giles, 201.
Tompkinson, John, 45.
Trafford, Francis, 246.
Trew, Richard, 89.
Truman, Thomas, 21.
Turner, Arthur, 78, 98, 99, 195, 196, 215.
Turner, Richard, 134.
Twotley, Andrew, 191.

U

Underwood, Anthony, 160.
Utie, George, 194.
Utie, Mary, 119-121.
Utie, Nathaniel, 1, 62, 119, 121.

V

Vanderdonck, Adrian, 251.
Vanderdonck, Mary, 251.
Vanhack, John, 2.
Vaughan, Anne, 139.
Vaughan, Robert, 6-8, 97, 105, 179.
Vaughan, Thomas, 117, 256.
Vaughan, Mrs. Thomas, 256.
Veitch, James, 162, 208.
Vicaris, John, 217.

W

Wade, John, 239.
Waerum, John, 248.
Walker, James, 171.
Wallcott, John, 249.
Ward, John, 98, 99.
Ward, Matthew, 16.
Ward, Thomas, 100, 222, 236.
Ward, Mrs. Thomas, 100.
Warren, Humphrey, 118.
Warren, Susan, 175, 176.
Washington, George, 172.
Washington, John, 172, 173.
Watson, Richard, 199, 200, 248, 249.

Watson, Thomas, 102-104, 127, 234.
Watson, Thomas, 247.
Webb, Martha, 175.
Wedge, John, 216, 217.
Wells, Swithen, 5.
Wennam, William, 187, 235.
Wharton, Elizabeth, 191.
White, Andrew, 150, 233.
White, Dennis, 73, 192.
White, John, 102, 103, 112.
White, Nicholas, 114, 115.
Whyniard, Thomas, 122, 125.
Wickes, John, 216, 217.
Wickes, Joseph, 8, 9, 74, 89, 112, 179, 215, 216, 235, 236.
Wild, Abraham, 31.
Wild, Elizabeth, 240, 241.
Wilkinson, Thomas, 213, 214.
Wilkinson, William, 205.
Willan, Richard, 11.
Willey, Humphrey, 156.
Williams, Anne, 197.
Williams, John, 52, 53, 221, 222.
Williams, Olive, 221.
Williams, Mary, 52, 53.
Willkins, Mary, 217.
Willson, George, 142.
Wilson, James, 104, 234.
Wilson, Robert, 52.
Winchester, John, 6, 254.
Winchester, Mrs. John, 254, 255.
Wintour, Robert, 41, 81, 150, 239.
Winwood, Edward, 194.
Withrington, Elizabeth, 54.
Wollman, Richard, 78, 213, 214.
Wood, John, 207.
Wood, Stephen, 137, 138.
Woosey, James, 93.
Wright, John, 142.
Wright, Richard, 44.
Wroth, Mrs., 194.
Wyatt, John, 246.
Wynne, Thomas, 95, 96.
Wynne, Mrs. Thomas, 95, 96.

Y

York, Anne, 192.
Young, Jacob, 234.
Young, William, 12.
Youngman, Samuel, 238.

TOPICAL INDEX

A

ABORTION. Drugs used to produce miscarriages, 240; case where chirurgeon is accused of being an abortionist, 240, 241.

ACCIDENTAL DEATH. By drowning, etc., 137, 138. See SUICIDE.

ACCIDENTAL KILLING. Case of, 135, 136. See HOMICIDE.

ACT OF TOLERATION. See BLASPHEMY.

ADULTERY. Extent of, 174; provincial officials accused of this offense or of fornication, 174-178; other colonists accused of the same offenses, 178-180; punishment of men and women compared, 181-186. See BASTARDS.

ALCOHOLIC BEVERAGES. Kind drunk, including beer, wine, cider, perry, brandy, etc., 149-151; amount a gentleman could consume a year, 149, 150. See DRUNKENNESS and ORDINARIES.

ASSAULT AND BATTERY. Cases of this which came before provincial, county and manor courts, 141-144.

ATTORNEYS. Training, 16; their appointment, 16-17; who could act as such, 17, 18; fees and their limitation, 18, 19; disbarment case, 19.

B

BASTARDS. Punishment of maidservants and of other women who gave birth to bastards, 187-197; punishment of their seducers, 187-197. See MARRIAGES.

BATTERY. See ASSAULT AND BATTERY.

BIGAMY. Punishment of those guilty of this offense, 205; cases of, 205, 206.

BENEFIT OF CLERGY. History of this privilege and who could claim it, 27, 28.

BLASPHEMY. Provisions of Act of Toleration of 1649 in re this offense, 165, 166; trial of a chirurgeon under this act, 166, 167. See PROFANITY.

BLOOD TEST. Theory that dead body would bleed if touched by murderer, 126; cases where this test was applied, 126, 127.

BRANDING IRONS. Their use, 35, 69, 70.

BRICKS AND BRICKMAKING. See HOUSES.

BURGLARY. Definition and case of, 56, 57. See HOUSEBREAKING and FORCIBLE ENTRY.

BURIALS. Customs and practices observed at funerals and burials, 256-258.

C

CHIRURGERY. Purges and blood-letting the most usual treatment for sick persons, 238, 239; drugs used, 239, 240; suits entered by chirurgeons to recover for their services, 241-246; cases of sick persons who could not afford the services of a chirurgeon, 247-249; chirurgeons accompany military expeditions, 249; chirurgeon disqualified, 241; lay practitioners, 246. See ABORTION, NURSES and MIDWIVES.

CLERK OF COURT. See TRIAL.

CLOTHING. Worn by colonists and their servants, 41-43.

CONSTABLES. Their appointment and duties, 14; cases where they refused to take oath of office or to perform the duties required of them, 14-15; colonists must assist them in the performance of their duties, 11, 12. See SHERIFFS.

CONTEMPT OF COURT. Attempt to uphold dignity of court by wearing of distinctive ribbon and medal, 3; litigants must stand bare-headed in presence of judges, 4; Quakers objected to doing this, 4; punishment of those guilty of drinking or swearing in presence of court, 4; other ways of showing contempt of court, 15, 16; cases where individuals were guilty of contempt of court, 5-9. See OFFICIAL DIGNITY.

CORONERS JURY. Verdicts of, 236-238.

CORPORAL PUNISHMENT. Could not be inflicted on a gentleman, 39. See PUNISHMENT.

COUNTY COMMISSIONERS. See JUSTICES OF THE PEACE.

COUNTY COURT HOUSES. Provision made for their erection, 35, 36.

CRIER. Duties of provincial court crier, 34. See TRIAL.

D

DEBTORS, Imprisonment of, 34, 36.

DEFAMATION. Reason for considering such cases in a book about crime, 207; instances of where persons were accused of various offenses, including dishonesty, 209, 210; perjury, 208, 209; theft, 211-213; hog stealing, 213-216; and immorality, 216-231.

DEODAND. Definition of, 136; cases where applied or not applied, 136, 137.

DEPUTY SHERIFF. See SHERIFF.

DIGNITY OF COURT. See CONTEMPT OF COURT.

DISEASES. See SICKNESS.

DIVORCE. Although probably no cases of absolute divorce, there were cases of where parties were allowed to separate, 201-205.

DOCTORS. See CHIRURGERY.

DOGS. Used in colony as protection of homes and for hunting, 77; cases where dogs attacked strangers or friends, 77-79.

DUCKING STOOLS. At St. Mary's and in counties, 33, 35.

DRUNKENNESS. Cases of drunkenness among the provincial officials, 145-147; views held by a proprietor and governor of Maryland regarding excessive drinking, 147; amount of drinking done, 147; deaths from drinking too much, 147, 148; statutes passed to punish drunkards, 148, 149; cases of those punished under these laws, 149; constables duty to supress drunkenness, 14. See ALCOHOLIC BEVERAGES and ORDINARIES.

E

EXCUSABLE HOMICIDE. See HOMICIDE.

F

FELO-DE-SE. See SUICIDE.

FELONIOUS HOMICIDE. See HOMICIDE.

FERRIES. Provision for, 47, 48; on which rivers, 48; duties of ferrymen, 48; charges for passengers and horses, 48; regulations regarding transport of servants, 48, 49.

FORGERY. Punishment of, 31, 32.

FORCIBLE ENTRY. Case of, 57. See also BURGLARY and HOUSEBREAKING.

FORNICATION. See ADULTERY.

FUNERALS. See BURIALS.

FURNITURE. See HOUSES.

G

GRAND JURY. Appointment and duties, 21-24. See TRIAL JURY.

GRAND LARCENY. See LARCENY.

H

HANGING. Sentences of, 119, 122.

HIGHWAYS. Roads and paths provided for, 46; where located, 46, 47; duties of overseers of highways, 46, 47; opinion of visitors about roads in the colony, 47.

HOG STEALING. See LIVESTOCK.

HOMICIDE. Definition of felonious, justifiable and excusable homicide, 119; felonious homicide is either murder or manslaughter, 119; Cases where persons are condemned for murder, 119-128; cases where they are accused of murder, but not convicted, 128-131; cases of manslaughter, 131-133; excusable homicide, 133-135; and accidental killing, 135, 136. See DEODAND, ASSAULT AND BATTERY, and SUICIDE.

HORSES. Part they played in economic life of colonists shown by uses to which they were put, 71, 72; their equipment, 72; horse races, 72; sale of and litigation about them, 72-74; horse traders, 72, 73; colony's policy in regard to importation of horses underwent a change, 75, 76; rangers licensed to round up wild horses, 76; dispute about limiting supply of horses, 76, 77; provision for horses at ordinaries, 154.

HOUSES. Size and type, 43-45; chimneys, 44; windows, 44; locks, 44; use of brick, 44; brickmakers, 44; typical houses, 44, 45; furniture in houses, 45; tableware and kitchenware, 45.

HOUSEBREAKING. Definition of, 56; case of, 57. See BURGLARY and FORCIBLE ENTRY.

I

INNS. See ORDINARIES.

J

JAILERS. See PRISONS.
JURY. See GRAND, PETIT and CORONER'S JURY.
JUSTICES OF THE PEACE. Punishment of those guilty of striking one, 9; their appointment and oath of office, 19; regulations governing meetings of county justices, 19, 20. See CONTEMPT OF COURT.
JUSTIFIABLE HOMICIDE. See HOMICIDE.

K

KITCHENWARE. See HOUSES.

L

LAWBOOKS. Names of those used by county courts, 36.

LAWYERS. See ATTORNEYS.
LARCENY. Distinction between grand and petit larceny, 26, 27; punishment for larceny under various statutes, 55, 56.

LIVESTOCK. Part played in economic life of colonists shown by uses to which they were put, 58; livestock bartered or exchanged for many other things, 59; litigation regarding cattle and swine, 59; where first livestock came from, 59; protection of cornfields against cattle, 59, 60; rangers allowed to round up wild herds, 60, 61; cattle markings must be recorded, 61; sale of cattle, 61-63; at first licenses were issued to hunt and kill swine, 63; latter revoked and laws passed against hog stealing, 64, 69, 70; punishment of servant who killed his master's cattle, 70; hogs in village of St. Mary's, 71. See also HORSES and DOGS.

M

MAGISTRATES. See JUSTICES OF THE PEACE.
MANSLAUGHTER. See HOMICIDE.
MARRIAGES. Regulations governing marriage, 197-199; the ceremony, 199; witnesses, cost, 199; Maryland a Gretna Green for Virginians, 201. See BIGAMY AND DIVORCE.
MIDWIVES. Cases of, 256.
MISCARRIAGES. See ABORTION.
MURDER. See HOMICIDE.

N

NURSES. Women acted as nurses or healers of the sick, 251-256.

O

OFFICIAL DIGNITY. Attempt of governor, councillor and burgess to uphold, 1-3.
ORDINARIES. Need of, 151, 152; where they were located, 152; license to keep one necessary, 152; and order must be preserved in them, 152, 153;

regulation of price of beers, wines and liquors, 153-154; cost of a night's lodging, 154; collection of ordinary keeper's accounts, 155; violation of regulations, 155, 156; ordinaries at St. Mary's, in the counties, and names of some of their keepers, 156-161. See SUNDAY OBSERVANCE.

P

PASSES. One was needed in order to leave the colony, 48, 49.

PETIT JURY. Sometimes called the trial jury. Duties, 24, 30; service on compulsory, 30; pay of jurors, 30. See GRAND JURY.

PETIT LARCENY. See LARCENY.

PERJURY. Punishment for, 30, 31.

PHYSICIANS. See CHIRURGERY.

PILLORY. At St. Mary's and in counties, 33, 35. Offenders ears nailed to, 30.

PRISONS. First in colony, 32; and at St. Mary's, 33; construction of state house and prison in 1674, 33, 34; criminals and debtors confined in prisons, 33; prisons in counties, 34-36; location of one in Baltimore county, 36; punishment of escaped prisoner, 37; prison on plantation of Augustine Herrman, 37; prison keepers, or jailers, 36; conditions in prisons, 37; prisoners liable for cost of own imprisonment, 37, 38; question of who should pay for servant's imprisonment, 38.

PROCEDURE. See TRIAL.

PROFANITY. Amount of, 161; punishment of those guilty of this offense, 161-163. See BLASPHEMY and SUNDAY OBSERVANCE.

PUNISHMENT. The more usual and some unusual ways of punishing criminals, 38-40.

Q

QUAKERS. They object to complying with certain regulations, 4; steps taken to drive them out of the colony, 4.

S

SERVANTS. Duties they performed, 80, 83; cost of transporting and keeping one, 80; profit derived from servant's labor, 80; number of servants owned by prominent colonists, 80, 81; character and type of servant class, 81; some worked for wages, some indented, 81, 82; meaning of indenture, 82; terms of indenture might be modified, 82; Cecil Calvert employed wage servants, 82, 83; most servants employed in raising tobacco, 83; must work Saturday afternoons, 83, 84; overseers of servants, 84; if a skilled workman, servant might serve a shorter time than if unskilled, 84; reason for requiring a servant to serve a certain number of years, 84; entering into the agreement voluntarily or involuntarily, 84; kidnapping, 84; Act of 1682 to prevent that practice, 85; Act of 1666 governing time servant must serve, 85; servants brought before provincial courts to have their ages determined, 85-88; cases where this was done, 87, 88; average age of servants, 87; destitute widows sometimes compelled to hire their children out, 88, 89; sale and assignment of children, 89; new master bound by terms of original indenture, 90; ways by which a servant could obtain his freedom, 90, 91; clothing, corn, etc., furnished servant at end of term of indenture, 91, 92; special agreement might give servant more than usual allowance, 92; servants bring suits to recover clothing, corn, etc., 92, 93; by Act of 1654 servant no longer allowed fifty acres at end of period of indenture, 92; servant sometimes maltreated, 93; must be respectful to masters, 93; must not complain of food, 94; disciplined by master, 93; master not always able to prove insubordination on part of servants, 94, 95; in some cases badly treated servant received justice at hands of the law, 95, 96; usually freed or sold to another master, 95; in other cases, however, the master or mistress who had maltreated a servant is either not adequately punished for their brutality, or sometimes not punished at all, 96-110; chances of a runaway servant to escape were slight, 110; reward for capture, 110; aid of Indians and of other colonies sought in capture of runaways, 110, 111; prison erected in Baltimore county to confine runaways, 111; servants running way from their Virginia masters would be returned to the latter, 111; no one could help a runaway servant escape, 111, 112; or harbor one after his escape, 111, 113, 114; cases where colonists were accused of these offenses, 112-115; the runaway servant if captured might have to serve additional time, 116, 117; he was sometimes whipped, 117, 118; constables must pursue runaway servants, 110; servants could not be transported on ferries, 48, 49. For cases where masters are accused of murdering their servants by beating them to death, see HOMICIDE. All suicides were among the servant class. See SUICIDE. For cases where unmarried women servants are accused of having children, see BASTARDS.

SHERIFFS. Colonists must assist them in performance of their duties, 9-11; how appointed and provisions of oath they must take, 12, 13; length of service, 12; must properly perform their duties, 13; liability of sheriff, 13, 14; if he allowed prisoner to escape, 36, 37; sheriff could collect fees for his services, 37.

SICKNESS. Nature and amount of sickness, 232, 233; ailments included ague, fever, dropsy, flux, scurvy, ulcers, venereal diseases, insanity, etc., etc., 233-236; knowledge shown of human anatomy by findings of coroner's juries, 236-238. See CHIRURGERY.

SLANDER. See DEFAMATION.

STATE HOUSE. Early and later ones at St. Mary's, 33, 34.

STOCKS. At St. Mary's and in the counties, 33, 35.

SUICIDE. When a person took his own life intentionally he was guilty of a felony, 138, 139.

SUNDAY OBSERVANCE. Regulations governing the observance of this day, 163-165; ordinaries came under these regulations, 165. See ORDINARIES.

SWEARING. Punishment of those guilty of this offense, 161-163. See PROFANITY and CONTEMPT OF COURT.

T

TABLEWARE. See HOUSES.

TAVERNS. See ORDINARIES.

THEFT. Among reasons so few cases of theft, was that there was little of value to steal, 41, 43; few paths and roads by which to escape, 46; few ferries, 48; necessity of having pass to leave colony, 48, 49; difficulty of disposing of article in province, 49; risk one ran in buying from a thief if he was a servant, 49, 50; act passed in 1663 to prevent servants stealing and selling their master's goods, 51, 52; cases under this act, 51; cases in which servants and others were accused of theft, 50-54. See LARCENY.

THEFTBOTE. Definition of and case where this occurred, 53, 54.

TOBACCO. Punishment of those who altered mark or quality of tobacco 32.

TOLERATION ACT OF 1649. See BLASPHEMY.

TREASON. Definition of grand and petit or petty treason, and case in which the latter accusation was made, 120, 121.

TRIAL. Procedure followed in the trial of a criminal described and explained, 21-29.

TRIAL JURY. See PETIT JURY.

W

WITCHCRAFT. Extent of belief in this, 167; punishment of witches, 167, 168; cases of those either accused or hanged for witchcraft, 168-173.

WITNESSES. Duty to testify when summoned, otherwise fined, 30, 31. See PERJURY.

WHIPPINGS. See PUNISHMENT and CORPORAL PUNISHMENT.